FERRARI
DINO 308 & MONDIAL
Gold Portfolio
1974-1985

Compiled by
R.M.Clarke

ISBN 1 85520 3588

BROOKLANDS BOOKS LTD.
P.O. BOX 146, COBHAM,
SURREY, KT11 1LG. UK

Brooklands Books

MOTORING

BROOKLANDS ROAD TEST SERIES

Abarth Gold Portfolio 1950-1971
AC Ace & Aceca 1953-1983
Alfa Romeo Giulietta Gold Portfolio 1954-1965
Alfa Romeo Giulia Berlinas 1962-1976
Alfa Romeo Giulia Coupés 1963-1976
Alfa Romeo Giulia Coupés Gold P. 1963-1976
Alfa Romeo Spider 1966-1990
Alfa Romeo Spider Gold Portfolio 1966-1991
Alfa Romeo Alfasud 1972-1984
Alfa Romeo Alfetta Gold Portfolio 1972-1987
Alfa Romeo Alfetta GTV6 1980-1986
Allard Gold Portfolio 1937-1959
Alvis Gold Portfolio 1919-1967
AMX & Javelin Muscle Portfolio 1968-1974
Armstrong Siddeley Gold Portfolio 1945-1960
Aston Martin Gold Portfolio 1972-1985
Aston Martin Gold Portfolio 1985-1995
Audi Quattro Gold Portfolio 1980-1991
Austin A30 & A35 1951-1962
Austin Healey 100 & 100/6 Gold P. 1952-1959
Austin Healey 3000 Gold Portfolio 1959-1967
Austin Healey Sprite 1958-1971
Barracuda Muscle Portfolio 1964-1974
BMW 1600 Collection No.1 1966-1981
BMW 2002 Gold Portfolio 1968-1976
BMW 316, 318, 320 (4 cyl.) Gold P. 1975-1990
BMW 320, 323, 325 (6 cyl.) Gold P. 1977-1990
BMW M Series Performance Portfolio 1976-1993
BMW 5 Series Gold Portfolio 1981-1987
Bricklin Gold Portfolio 1974-1975
Bristol Cars Gold Portfolio 1946-1992
Buick Automobiles 1947-1960
Buick Muscle Cars 1965-1970
Cadillac Allanté 1986-1993
Cadillac Automobiles 1949-1959
Cadillac Automobiles 1960-1969
Charger Muscle Portfolio 1966-1974
Checker ☆ Limited Edition
Chevrolet 1955-1957
Impala & SS Muscle Portfolio 1958-1972
Chevrolet Corvair 1959-1969
Chevy II & Nova SS Muscle Portfolio 1962-1974
Chevy El Camino & SS 1959-1987
Chevelle & SS Muscle Portfolio 1964-1972
Chevrolet Muscle Cars 1966-1971
Chevy Blazer 1969-1981
Chevrolet Corvette Gold Portfolio 1953-1962
Chevrolet Corvette Sting Ray Gold P. 1963-1967
Chevrolet Corvette Gold Portfolio 1968-1977
High Performance Corvettes 1983-1989
Camaro Muscle Portfolio 1967-1973
Chevrolet Camaro Z28 & SS 1966-1973
Chevrolet Camaro & Z28 1973-1981
High Performance Camaros 1982-1988
Chrysler 300 Gold Portfolio 1955-1970
Chrysler Valiant 1960-1962
Citroen Traction Avant Gold Portfolio 1934-1957
Citroen 2CV Gold Portfolio 1948-1989
Citroen DS & ID 1955-1975
Citroen DS & ID Gold Portfolio 1955-1975
Citroen SM 1970-1975
Cobras & Replicas 1962-1983
Shelby Cobra Gold Portfolio 1962-1969
Cobras & Cobra Replicas Gold P. 1962-1989
Cunningham Automobiles 1951-1955
Daimler SP250 Sports & V-8 250 Saloon Gold P. 1959-1969
Datsun Roadsters 1962-1971
Datsun 240Z 1970-1973
Datsun 280Z & ZX 1975-1983
DeLorean Gold Portfolio 1977-1995
Dodge Muscle Cars 1967-1970
Dodge Viper on the Road
ERA Gold Portfolio 1934-1994
Excalibur Collection No.1 1952-1981
Facel Vega 1954-1964
Ferrari Dino 1965-1974
Ferrari Dino 308 & Mondial Gold Portfolio1974-1985
Ferrari 328 • 348 • Mondial Gold Portfolio 1986-1994
Fiat 500 Gold Portfolio 1936-1972
Fiat 600 & 850 Gold Portfolio 1955-1972
Fiat Pininfarina 124 & 2000 Spider 1968-1985
Fiat X1/9 Gold Portfolio1973-1989
Fiat Abarth Performance Portfolio 1972-1987
Ford Consul, Zephyr, Zodiac Mk.I & II 1950-1962
Ford Zephyr, Zodiac, Executive, Mk.III & Mk.IV 1962-1971
Ford Cortina 1600E & GT 1967-1970
High Performance Capris Gold Portfolio 1969-1987
Capri Muscle Portfolio 1974-1987
High Performance Fiestas 1979-1991
High Performance Escorts Mk.I 1968-1974
High Performance Escorts Mk.II 1975-1980
High Performance Escorts 1980-1985
High Performance Escorts 1985-1990
High Performance Sierras & Merkurs Gold Portfolio 1983-1990
Ford Automobiles 1949-1959
Ford Fairlane 1955-1970
Ford Ranchero 1957-1959
Ford Thunderbird 1955-1957
Ford Thunderbird 1958-1963
Ford Thunderbird 1964-1976
Ford GT40 Gold Portfolio 1964-1987
Ford Bronco 1966-1977
Ford Bronco 1978-1988
Goggomobil ☆ Limited Edition
Holden 1948-1962
Honda CRX 1983-1987
International Scout Gold Portfolio 1961-1980
Isetta 1953-1964
ISO & Bizzarrini Gold Portfolio 1962-1974
Jaguar and SS Gold Portfolio 1931-1951
Jaguar XK120, 140, 150 Gold P. 1948-1960
Jaguar Mk.VII, VIII, IX, X, 420 Gold P. 1950-1970
Jaguar Mk.1 & Mk.2 Gold Portfolio 1959-1969
Jaguar C-Type & D-Type ☆ Limited Edition
Jaguar E-Type Gold Portfolio 1961-1971
Jaguar E-Type V-12 1971-1975
Jaguar S-Type & 420 ☆ Limited Edition
Jaguar XJ12, XJ5.3, V12 Gold P. 1972-1990
Jaguar XJ6 Series I & II Gold P. 1968-1979
Jaguar XJ6 Series III Perf. Portfolio 1979-1986
Jaguar XJ6 Gold Portfolio 1986-1994
Jaguar XJS Gold Portfolio 1975-1988
Jaguar XJS Gold Portfolio 1988-1995
Jeep CJ5 & CJ6 1960-1976
Jeep CJ5 & CJ7 1976-1986
Jensen Cars 1946-1967
Jensen Cars 1967-1979
Jensen Interceptor Gold Portfolio 1966-1986
Jensen Healey 1972-1976
Lagonda Gold Portfolio 1919-1964
Lamborghini Countach & Urraco 1974-1980
Lamborghini Countach & Jalpa 1980-1985
Lancia Aurelia & Flaminia Gold Portfolio 1950-1970
Lancia Fulvia Gold Portfolio 1963-1976
Lancia Beta Gold Portfolio 1972-1984
Lancia Delta Gold Portfolio 1979-1994
Lancia Stratos 1972-1985
Land Rover Series I 1948-1958
Land Rover Series II & IIa 1958-1971
Land Rover Series III 1971-1985
Land Rover 90 110 Defender Gold Portfolio 1983-1994
Land Rover Discovery 1989-1994
Land Rover Story Part One 1948-1971
Lincoln Gold Portfolio 1949-1960
Lincoln Continental 1961-1969
Lincoln Continental 1969-1976
Lotus Sports Racers Gold Portfolio 1953-1965
Lotus Seven Gold Portfolio 1957-1974
Lotus Caterham Seven Gold Portfolio 1974-1995
Lotus Elite & Eclat 1974-1982
Lotus Elan Gold Portfolio 1962-1974
Lotus Elan Collection No. 2 1963-1972
Lotus Elan & SE 1989-1992
Lotus Cortina Gold Portfolio 1963-1970
Lotus Europa Gold Portfolio 1966-1975
Lotus Elite & Eclat 1974-1982
Lotus Turbo Esprit 1980-1986
Marcos Cars 1960-1988
Maserati 1965-1975
Mazda Miata-MX-5 Performance Portfolio 1989-1996
Mazda RX-7 Gold Portfolio 1978-1991
Mercedes 190 & 300 SL 1954-1963
Mercedes G Wagen 1981-1994
Mercedes S & 600 1965-1972
Mercedes S Class 1972-1979
Mercedes SLs & SLCs Gold Portfolio 1963-1971
Mercedes SLs & SLCs Gold Portfolio 1971-1989
Mercedes SLs Performance Portfolio 1989-1994
Mercury Muscle Cars 1966-1971
Messerschmitt Gold Portfolio 1954-1964
MG Gold Portfolio 1929-1939
MG TA & TC Gold Portfolio 1936-1949
MG TD & TF Gold Portfolio 1949-1955
MGA & Twin Cam Gold Portfolio 1955-1962
MG Midget Gold Portfolio 1961-1979
MGB Roadsters 1962-1980
MGB MGC & V8 Gold Portfolio 1962-1980
MGB GT 1965-1980
MG Y-Type & Magnette ZA/ZB ☆ Limited Edition
Mini Gold Portfolio 1959-1969
Mini Gold Portfolio 1969-1980
High Performance Minis Gold Portfolio 1960-1973
Mini Cooper Gold Portfolio 1961-1971
Mini Moke Gold Portfolio 1964-1994
Mopar Muscle Cars 1964-1967
Morgan Three-Wheeler Gold Portfolio 1910-1952
Morgan Plus 4 & Four 4 Gold P. 1936-1967
Morgan Cars 1960-1970
Morgan Cars Gold Portfolio 1968-1989
Morris Minor Collection No. 1 1948-1980
Shelby Mustang Muscle Portfolio 1965-1970
High Performance Mustang IIs 1974-1978
High Performance Mustangs 1982-1988
Nash-Austin Metropolitan Gold P. 1954-1962
Oldsmobile Automobiles 1955-1963
Oldsmobile Muscle Cars 1964-1971
Oldsmobile Toronado 1966-1978
Opel GT Gold Portfolio 1968-1973
Packard Gold Portfolio 1946-1958
Pantera Gold Portfolio 1970-1989
Panther Gold Portfolio 1972-1990
Plymouth Muscle Cars 1966-1971
Pontiac Tempest & GTO 1961-1965
Pontiac Muscle Cars 1966-1972
Pontiac Firebird & Trans-Am 1973-1981
High Performance Firebirds 1982-1988
Pontiac Fiero 1984-1988
Porsche 356 Gold Portfolio 1953-1965
Porsche 911 1965-1969
Porsche 911 1970-1972
Porsche 911 1973-1977
Porsche 911 Turbo 1975-1984
Porsche 911 SC & Turbo Gold Portfolio 1978-1983
Porsche 911 Carrera & Turbo Gold P. 1984-1989
Porsche 924 Gold Portfolio 1975-1988
Porsche 928 Performance Portfolio 1977-1994
Porsche 944 Gold Portfolio 1981-1991
Range Rover Gold Portfolio 1970-1985
Range Rover Gold Portfolio 1986-1995
Reliant Scimitar 1964-1986
Riley Gold Portfolio 1924-1939
Riley 1.5 & 2.5 Litre Gold Portfolio 1945-1955
Rolls Royce Silver Cloud & Bentley 'S' Series Gold Portfolio 1955-1965
Rolls Royce Silver Shadow Gold P. 1965-1980
Rolls Royce & Bentley Gold P. 1980-1989
Rover P4 1949-1959
Rover P4 1955-1964
Rover 3 & 3.5 Litre Gold Portfolio 1958-1973
Rover 2000 & 2200 1963-1977
Rover 3500 1968-1977
Rover 3500 & Vitesse 1976-1986
Saab Sonett Collection No.1 1966-1974
Saab Turbo 1976-1983
Studebaker Gold Portfolio 1947-1966
Studebaker Hawks & Larks 1956-1963
Avanti 1962-1990
Sunbeam Tiger & Alpine Gold P. 1959-1967
Toyota MR2 1984-1988
Toyota Land Cruiser 1956-1984
Triumph Dolomite Sprint ☆ Limited Edition
Triumph TR2 & TR3 Gold Portfolio 1952-1961
Triumph TR4, TR5, TR250 1961-1968
Triumph TR6 Gold Portfolio 1969-1976
Triumph TR7 & TR8 Gold Portfolio 1975-1982
Triumph Herald 1959-1971
Triumph Vitesse 1962-1971
Triumph Spitfire Gold Portfolio 1962-1980
Triumph 2000, 2.5, 2500 1963-1977
Triumph GT6 Gold Portfolio 1966-1974
Triumph Stag Gold Portfolio 1970-1977
TVR Gold Portfolio 1959-1986
TVR Performance Portfolio 1986-1994
VW Beetle Gold Portfolio 1935-1967
VW Beetle Gold Portfolio 1968-1991
VW Beetle Collection No.1 1970-1982
VW Karmann Ghia 1955-1982
VW Bus, Camper, Van 1954-1967
VW Bus, Camper, Van 1968-1979
VW Bus, Camper, Van 1979-1989
VW Scirocco 1974-1981
VW Golf GTI 1976-1986
Volvo PV444 & PV544 1945-1965
Volvo Amazon-120 Gold Portfolio 1956-1970
Volvo 1800 Gold Portfolio 1960-1973
Volvo 140 & 160 Series Gold Portfolio 1966-1975

Forty Years of Selling Volvo

BROOKLANDS ROAD & TRACK SERIES

Road & Track on Alfa Romeo 1964-1970
Road & Track on Alfa Romeo 1971-1976
Road & Track on Aston Martin 1962-1990
R & T on Auburn Cord and Duesenburg 1952-84
Road & Track on Audi & Auto Union 1952-1980
Road & Track on Audi & Auto Union 1980-1986
Road & Track on Austin Healey 1953-1970
Road & Track on BMW Cars 1966-1974
Road & Track on BMW Cars 1975-1978
Road & Track on BMW Cars 1979-1983
R & T on Cobra, Shelby & Ford GT40 1962-1992
Road & Track on Corvette 1953-1967
Road & Track on Corvette 1968-1982
Road & Track on Corvette 1982-1986
Road & Track on Corvette 1986-1990
Road & Track on Ferrari 1975-1981
Road & Track on Ferrari 1981-1984
Road & Track on Ferrari 1984-1988
Road & Track on Fiat Sports Cars 1968-1987
Road & Track on Jaguar 1950-1960
Road & Track on Jaguar 1961-1968
Road & Track on Jaguar 1968-1974
Road & Track on Jaguar 1974-1982
Road & Track on Jaguar 1983-1989
Road & Track on Lamborghini 1964-1985
Road & Track on Lotus 1972-1981
Road & Track on Maserati 1975-1983
R & T on Mazda RX7 & MX5 Miata 1986-1991
Road & Track on Mercedes 1952-1962
Road & Track on Mercedes 1963-1970
Road & Track on Mercedes 1971-1979
Road & Track on Mercedes 1980-1987
Road & Track on MG Sports Cars 1949-1961
Road & Track on MG Sports Cars 1962-1980
Road & Track on Mustang 1964-1977
R & T on Nissan 300-ZX & Turbo 1984-1989
Road & Track on Pontiac 1960-1983
Road & Track on Porsche 1951-1967
Road & Track on Porsche 1968-1971
Road & Track on Porsche 1972-1975
Road & Track on Porsche 1975-1978
Road & Track on Porsche 1985-1988
R & T on Rolls Royce & Bentley 1950-1965
R & T on Rolls Royce & Bentley 1966-1984
Road & Track on Saab 1972-1992
R & T on Toyota Sports & GT Cars 1966-1984
R & T on Triumph Sports Cars 1953-1967
R & T on Triumph Sports Cars 1967-1974
R & T on Triumph Sports Cars 1974-1982
Road & Track on Volkswagen 1951-1968
Road & Track on Volkswagen 1968-1978
Road & Track on Volkswagen 1978-1985
Road & Track on Volvo 1957-1974
Road & Track on Volvo 1977-1994
R&T - Henry Manney at Large & Abroad
R&T - Peter Egan's "Side Glances"
R&T - Peter Egan "At Large"

BROOKLANDS CAR AND DRIVER SERIES

Car and Driver on BMW 1955-1977
Car and Driver on BMW 1977-1985
C and D on Cobra, Shelby & Ford GT40 1963-84
Car and Driver on Corvette 1978-1982
Car and Driver on Corvette 1983-1988
C and D on Datsun Z 1600 & 2000 1966-1984
Car and Driver on Ferrari 1955-1962
Car and Driver on Ferrari 1963-1975
Car and Driver on Ferrari 1976-1983
Car and Driver on Mopar 1956-1967
Car and Driver on Mopar 1968-1975
Car and Driver on Mustang 1964-1972
Car and Driver on Pontiac 1961-1975
Car and Driver on Porsche 1955-1962
Car and Driver on Porsche 1963-1970
Car and Driver on Porsche 1970-1976
Car and Driver on Porsche 1977-1981
Car and Driver on Porsche 1982-1986
Car and Driver on Saab 1956-1985
Car and Driver on Volvo 1955-1986

BROOKLANDS PRACTICAL CLASSICS SERIES

PC on Austin A40 Restoration
PC on Land Rover Restoration
PC on Metalworking in Restoration
PC on Midget/Sprite Restoration
PC on MGB Restoration
PC on Sunbeam Rapier Restoration
PC on Triumph Herald/Vitesse
PC on Spitfire Restoration
PC on Beetle Restoration
PC on 1930s Car Restoration

BROOKLANDS HOT ROD 'MUSCLECAR & HI-PO ENGINES' SERIES

Chevy 265 & 283
Chevy 302 & 327
Chevy 348 & 409
Chevy 350 & 400
Chevy 396 & 427
Chevy 454 thru 512
Chrysler Hemi
Chrysler 273, 318, 340 & 360
Chrysler 361, 383, 400, 413, 426, 440
Ford 289, 302, Boss 302 & 351W
Ford 351C & Boss 351
Ford Big Block

BROOKLANDS RESTORATION SERIES

Auto Restoration Tips & Techniques
Basic Bodywork Tips & Techniques
Camaro Restoration Tips & Techniques
Chevrolet High Performance Tips & Techniques
Chevy Engine Swapping Tips & Techniques
Chevy-GMC Pickup Repair
Chrysler Engine Swapping Tips & Techniques
Engine Swapping Tips & Techniques
Ford Pickup Repair
How to Build a Street Rod
Land Rover Restoration Tips & Techniques
MG 'T' Series Restoration Guide
MGA Restoration Guide
Mustang Restoration Tips & Techniques
Performance Tuning - Chevrolets of the '60s
Performance Tuning - Pontiacs of the '60s

MOTORCYCLING

BROOKLANDS ROAD TEST SERIES

AJS & Matchless Gold Portfolio 1945-1966
BSA Twins A7 & A10 Gold Portfolio 1946-1962
BSA Twins A50 & A65 Gold Portfolio 1962-1973
Ducati Gold Portfolio 1960-1974
Ducati Gold Portfolio 1974-1978
Laverda Gold Portfolio 1967-1977
Norton Commando Gold Portfolio 1968-1977
Triumph Bonneville Gold Portfolio 1959-1983

BROOKLANDS CYCLE WORLD SERIES

Cycle World on BMW 1974-1980
Cycle World on BMW 1981-1986
Cycle World on Ducati 1982-1991
Cycle World on Harley-Davidson 1962-1968
Cycle World on Harley-Davidson 1978-1983
Cycle World on Harley-Davidson 1983-1987
Cycle World on Harley-Davidson 1987-1990
Cycle World on Harley-Davidson 1990-1992
Cycle World on Honda 1962-1967
Cycle World on Honda 1968-1971
Cycle World on Honda 1971-1974
Cycle World on Husqvarna 1966-1976
Cycle World on Husqvarna 1977-1984
Cycle World on Kawasaki 1966-1971
Cycle World on Kawasaki Off-Road Bikes 1972-1979
Cycle World on Kawasaki Street Bikes 1972-1976
Cycle World on Norton 1962-1971
Cycle World on Suzuki 1962-1970
Cycle World on Suzuki Off-Road Bikes 1971-1976
Cycle World on Suzuki Street Bikes 1971-1976
Cycle World on Triumph 1967-1972
Cycle World on Yamaha 1962-1969
Cycle World on Yamaha Off-Road Bikes 1970-1974
Cycle World on Yamaha Street Bikes 1970-1974

MILITARY

BROOKLANDS MILITARY VEHICLES SERIES

Allied Military Vehicles No.2 1941-1946
Complete WW2 Military Jeep Manual
Dodge Military Vehicles No.1 1940-1945
Hail To The Jeep
Land Rovers in Military Service
Military & Civilian Amphibians 1940-1990
Off Road Jeeps: Civ. & Mil. 1944-1971
US Military Vehicles 1941-1945
US Army Military Vehicles WW2-TM9-2800
VW Kubelwagen Military Portfolio 1940-1990
WW2 Jeep Military Portfolio 1941-1945

1986

CONTENTS

5	Dear Santa Claus	Car	Nov	1974
10	Dino 308 GT4	Road & Track	Sept	1974
13	The Ferrari Dino 308 GT4 2+2 Road Test	Motor Sport	Oct	1974
16	Ferrari Dino 308 GT4 Road Test	Autosport	Dec 12	1974
19	Dino is Newer Better?	Autocar	Aug 16	1975
22	Fantasy Cars for Everyman Ferrari Dino vs. Maserati Merak Comparison Test	Car and Driver	June	1975
28	Ferrari 308 GTB European Test	Road & Track	Mar	1976
31	'Signor, you go more slow, eh?'	Car	July	1976
34	Ferrari 308 GTB	Car South Africa	June	1976
36	Ferrari 308 GTB: Bravura! Road Test	Autosport	July 29	1976
38	The Ferrari 308 GTB	Motor Sport	Dec	1976
41	Ferrari 308 GTB Road Test	Autocar	Oct 23	1976
47	Fazzaz on the Fly 308 GTB	Road & Track	Dec	1976
49	Ferrari 308 GTB Road Test	Motor Trend	April	1977
53	Ferrari 308 GTB Road Test	Car and Driver	Mar	1977
58	Ferrari's Little Red Rocket 308 GTB Road Test	Modern Motor	July	1977
65	You can Afford a Ferrari 308 GTS	Motor Trend	June	1978
69	Ferrari 308 GTS Road Test	Car and Driver	June	1978
75	Tale of Two Targas Ferrari 308 GTB vs. Porsche 911 SC Comparison Test	Motor	Aug 26	1978
80	Ferrari Dino 308 GT4 Road Test	Road & Track	Nov	1979
83	Ferrari 308 GTBi Road Test	Car and Driver	Oct	1980
87	Dolce as well as Presto Mondial 8	Autocar	April 24	1980
90	Spin in a Spyder 308 GTS	Motor Manual	Dec	1980
94	Ferrari 308 GTSi Road Test	Road & Track	Mar	1981
97	308GTBi, 308GTSi & Mondial 8 Buyer's Guide	Road & Track Special		1981
98	Mondial la Magnifica! Mondial 8	Car	July	1981
104	Maranello's Mondial 8 Road Test	Motor Sport	Aug	1981
107	Ferrari Mondial Road Test	Motor Trend	Nov	1981
112	Ferrari Mondial 8 Road Test	Car and Driver	Nov	1981
118	Four Seats and the Best Handling Mondial 8	Motor	Dec 5	1981
121	Ferrari with Fight	Car	Oct	1981
124	Silver Dream Machine Mondial Quattrovalvole Road Test	Motor	Oct 30	1982
130	Ferrari 308 GTSi Road Test	Road & Track Special		1983
133	Ferrari 208 GTBi Turbo	Car and Driver	May	1983
134	Ferrari 308 Owner Survey	Road & Track	June	1983
136	Ferrari 308 GTBi Quattrovalvole	Motor Sport	Mar	1983
139	Ferrari 308 Quattrovalvole Road Test	Car and Driver	Aug	1983
144	Ferrari 308 GTBi Quattrovalvole Road Test	Motor	Oct 29	1983
148	Return Match Ferrari 308 GTB Quattrovalvole vs. Porsche 911 Carrera Comparison Test	Motor	Feb 4	1984
153	Four-Value Flash 308 GTB	Sports Car World	June	1984
156	Ferrari Mondial Cabriolet Quattrovalvole Road Test	Road & Track	May	1984
160	Ferrari 308 GTB QV Road Test	Fast Lane	Feb	1985
164	Twin-Turbo Ferrari 308	Car and Driver	Feb	1985
165	Ferrari 308 GTB	Classic & Sportscar	April	1993
172	308 GT4 Owner's View	Classic & Sportscar	Oct	1985

ACKNOWLEDGEMENTS

These V8-engined Ferraris of the Seventies and Eighties continue to fascinate enthusiasts, and our earlier Road Test books on them, *Ferrari Dino 308 1974-1979* and *Ferrari 308 and Mondial 1980-1984* have sold right out. We decided to combine both books, include some recent material and issue them as one of our new enlarged Gold Portfolios. Brooklands Books are a living archive service for motoring enthusiasts, republishing valuable material which is no longer available about cars of interest. For volumes like this one, we depend on the generosity and understanding of those who originally published the copyright material we reproduce. So we are pleased to acknowledge our debt for material in the present volume to the owners of *Autocar, Autosport, Car, Car and Driver, Car South Africa, Classic and Sportscar, Fast Lane, Modern Motor, Motor, Motor Manual, Motor Sport, Motor Trend, Road & Track* and *Sports Car World*.

R M Clarke

The story of these V8-engined Ferraris began in 1973, when the 308 GT4 was unveiled at the Paris Motor Show. As a replacement for the original 246 series Dino, it once again wore Dino badges, and it shared the 246's transverse mid-engined layout. However, the 308 GT4 was a 2+2 rather than an uncompromising two-seater, and it broke new ground in having a body styled by Bertone.

The new styling was not as widely appreciated as Ferrari had hoped, although few people complained about the glorious new twin-cam 3-litre V8, which pushed out 255bhp and gave the car a 155mph top speed. At least, it did for most countries; the emissions-controlled version homologated for the USA for 1975 mustered only 240bhp, and for the Italian market (where petrol engines bigger than 2 litres are heavily taxed) there was a 1.991cc V8 with just 155bhp in what was badged a 208 GT4. Sales were steady but not spectacular. The Dino badging was dropped in May 1976, and production began to tail off during 1978. The last of 2,826 308 GT4s was built in January 1980.

Its replacement was the Mondial 8, announced at the Geneva Show in March 1980. This car had much more attractive Pininfarina styling and a longer wheelbase which gave both more passenger space and more luggage space. Power came once again from the 3-litre V8, which now benefited from fuel injection and electronic ignition, but with only 214bhp (205bhp for the USA), this heavier car was also a disappointment to Ferrari fans. It survived until 1982, by which time just 703 examples had been built.

Ferrari had been quite clear about the problems with these cars; the 308 GT4 was not sufficiently attractive, and the Mondial 8 not fast enough. So the new car which arrived in summer 1982 retained the body of the Mondial but substituted a 240bhp 32-valve engine (which lost 10bhp in emissions-controlled US form). This was much more like it, and sales of the Mondial Quattrovalvole showed that Modena had got the formula right this time. When a gorgeous cabriolet version joined the original coupe in 1984, sales practically doubled overnight. By the time the 3-litre engine gave way to a 3.2-litre edition for 1986, a total of 1,145 Mondial Quattrovalvole coupes and 629 cabriolets had been made.

To the buyer of a new Ferrari, it matters a great deal whether his (or her) car offers more performance and better looks than other sporting exotics on offer at the time. That is why the 308 GT4 and Mondial 8 lost out to rival makes when they were new. But to the enthusiast today, all that matters is whether the car is attractive in its own right. There are plenty of Ferrari enthusiasts who number the models in this book among their most coveted cars, and they will find plenty to feed their passion in its pages.

James Taylor

DEAR SANTA CLAUS...
Whereby Mel Nichols spends two days with Ferrari's new 308GT4—a Dino with REAL sting!

SO THE JEWEL CALLED THE DINO 246GT has come and gone, and in its place we now have a bigger and heavier car of quite different shape, powered not by the traditional Dino V6 but by a brand-new V8. A car that, if you are to believe a self-appointed critic who accosted us in a garage, lacks the character and instant appeal of the 'proper Dino'. The snarl of the 3.0litre V8 firing up drowned whatever else he had to say, but he did at least turn and listen appreciatively as the car accelerated our ears beyond his clutches. His reaction was understandable, of course—the Dino 246GT, the car that has come closer than most to being all things to all men, is a tough act to follow.

It isn't clear why Ferrari killed the V6. It was popular, and priced for relatively high sales volumes. On the other hand, it was strictly a two-seater and potential buyers were asking for more room—if only for oddments stowage—and some wanted more power. Ferrari's answer was to create a 2+2 Dino and endow it with a V8. But most people thought this new car would supplement the Dino V6, not replace it, and that seems to be the main reason why the Dino 308GT4 has come onto the market rather innocuously. No doubt its styling has contributed, too. There is, indeed, great similarity between the Dino's lines and those of the two-year-old Lamborghini Urraco, also Bertone styled. There are few ways of tackling the problem of styling a mid-engined 2+2 with good visibility and practicality.

So styling *is* the Dino 308GT4's most controversial point. From the front and sides it is well-proportioned, meaty and aggressive. From the rear it is fiddly and unattractive, rather like a woman with a narrow, pinched backside. Overall, it is neat and effective with just enough meanness lurking in its lines to make it identifiably Ferrari. And it's a car that's proving very popular: within two months of the car becoming available in Britain, 30 Dinos have been sold.

Their owners, like CAR, have no doubt discovered that this new Ferrari is *not* the Dino with middle-aged spread. It may have come about with little impact—but it certainly has plenty when you drive it. Enough, in fact, to suggest that it is probably closer to a modern-day Miura than anything else! Its V8 engine and very thorough design and development are the keys to it.

Like most of its Ferrari sisters, the 308 has a tubular steel chassis with the steel body bolted onto it. After that, it departs markedly from Maranello tradition: it is styled by Bertone rather than Pininfarina, it is the first 2+2 mid-engined Ferrari, and it carries Ferrari's first roadgoing V8. Compared with the Dino 246, it is 8in longer in the wheelbase (at 100.4in) and 6in longer overall. The 57.5in track is just over an inch wider, the 71in width is 4in up and the 46.5in height makes the 308 2.6in taller than the old Dino., Weight? Ferrari's brochure puts it at 2536lb, but the owners' handbook says 3009lb, which is the all-up total for a fully optioned car with air conditioning.

The main connection between the new and old Dinos—name apart—is in the general mechanical layout. The engine/transmission unit sits sideways across the car just ahead of the rear wheels; there is a front-mounted radiator, a little room under the nose to carry oddments as well as the emergency spare wheel, and a boot in the section aft of the engine for the real luggage. Suspension is similar for both cars, with two wishbones, coil springs and Koni dampers doing the work at all four wheels. The rear dampers are in unit with the springs, mounted on the upper wishbones (which are fabricated steel, not cast alloy as in the bigger Ferraris). Anti-roll bars are used front and rear, and the wheels and tyres are the same on the new Dino as they were on the old: cast Cromodora wheels, 14 by 6.5in, wearing 205/70VR Michelin XWX radials.

The engine, however, has few common points with the old V6: Ferrari started with a clean sheet of paper. The banks of its V are at 90degrees, not 65degrees (so it takes up more space in the engine compartment) and the bore and stroke are rather less oversquare at 81 by 71mm. The total capacity is 2926cc, and the compression

Hard-cornering 308 (opposite) reveals barest trace of roll despite throttle-provoked oversteer. Transverse V8 is reasonably accessible (right) but shrouded by air cleaner. Extensive instrument display and minor controls (below) are housed in wrap around section of fascia

ratio inside all that magnificent alloy is 8.8 to one, a little down on the V6's. The V8 has four cams of course, with delicious black crackle covers hiding their operation and toothed belts rotating them; Ferrari now seem content with belts rather than the chains used previously. (You change them every 24,000 miles, but they don't need adjusting in the meantime). Four twin-choke down-draught Webers pour in the fuel, gulping their air from pipes connected to the inlets just behind the rear side window. The five-speed transmission snuggles beside the alloy wet sump, with the limited-slip differential coming right in behind it, so that the axles, as they spear out to the wheels, are just about in line with the outer edge of the rearmost cylinder bank: the power unit is still *that* compact. Power from it all is 255DIN bhp at 7700 rpm.

Getting into the Dino is initially daunting because the driving position is well forward and appears to offer little room. That is an erroneous impression, however. The seat may be well forward, but so are the steering wheel and the pedals. It is a good driving position, a little long in reach to the wheel, in the Italian supercar tradition, and fractionally offset to the left, but it is one into which you settle rapidly and easily, encouraged by the superb vision (the bonnet drops away so sharply you see very little of it). The only problem is the incredibly long throw of the clutch; one must really stretch the left leg to depress it fully. It is tiresome while the gearbox is cold for it must then be pushed all the way down. When everything is warm and sweet, you can get by perfectly with a three-quarter depression.

Even if it has been standing all night, firing up the engine is simplicity itself. Just a little choke, a good twist of the key and a hearty prod on the throttle brings the V8 snapping into life. Re-adjust the choke, feel the engine smooth out beneath your right toes and then move your hand across to the slim metal lever for first, which is over and back in the shiny metal gate. The top four rest in the H-pattern. A business-like change with reverse (opposite first) locked out by a little tit that requires depression of the lever to overcome it. The clutch is strong, but the take-up is sweet and easy, balanced perfectly against the revs which are, in their turn, controlled beautifully by one of those throttles that tells you everything and brings from the engine a response precisely in accordance with your wishes. Nothing less, nothing more.

As the gearbox warms and the need for that long clutch push fades, you start to feel very self-confident in the Dino, assisted by the marvellous lack of fuss from the engine, the ease of the gearchange and the excellent vision. The car feels tight and neat about you; it is very easy to place in traffic and in tight areas. Within a few miles, you're using the power of the engine and the responsive handling to carve through the traffic, nipping and darting, as if you're in a 1275 Cooper.

And what power there is in that engine! You expect performance in a Ferrari, but this V8 comes as a surprise because not only is it mail-fisted but its power is spread over an enormous rev range. There are no flat spots, not the slightest trace of camminess; the engine just gives more and more and more power as it revs. It is just like turning up a dial. The answer is the torque that just wasn't there in the V6—210lb/ft at

Rear seats (below) will carry adults, but are best with children. Front compartment (right) is comfortable and practical, but intrusive wheels arches necessitate offset pedals. Small, soft bags can be tucked around emergency spare (bottom) but main luggage area is in aftmost section of tail (bottom right)

5000rpm, compared with 166lb/ft at 5500rpm and all in a car that weighs only 250lb more.

This means that compared with the V6, the Dino 308 feels a much meaner, more potent car. Meaner isn't really the correct word because there is nothing intimidating about the thing. It's just that you *know* it has real go, and you delight in the swift blast of using it, unleashing it. In the 246 the enjoyment came more from sweetness and not the sheer sting. This new Dino has suddenly become a very much alive and exciting car. Mentally, you applaud the Ferrari engineers for the masterful job they've done with the V8. Its flexibility is outstanding—it pulls happily from 1000rpm and 22mph in fifth—and its smoothness is remarkable. An excellent device for city driving, whether blasting past slower cars into the gap you've spotted way up ahead, or simply for pottering in fifth as you talk with your passenger.

Out on the open road it is even more enjoyable as you exploit the excellence that is embodied in the car. At the 7700rpm redline, the gear ratios give 45, 65, 91 and 123mph, with fifth running out to 155 at just under 7400rpm. Performance is there aplenty: anytime you like you can reach 60mph from a standstill in 6.2seconds and 100mph in 16 seconds. That, of course, leaves the Dino's Merak and Urraco opponents gasping in its wake (although the forthcoming quad-cam Urraco should go a long way towards levelling things up). Among 2+2s at this money, only the Porsche Carerra is quicker in acceleration, although it isn't as fast at the top end. And the Dino is quicker than a Bora, and not far behind a Pantera.

More important than the single fact that the Dino has a lot of performance is the point that it can be used a great deal of the time. The gear ratios are thoroughly well suited both to cut and thrust in traffic and to very fast coverage of country roads. On paper, second might look a little low; in practice, the torque of the engine is so great you never feel that there is a gap between second and third.

Just as one expects a Ferrari to go hard, one also expects it to handle—especially if it is a mid-engined Ferrari. The Dino handles as easily as its engine performs; there is the characteristic touch of understeer, but the faster you go the more it is neutralised. There's medium weight in the steering at town paces, but with speed it becomes perhaps a fraction too light and, even though it is rack-and-pinion, it manages to have the slight lack of feel that is common to most

Ferraris. As it stands there is just enough feel to give adequate warning when the front wheels are losing their grip coming into those tight, wet bends that always catch out mid-engined cars. Into sharp corners, the 3.2turns lock-to-lock steering feels slightly low-geared, but it is fine for fast roads with sweeping bends. Along such roads, the Dino is sighted tight into the apex and then the power is squeezed on—hard. But herein may lie a trap for the unwary if they hark back too much to the old Dino. The V8 has such potency that power oversteer is *very* readily available; it can be induced in a way that it never could in the V6. It can come unintentionally until you realise the extent of power reaching the wheels in second or third gears. This may not be so good for those who buy Ferraris for the name rather than the pleasure; for those who buy the cars to drive, it is excellent. You're on your mettle, because you have a machine that will betray if you're careless, but it is a machine that you can use to delightful ends because of it. It is thus a more sporting car than the late lamented Dino 246.

Even in strong crosswinds on motorways, the 308 is extremely stable. Similarly, it is unaffected by mid-corner bumps on backroads. It is thus a very refined, most enjoyable car to drive as fast as possible as often as possible. Brakes are excellent, needing a good firm push to get them working but never fading.

Ride? Apart from the engine, the ride is probably the Dino's most outstanding feature. The moment you move off you note the remarkable lack of low-speed harshness and noise. Cars of such high dynamic qualities usually have both, and are accepted for it. That the Dino soaks up bumps so magnificently without the occupants ever being aware of the powerful damping, shows just how thorough Ferrari's design and development has been. Progress over the earlier Dino, and opposition makes, is great. Unfortunately, suppression of wind noise isn't so good: it starts to intrude above 130mph with some buffeting around the window frames. Engine noise is there all the while, of course. It is pleasant, with just a touch of savagery to remind you of the power creating it, but it would be uncomfortable if it were louder.

Seats in the Dino are comfortable, complementing the ride very well. Cloth covering is standard; leather is a £315 option. With the front passengers seat run forward—generally uncomfortably so—it is possible to carry an adult in the back. Mostly, however, the rear seats are for children unless an adult sits across the back rather than in one seat. Most owners will probably use the seats not so much for extra people as for oddments and extra baggage; a cure for the most serious Dino 246 deficiency. For the driver, once the long clutch throw has been accepted as a fact of the car, which is overcome when the engine is hot anyway, the driving position is very good. Comfortable, relaxing, it sets him up well to drive the car as hard as it will go. Lateral location in the seat is adequate, and left foot comfort is looked after by the platform screwed to the small central tunnel. The good all-round vision is the best yet in a mid-engined car, the detachable top Fiat X1/9 (also a Bertone design) apart.

The gearchange is crisp and positive, among the best we've sampled in a mid-engined car and there are few faults with the minor control layout. Main functions are looked after by the three steering column stalks, although pulling one of them back to wash the screen doesn't bring on the wipers simultaneously, as it should do. For Britain, Maranello Concessionaires wire the fog lights up so that they can be flashed in daylight without you having to worry about bringing up the headlights. The instrument panel wraps around, rather like the new Espada's, so that the rest of the switchgear can be set into its extremities and is thus easily within the driver's reach. Instruments —electronic tachometer AND speedo, water and oil temperature, oil pressure, fuel gauge and a small clock—are carefully arranged so that none is obscured by the wheel, a nasty fault in the old Dino, and in the BB. The care embodied here seems to typify the thought and attention that has been put into the entire car.

The vast windscreen means that the cabin becomes hot even in days of minimal sun, and the ventilation is not good unless the optional air conditioning (£351) is fitted. With this, the car is pleasant; but switches for the temperature and fan controls are just a little too far forward on the central tunnel. Electric windows and tinted glass are standard features, and the windows work surprisingly rapidly.

Boot space? Not bad; about as good as in the Urraco. The rear one won't take large bags because it isn't deep enough, but it will take a goodly amount of smaller ones. There is room under the bonnet for a little soft luggage so long as you don't have a puncture. The road wheel, substituted for the narrow section spare, would then take up all the room. Fuel tank is a useful 17.6gallons—but since the consumption is a relatively hefty 15mpg most of the time, the range is a less impressive 264 miles. With very hard driving, the fuel consumption will increase to around 13mpg. On the other hand, gentle usage doesn't seem to take the figure much beyond 16.5. So Ferrari's quoted 15 to 16mph is quite correct, even if disappointing for owners. Or is it, considering the performance?

Other pleasant details about the Dino are the fact that the windscreen wipers have been sited for rhd, and the engine and boot compartment release levers have been shifted across to the right-hand door pillar. Engine accessibility cannot be expected to be an optimum feature, but in the Dino it is better than average among mid-engined cars. The twin coils, the distributor, the oil filter and the carburettors are all easily reached. The only hard to get at items for routine servicing are the plugs in the forward cylinder bank. Faults include boot and bonnet stays that require manual releasing, rather than self-tensioning struts. And the slats in the engine cover permit water, in a heavy downpour, to drip down onto the rear row of spark plugs, getting past the rubber insulating meant to be protecting them and making starting difficult and running erratic afterwards. A cure for this is on the way.

The turning circle, a terrible 41ft, cannot be cured and parking and low-speed manoeuvring are ponderous.

And such are the faults of the Dino; a scant few. This is a thoroughly impressive and oh-so-desirable car, a car that may have come quietly but is certainly going to make its mark now it's here. It is distinguished by exceptional ride, thoughtful design and careful development, by the engine's remarkable flexibility and of course its solid performance. At £8000, offering what it does, it seems like remarkably good value. You can indeed think of it as truly worthy replacement for the Dino, one that is even more enjoyable to drive. Or you can think of it as a more sophisticated and thoroughly refined alternative to outright street-racer machines like the Porsche Carrera. You could think of it as delightful alternative to things like the BMW CS range, which it outperforms, out-handles, damned-near betters for fuel consumption and the cost is near-enough the same.

You can even think of it, perhaps, as a mid-'seventies version of the Miura: not quite as potent, not quite as ferocious and demanding, but a good deal more refined and practical.

Dino 308GT4 performance

Gear maximums	(7700rpm)
First	45mph
Second	65
Third	91
Fourth	123
Fifth	155 (7300rpm)

0-30	2.3secs
40	3.4
50	5.0
60	6.2
70	8.1
80	10.2
90	12.8
100	16.0

Speedo 8% inaccurate at 100mph.

DINO 308 GT4

A worthy addition to the long line of great GT cars from Ferrari

BY PAUL FRÈRE

Dino BELIEVE IT OR NOT, but it is a fact that Ferrari does not keep any sort of demonstrator, and trying to get hold of a new model is invariably a major problem. Past experience indicates that this is not caused by lack of good will, as every time I have been lent a car by the factory it came straight from the assembly line—not the best way to achieve first class performance figures—after which it was supposed to serve as a demonstrator for some dealer.

The same old problem cropped up again when it came to trying Maranello's latest creation, the Dino 308 GT4. Then I was informed that Jacques Swaters, the Belgian importer who runs Ecurie Francorchamps that has done so well so often at Le Mans, had one available and was willing to lend it to me for two days. This was time enough to add 1000 miles to the odometer of a car that had the further advantage that it could not be suspected by anyone of being a specially prepared sample. Germany is (I hope only for the time being) the only country in Europe where a car of this sort can be extended without risking huge fines, so I made the test coincide with a visit to Wolfsburg. I am indebted to Anton Conrad, the Volkswagen press officer, for offering me the facilities of VW's magnificent high-speed track. All the performance testing was done there with a minimum of fuss; I even used their test equipment: speed traps, fifth wheel with recording apparatus, etc.

Enzo Ferrari once told me, "A Ferrari is a 12-cylinder car." Consequently, the 308 GT4 isn't a Ferrari and that's why it's called a Dino. It has only eight cylinders, but by any other standard, including performance, it is a Ferrari. In fact, it's a lot quicker than the short-lived 365 GTC4 and it has no trouble keeping up with any other fully road-equipped exotic supercar I have tried except a Daytona, the fastest Corvette, and a lightweight Carrera which the Dino matches in speed but not in acceleration. But then the lightweight Carrera cannot be fairly compared with the luxuriously equipped Dino.

In many ways the Dino departs from the Ferrari tradition, the most obvious novelties being the use of an 8-cyl engine for the first time in a production Ferrari and that Bertone has been entrusted with the design of the body rather than Pininfarina. It is also one of the rare central-engine cars featuring rear occasional seats. The emphasis is very much on the word occasional, however, and though there may be room for two small children or to take an adult seated across the car to the movies, it requires a good deal of optimism (and bad faith) to call the Dino a 2+2. But certainly the so-called rear seats are a good answer to the justified criticism that mid-engine cars are infuriating for lack of oddment space, and they might be quite useful to accommodate any luggage that cannot be swallowed by the smallish rear luggage compartment.

In its general layout the 8-cyl Dino is very similar to its smaller 6-cyl stablemate with the engine and transmission unit set across the car just ahead of the rear wheels. Being a 90° engine, it takes up a little more longitudinal room than the 65° V-6; Ferrari now seems satisfied that cog-belts are a reliable and more silent alternative to chains for driving the four overhead camshafts. The running gear, too, is very similar to its forerunner: unequal-length wishbones front and rear and the same 6½-in.-wide alloy wheels shod with Michelin XWX high-speed radials.

Whatever you think of the body shape, it certainly provides a good driving position, keeps the car firmly down at speed, and produces little wind noise. First impressions at the wheel, apart from the comfortable position, are of a heavy clutch (which, I understand, is being attended to), an excellent and precise gear change by a thin lever moving in a visible gate, and an amazingly flexible and docile engine. This flexibility, combined with the respectable output of 83 bhp per liter, is perhaps the most remarkable and endearing feature of the car. There isn't the slightest problem in driving the Dino in heavy town traffic, and it makes life so easy on motorways where, after having had to slow down to 80-90 mph, all you have to do is stay in 5th gear and push your right foot down to zoom past the car that got in your way. In fact the engine will accelerate in 5th from 1000-1100 rpm (25 mph) and soar right up to its maximum speed of 152 mph without flat spots or cam effects being felt. Maximum speed is reached at an

The 2.9-liter V-8 has four camshafts (driven by toothed belts) and four Weber carburetors, is mounted transversely with the 5-speed transaxle.

indicated 7300 rpm (in fact a calculated 6700) and the engine is safe up to 7600, providing an exceptionally wide useful range.

Obviously, though, some advantage is gained by changing down and the ratios are well chosen, giving useful maxima of 44, 63, 87 and 119 mph in the lower gears, which were the change-up points used for getting the acceleration figures. First gear is near the driver and back, the other four being in the usual H pattern, and the change from first into second is rather slow in contrast to the others.

Handling is typically Ferrari and somehow, even though their central-engine cars use rack-and-pinion steering rather than the worm gear of the front-engine models, the people in Maranello have managed to retain that somewhat dead feel that has been a characteristic of all Ferraris for more than two decades. You may like it or not, but the friction damping introduced certainly reduces kick-back more than is usual with a rack-and-pinion mechanism, and the control is very precise with good self centering as soon as the car gathers speed. Though it's far from finger-light, the steering never becomes very heavy, but maneuvering is really hard work because of the appalling steering lock. I would also favor higher gearing because the 3.5 turns from lock-to-lock would correspond to 4.5 or more with a decent turning circle. When cornering, the typical Ferrari understeer is evident, but the faster you go the less it becomes and there is a reassuring feeling of stability and safety in fast corners taken near the limit. Lifting off makes the car take a slightly tighter line just as it should, and straight-line stability is nearly perfect, the car showing little sensitivity to side winds. Quick changes of direction are effected without fuss, as you would expect with a low polar moment of inertia, but some of the inherent agility is lost by the low-geared steering. The expert deplores it, but it may be just as well for the normal consumer.

With four ventilated discs and a huge servo, the brakes seem to be well up to their task. They require a good push for an emergency stop—much better than to be over-sensitive in a fast car—but with hard pads, the lack of initial bite when cold can be a bit disconcerting. Unfortunately, we had no chance to drive the car on a race track to find out how long the brakes would remain fade-free, but on the road there wasn't the slightest trace of fade, though heavy rain increased the response delay. In the wet, the grip on the driving wheels remains very good, and despite the rear weight bias the car is not particularly prone to aquaplaning.

Apart from the performance, which you take for granted in a Ferrari, and the aforementioned remarkable flexibility of the engine, perhaps the most outstanding feature of the Dino 308 is the excellent ride it provides. There is no low-speed harshness and at speed the road irregularities are beautifully smoothed out, damping being so good that oscillations are virtually non-existent, though passengers never become aware of the powerful dampers and praise must go to Koni as well as to Ferrari. On rough roads and more particularly cobblestones, the impression of comfort is enhanced by the commendably low level of road noise despite the high-speed Michelins, which are not renowned for their sound-absorbing properties on such surfaces. The progress, compared to earlier Ferraris, is enormous. Noise from other parts could still be lowered, however. Though the body is obviously well shaped to reduce wind noise, the latter increased noticeably above 125 mph when the window frame on the driver's side was pulled away from the rubber seals by the low pressure on the outside. This was probably a matter of adjustment, as it didn't happen on the other side, and generally speaking I was told the body was still very "prototype" and several details would be changed or already had been modified in current production. One point that is certainly worth directing some attention to in the course of future development is the intrusion of engine noise into the interior. Conversation becomes difficult at speed and the radio completely useless; the Maserati Bora proves that a mid-engine car doesn't have to be noisy.

Except for the console-mounted radio, which is too far from the driver for comfort, all controls and instruments are well laid out. Electric window lifts are standard, the pedals allow easy heel-and-toeing and at last here is a car with pop-up headlights (other than the Opel GT) which is fitted with additional long-range lights for daylight flashing. It seems something of a waste, though, that they can't be switched on to supplement the otherwise excellent twin driving lights.

About the worst feature is the heating and ventilating system. Warm-weather ventilation should be all right, but in cold or cool weather there is just no alternative but to be cooked or frozen. I tried every combination for regulating the flow of water through the heater element and the flow of air without ever achieving the desired result. Surely it is a shame that in a car of this class the heater temperature is "adjusted" by modifying the flow of water rather than by mixing warm and cold air. I suppose the only way to achieve real comfort is to buy a car with the optional air conditioning, though an acceptable makeshift solution is to have the heat on and open a window a few inches. Luckily this is one of the few cars in which the windows can be opened quite wide without increasing wind roar noticeably or creating annoying drafts.

Safety belts are a disappointment too. The anchorage points of the 3-point belts are so badly chosen as to make the belts practically useless, and in a car of this class and price you'd expect automatic inertia reel belts rather than the manually adjustable type that get into people's way. There are lap belts for the rear seats but they are really only for children; adults will find the seating extremely uncomfortable when fastened in properly.

Engine accessibility is never a strong point in central-engine cars, but routine servicing is reasonable in the Dino except for the sparkplugs on the bulkhead side cylinder head, and I would not like to be the mechanic entrusted with adjusting the valves on that part of the engine. The exhaust system is a cleverly designed bunch of spaghetti-like tubes leading into an enormous transverse silencing box neatly fitting behind the power unit. Opening the front lid discloses the radiator, the light-alloy pedal carrier to which the brake servo is bolted, the generous tool kit and the emergency, thin-rimmed spare wheel fitted with a thin, high-pressure tire. Thanks to this shallow wheel, there is some space left for additional flat luggage.

Making up the sums after driving the Dino for some 1000 miles, mostly on motorways and as fast as circumstances permitted, the fuel consumption turned out to be rather a good surprise at 12.5 miles to the U.S. gallon, including the performance tests. Relating this to weight and performance, the people in Le Mans would call the Dino a likely efficiency-index winner. U.S. drivers should get better mileage.

PRICE	
List price	est $19,000

ENGINE & DRIVETRAIN	
Type	dohc V-8
Bore x stroke, mm	81.0 x 71.0
Displacement, cc/cu in.	2926/179
Compression ratio	8.8:1
Bhp @ rpm, net	242 @ 7700
Torque @ rpm, lb-ft	na
Fuel requirement	premium
Transmission	5-sp manual
Gear ratios: 5th (0.89)	3.14:1
4th (1.20)	4.20:1
3rd (1.62)	5.67:1
2nd (2.23)	7.82:1
1st (3.23)	11.30:1
Final drive ratio	3.50:1

CHASSIS & BODY	
Body/frame	tubular steel chassis, steel panels
Brake system	vented disc, vacuum assisted
Wheels	cast alloy, 14 x 6½J
Tires	Michelin XWX, 205/70VR-14
Steering type	rack & pinion
Turns, lock-to-lock	3.5
Suspension, front & rear: unequal-length A-arms, coil springs, tube shocks, anti-roll bar	

GENERAL	
Curb weight, lb	2930
Wheelbase, in.	100.4
Track, front/rear	57.9/57.5
Length	170.0
Width	71.0
Height	46.5
Fuel capacity, U.S. gal.	22.0

CALCULATED DATA	
Lb/bhp (test weight)	13.3
Mph/1000 rpm (5th gear)	22.2
Engine revs/mi (60 mph)	2700

ROAD TEST RESULTS ACCELERATION	
Time to distance, sec:	
0-100 ft	3.0
0-500 ft	8.0
0-1320 ft (¼ mi)	14.6
Speed at end of ¼ mi, mph	91.5
Time to speed, sec:	
0-30 mph	2.2
0-50 mph	5.1
0-60 mph	6.4
0-80 mph	10.5
0-100 mph	16.5
0-120 mph	24.6

SPEEDS IN GEARS	
5th gear (6700 rpm)	152
4th (7000)	118
3rd (7000)	87
2nd (7000)	63
1st (7000)	44

FUEL ECONOMY	
Hard driving, mpg	12.5

SPEEDOMETER ERROR	
100 km/h indicated is actually	93
200 km/h	183

ROAD TEST

The Ferrari Dino 308 GT4 2+2

THE FERRARI DINO 308 GT4, introduced last Autumn, is more than a big brother to the familiar, beautiful little 246GT. It is an entirely new car, the few things in common being restricted to the steering wheel and alloy road wheels and it is particularly significant in Ferrari folklore in being the first Maranello production car to adopt a V8 engine. British journalists have been waiting eagerly and frustratedly for the opportunity to try this 156 m.p.h. Dino, so we are especially grateful to David Grayson, the Managing Director of Maranello Concessionaires Ltd., for ending our personal frustrations by loaning us his one and only 308 demonstrator for an afternoon.

Power unit and performance apart, the 308 will be welcomed by Ferrari customers, particularly those whose families have outgrown their 246 GTs, for having a very genuine 2+2 seating configuration within almost as compact a total area as the definitely two-seater 246 GT. In addition the tail area behind the mid-mounted, transversely-directed, all-aluminium V8 contains a thoroughly practical luggage boot (wide enough and deep enough for several sets of golf clubs!) to obviate the holdall and toothbrush image of the 246.

As with the 246 (2.4 litres, 6-cylinders), the 308 title indicates the capacity and number of cylinders. The former's 65 degree V6 engine has very little in common with the new 3-litre V8 which has its ancestry in the 4.4-litre V12 engine contemporarily best known in its Daytona application. However, whereas the V12 is of the usual 60 degree vee-slant, the arrow has been broadened in the new engine to the 90 degrees most suitable for a V8. Bore and stroke dimensions are identical to those of the V12 at 81 mm. x 71 mm. (92.5 mm. x 60 mm. are the even more violently oversquare dimensions of the V6), presenting a capacity of 2,926.9 c.c. Wet liners are used in the alloy cylinder block, the crankshaft runs in five main bearings and is lubricated from a wet sump system, while the transaxle upon which the engine is mounted contains five forward and one reverse gears, a limited slip differential and its own lubricant.

Cylinder head design follows similar practice to the V12, each head carrying two overhead camshafts operating two valves (vee-slanted at 46 degrees) per hemispherical combustion chamber via thimble tappets. Rubber, toothed timing belts received the ultimate blessing when Ferrari adopted them instead of his usual chains for the Berlinetta Boxer flat-12 engine and the 308's four camshafts are similarly driven to eradicate the familiar Ferrari valve-gear thrash. Ignition is by a separate Marelli distributor on each bank, driven by the uppermost (in terms of the angled heads) camshafts, and twin coils. A modest 8.8:1 compression ratio is used and 98 to 100 octane rating fuel specified, fed to the engine via a single Corona fuel pump and four, twin-choke, downdraught Weber 40 DCNF carburetters from a 17.3 gallon (including 3.3 gallon reserve) tank.

Undoubtedly the 308 must have gained weight in comparison with the 246, but I cannot believe the pound and kilogramme figures quoted in the respective official Ferrari handbooks I have in front of me, which when converted make the former a real heavyweight at 26 cwt. 97 lb., against 21 cwt. for the latter. As this would give the 308 a poorer power-to-weight ratio than the 246 and would consequently remove the credibility of Ferrari's claimed performance figures, I am inclined to prefer the Maranello Concessionaires' brief specification sheet claims of 23 cwt. for the new model and 21½ cwt. for the old. All the foregoing refer to dry weight. Similarly I have various power output claims in front of me: 255 b.h.p. at 7,600 r.p.m.; 255 b.h.p. at 7,700 r.p.m.; 250 b.h.p. at 7,600 r.p.m. 250 b.h.p. at 7,700 r.p.m.; quoted as DIN, SAE, net or gross in all the permutations. I leave the reader to take his pick, for any of these figures is outstanding and compares with the 195 b.h.p. (DIN or SAE?) at 7,600 b.h.p. of the

smaller V6. Torque figures for the new engine are equally impressive: 209.76 lb. ft. at 5,000 r.p.m., against the V6's 165.5 lb. ft. at 5,500 r.p.m., both figures quoted from the Ferrari handbooks and probably gross.

Even after several years of familiarity (and to date 700 have been sold in Britain) there are people who positively drool over the curvaceous, sensual lines of the Dino 246, a modern day classic and undoubtedly one of the most beautiful designs to be put into production in the history of the motor car. Alongside this Pininfarina masterpiece, Bertone's efforts on the 308 are slightly disappointing, the smooth, delicate mouldings having given way to the harsher, more acute, sharply-edged contours which seem to identify all designers' thoughts for the 70s. Nevertheless, the outcome is attractively, excitingly exotic, and those who regard it as an ugly duckling alongside its older stablemate can always console themselves with the thought that any attempt to adapt the lines of the 246 into a 2+2 would probably have been disastrous. On this subject of the Italian design war, it is interesting to note that Bertone was responsible for the sharp-edged 186 m.p.h. Lamborghini Countach, while Ferrari chose Pininfarina to shape the more rounded features of the 188 m.p.h. Berlinetta Boxer, these two cars disputing the war for the World's fastest production car.

The slab sides make the 308 look much wider than the 246 and present a comparatively vast amount of elbow room inside, yet the overall width is only 0.12 in. greater at 5 ft. 7.32 in. Those two rear seats have been accommodated by building the 308 with an 8 in. longer wheelbase, yet the sharply cropped nose and tail have kept down the overall length, at just over 14 ft., to 2½ in. more than the 246. Of necessity, for rear seat headroom, the 308 is some 3 in. higher, while the front and rear tracks have been increased by almost 2 in. to improve stability with the longer wheelbase. The suspension layout is identical to that of the 246 though there are detail changes in geometry and none of the parts appear to be interchangeable. In brief, both ends gain their impeccable handling and roadholding characteristics from pressed-steel wishbone arrangements, with coil springs/Koni telescopic damper units and anti-roll bars. The model was so new to the Concessionaires that there was uncertainty as to what improvements might have been made to the brakes on the 308, for those who regularly drive both models agree that those on the heavier 308 are much better. No disc sizes or swept areas have been published by Ferrari for the 308, though they have done so for the 246. However, ventilated outboard discs are fitted all round, a tandem master cylinder and vacuum servo-unit are mounted on the front bulkhead and the rear circuit incorporates a pressure limiting valve. The central handbrake lever operates on the normal rear pads, so is self-adjusting.

Maranello Concessionaires seems to have become a haven for retired racing drivers, for with such a high performance marque it is essential that the sales staff's driving ability and technical knowledge should be of impeccable calibre. Thus they have Mike Salmon as Sales Director (who this season has come out of retirement on several occasions to race the historic Aston Martin Project 212) and Mark Konig, he of Nomad sports-racing car fame, as Assistant Sales Manager. The latter gave me a brief introduction to the car along the Egham by-pass before handing over to me for an acclimatising canter round the lanes. There is no such thing as being let loose with one of Maranello's demonstrator Ferraris, if only because of insurance technicalities, so with Mark Konig returning to the sales room, Sue the receptionist had to be persuaded into the passenger seat to validate Maranello's policy, whilst we searched Surrey for photographic locations. (I might add that MOTOR SPORT has its own test car policy in any case). Unfortunately, such an arrangement isn't conducive to learning the car properly nor driving it in accustomed fashion to get a positive idea of how such a high performance car behaves when it is "performing highly". Impressions thus gained might

be improved upon or, highly unlikely in this instance, made worse with longer acquaintance.

The episode did serve to prove that the rear seats, or at least one of them, is suitable for adults, the photographer finding that with his knees splayed one either side of the passenger seat he was perfectly comfortable. The same would not have been true behind the driver's seat, without the driver choosing an uncomfortably short-armed position. For children even up to young teenage size they should be ideal. Unlike the Porsche's occasional seats, the Dino 308's rear seats are the real thing, luxuriously shaped and trimmed, with headrests on the rear bulkhead and with lap-strap seat belts as standard, stowed away in central and side open lockers when not in use. The front seats are a big improvement on those of the 246, mainly because the absence of a bulkhead immediately behind them has enabled them to have adjustable

FERRARI 308GT

back rests, accurately adjustable with a knurled knob, whilst a separate release lever allows them to be folded forwards for rear seat access. Adjustable headrests are included. Trimming of all the seating surfaces and the door panels is in realistic artificial suede. I found neither the seat nor the driving position quite so comfortable as the latest Porsche, the thin-rimmed, leather-covered wheel being less vertical and closer than I would have liked while having the seat adjusted to cope with the long travel of the initially heavy clutch pedal.

Dino 246 drivers will gain two immediate impressions on transferring to a 308. First the good one: the torque and flexibility of the V8 is quite wonderful. On the one hand this impressive engine will rev. to the definitive 7,700 r.p.m. red line on the Veglia tachometer, an utterly staggering number of revs. for a road-going production V8, and on the other it will crawl along and pick-up smoothly from just over 1,000 r.p.m. in fifth gear, equal to 22—25 m.p.h. The torque curve peaks at 5,000 r.p.m., but must be comparatively gentle in its shape all the way from 3,500 r.p.m., for mid-range acceleration is excellent. Thus it is not essential to stir the gear-lever incessantly for good performance in the manner one would have to do with the dynamo-like short-stroke V6. For out-and-out performance it pays to waggle the lever through its traditional Ferrari five-speed alloy gate, as maximum power doesn't occur till round about maximum revs—and through the gears the needle hits that in a couple of blinks of an eyelid. It would be wise to keep one of those eyelids open and glued to the mirror if so doing (and the mirror, stuck to the top of the steeply-raked screen is far too close to one's eye in any case), for even at 7,000 r.p.m. claimed speeds in the gears are 41 m.p.h., 59 m.p.h., 113 m.p.h. and 148 m.p.h. Another 400 r.p.m. are theoretically available in fifth to give the claimed 156 m.p.h.

Now for the bad impression. Ferraris of any cylinder configuration have been renowned for the beautiful noises they make, music to the ears whether in the cockpit or behind the exhausts. The ultimate example must be when *Motoring News'* Editor had a letter recently from an American asking whether it would be possible for him to drop a V12 Ferrari engine into his road-going Lola T70 because, "I just love that Ferrari noise"! The 308 sounds like . . . well, to be polite and accurate, exactly like a twin-cam Lotus-Ford engine, but with rather more buzziness within the cockpit than a twin-cam Lotus Europa. Even at its smooth tickover there is no hint of a V8 warble and the engine continues to sound like a small, well-tuned 4-cylinder unit, not at all in keeping with a very high performance, £8,000 car.

Instrumentation is clear on the typically Ferrari aluminium facia panel, but why is there not an ammeter or voltmeter? There are gauges for fuel, oil pressure and temperature, and water temperature grouped around the 180 m.p.h. speedometer and 10,000 r.p.m. tachometer and even a clock, which could have been replaced by an invaluable ammeter. Two-speed screen-wipers and washers are controlled by a stalk on the right of the steering column. A short stalk on the left controls the flashing indicators and a longer one looks after the operation of the four pop-up Halogen headlights, which rise much more rapidly than do Lotus ones and have a facility for emergency operation (with the car stationary) should the electric motors fail. Twin Carello auxiliary lamps are fitted under the radiator grille.

The gear-lever gate is contained in a slim central console. Gear-lever movement has been considerably reduced compared with the 246, though gearchanges can be notchy unless the clutch pedal is pushed right into the bulkhead. Also mounted on the console are the choke lever, cigar lighter and, in the test car's case, controls for the optional heated rear screen, electric windows and air-conditioning equipment, a long reach within the restrictions of the fixed belts. Normal heater switches are grouped on the left of the facia panel, matched by auxiliary light, hazard warning and heater fan switches on the right, all within flicking distance. There is a useful lockable cubby-hole containing the fuse boxes and an inspection lamp and deep pockets are formed in the doors by the armrests-cum-door pulls. Each door contains a courtesy light in its central panel (there is also a central one above the rear screen) and a warning light in its trailing edge.

Twin, lockable levers in the driver's door pillar release the self-supporting rear lids for the engine and fully-carpeted boot, the front "bonnet" is released by another lever under the facia and all three have separate emergency releases. Under the "bonnet" lie the radiator, from which hot air is ducted upwards through the steel panel, the emergency space-saver spare wheel, battery, servo, washer bottle and motors for the lights. I can't say that I would wish to drive very far on the narrow spare wheel, which carries a Michelin 105 R18X tyre, while the normal 6½J road-wheels are shod with Michelin 205/70 VR14 XWX tyres. A maximum speed of 90 m.p.h. is recommended when the spare wheel is fitted, though the handbook adds a disconcerting rider: "It is necessary to avoid any sharp braking because this special wheel will be the first wheel to lock or skid causing the tyre unnecessary damage"!

If the extra weight has made this new Dino marginally less agile than the 246, it has also improved its straight line and cornering stability—or at least the wider track and revised steering geometry has. As there are very few vehicles more stable than a 246 Dino it seems superfluous to add that the 308 has instantly become one the the world's best handling and roadholding road cars. The steering feels very little different to that of the 246—indeed, the same Cam Gears rack and pinion is used—and lock to lock takes an identical 3¼ turns. As those turns provide a poor turning circle of only 39.3 ft., it can be seen that the steering is not as high geared as one might expect. My own preference would be for a slightly higher ratio, more for low speed travel than for high speed work when the handling becomes neutral and less steering wheel movement is required. On the other hand the steering is superbly sensitive and responsive without becoming reactionary over bumps and is light enough to make parking easy. If the car is cornered on a trailing throttle it understeers noticeably, but any owner who complains generally about understeer should learn to drive: cornered under power, as much to compensate for the effect of the limited slip differential as to match general handling characteristics, the mild degree of understeer is exactly what is required to provide feel, stability and evasive qualities. Roadholding is phenomenal, there is minimal roll and, though suspension movement is small, the ride is surprisingly good. The improved cornering stability compared with the 246 is most noticeable on long, sweeping bends; the latter car tends to exhibit some diagonal pitching which has been eradicated on the 308.

Like Porsche, who dispense with a servo completely, Ferrari do not believe in over assisting the brakes at the expense of feel. Thus, high speed stops require a moderately heavy foot to assist the servo. The Dino then pulls up in ultra-short distances with hardly a trace of squat or nose-dive. In traffic the brakes are equally at home, braking effort being proportionally less than at speed and without any fierceness.

Ferrari claim a standing quarter-mile time of 14.4 seconds and a standing kilometre time of 26.2 seconds, with a terminal speed of 131 m.p.h., for this four-seater Dino. The kilometre figure is in fact only 0.6 seconds quicker than the Dino 246GT, but this marginal difference belies the improved ease with which the torquey V8 copes with such fierce acceleration.

This magnificent and practical new Dino, with its 3-litre V8 engine quite docile at one end of the scale and providing brilliant performance at the other, costs £8,339.76 from any of the 14 Dino dealers in Britain (including Maranello Concessionaires). The only sour note is the long list of extra-cost extras, most of which customers will regard as essential. The test car, for example, had the addition of metallic paint (£152), electric windows (£114), air-conditioning (£374), heated rear window (£30.42) and tinted glass (£44.46). A mixture of leather and cloth upholstery costs an extra £253, or all-leather upholstery £301. If the actual cost of such extras is inconsequential to a customer in this price bracket, the problem of choosing his personal permutation and awaiting the arrival from Italy of the right specification car could be annoying. It is unlikely to cause a shortage of customers, for this must be the car for which nearly every Dino 246GT owner has been waiting.

C.R.

The Ferrari Dino 3-litre V8 Engine.

Road test
by John Bolster

The pronounced wedge shape of the new 2+2 Bertone styled bodyshell is strikingly angular compared with the superseded 2-seater.

The latest from Maranello: Ferrari Dino 308 GT 2+2

The Ferrari Dino 246 GT was the car which most sporting drivers would prefer to own. In spite of a price which, though not high for a machine of this calibre, is quite a considerable sum, the Dino has been a best-seller among thoroughbred sports cars. The announcement of a new Dino is therefore a most important event and everybody has been asking whether or not it can really be better than the car we have all loved.

Let me put your minds at rest straight away! Having "owned" one for a week,

The bonnet is very short and the screen steeply raked. Visibility is excellent.

thanks to my trusting friends at Maranello Concessionaires, I can state that the new car is actually better in every important respect than the 246, unlikely though that may seem. This is not just a face-lift but an entirely new design, built on experience gained with the earlier car but far from being a copy.

Let us examine the specification. The steel frame is constructed of oval and square tubes, reinforced by the floor pan and the steel body shell, which owes its shape to Bertone's artistry. Here a great change is at once apparent, for while the earlier car was strictly a two-seater, the 308 is a 2 + 2. As the wheelbase has only been increased by 7.7 ins, the extra space has been found by moving the driver and front passenger forward. In spite of this, the body has a much shorter nose and tail so the length is only 3.3 ins greater, though there is still a useful little luggage boot behind the engine.

As for the new rear seats, the left one is quite practical if the front passenger sits well forward but the one behind the driver is strictly for a small child. The front boot is occupied by the radiator, fans, and narrow spare wheel and tyre—I suppose you either throw the muddy one away or drop it in mother-in-law's lap!

For real roadholding, wishbone suspension seems to be indispensable and the Ferrari has it both ends, with hefty pressed-steel links and anti-roll bars. The front uprights are on ball joints, steered by rack and pinion, the drive to the rear hubs going through Birfield Rzeppa constant-velocity driveshafts, the inner ones absorbing the plunging motion. Servo-assisted ventilated disc brakes are outboard-mounted all round, with 10¼ ins front and 10¾ ins rear discs; the Campagnolo magnesium alloy wheels carry 205/70 VR 14 tyres.

Like the 12-cylinder Ferraris, the V8 engine is of light alloy with wet liners. It has the same bore and stroke as its larger sisters, giving a capacity of 2926 cc, with cylinder banks set at 90 deg. With four twin-choke

Road test

downdraught Weber carburetters, the power output is 250 bhp DIN at 7,700 rpm, compared with 195 bhp at 7,600 rpm for the 246, so the slight weight increase does not matter. The two valves per cylinder are inclined at 46 deg and operated by twin overhead camshafts through piston-type tappets, which have their clearances set with Fiat interchangeable biscuit shims. The camshafts are driven by toothed belts, with a primary gear reduction to allow large driving pulleys to be used.

The clutch is mounted on the flywheel, driving the gearbox through what the French so artistically term a cascade of pinions. These three gears live in their own housing and drive the two-shaft gear cluster, which is situated in an extension of the engine sump but with a division to allow different types of oil to be used for power unit and transmission. The final drive is by helical-toothed spur wheels, the engine being mounted transversely just behind the rear seats and ahead of the driveshafts, which are parallel with the crankshaft. Air scoops for the engine are neatly blended into the body panels behind the rear windows, the twin fuel tanks being just ahead of the rear wheels.

Bertone's body, which is built by Scaglietti, has an entirely different shape from the Pininfarina shell on the 246. The very short bonnet and the far forward position of the steeply raked screen are unusual and the angular treatment contrasts with the curves of the 2-seater. Nevertheless, the general effect is both pleasing and functional, the sight of that low, red projectile in the mirror causing most people to move over smartly.

The car is easy to enter through the wide doors, the all-round view being excellent. The new steering wheel has no spoke that can obscure the rev-counter, so there is no excuse for over-revving. After two pumps on the accelerator pedal, the engine starts at once from cold, without using the choke, and soon idles evenly.

The character of the engine is completely different from that of the earlier V6, for whereas the six had little torque, the eight is all torque. The 246 relied on the driver to keep stirring the gearlever but the 308 is flexible beyond belief and can be driven all day in fourth and fifth gears if desired. Even at 2000 rpm there is plenty of punch, but the short-stroke unit will spin up to its peak revs of 7,700 whenever required.

The 308 feels more potent altogether than the 246 and its acceleration figures at high speeds are considerably better. Nevertheless, I found it quite difficult to beat the 0-60 time of the earlier car. I think that this was because first gear is out to the left opposite reverse, so one has to come through the gate and forward for second. I never got a first-to-second change that entirely satisfied me, the gearbox tending to resist very rapid handling though behaving well during normal driving. At this time of year, too, the roads are never really dry, wheelspin being a problem at the getaway though the traction was outstanding thereafter.

Whenever possible, I time cars in both directions over a measured distance, which is the most accurate way of determining the maximum speed. In this case, it was not convenient to follow that routine but I was able to reach 7,700 rpm in fifth gear, which is the red mark on the rev-counter dial. By calculation, it would appear that this is equivalent to just over 160 mph and I was able to check the rev-counter for accuracy. I think, therefore, that it would be fair to call the 308 a 160 mph car, or a little more if one were willing to over-rev briefly. It's very surprising how the car continues to accelerate in fifth gear when approaching its maximum speed, proving that the body has an efficient shape.

The Ferrari is marvellously steady at high speeds and is remarkably stable in gusty winds, which is by no means always the case with mid-engined cars. The cornering power

Blasting up a wooded hillside the Dino displays its precision handling.

seems even higher than that of the 246, the characteristic tending towards under-steering, though this is not excessive. The steering is never heavy and gives just the right amount of feel, though the turning circle on full lock is rather large. Perhaps the roadholding on wet surfaces is the most outstanding feature of the 308. It does not feel any larger than the previous Dino to handle, though the interior width gives room for the wildest driver to wave his elbows to his heart's content.

The brake pedal and accelerator are well arranged for heel and toe while the hand brake, operating in its own little drums, holds the car well on steep gradients. The ventilated disc brakes are powerful and stand up well to hard driving, with no sign of incipient fading. However, it is rather easy to lock the front wheels during hard braking on wet or gritty roads. The rear braking circuit has a control valve that is sensitive to suspension height and perhaps a little adjustment at this point was indicated. It's easy to tell if the front wheels are starting to lock because one sits so close to them that a characteristic "zip-zip-zip" sound can be heard; the brakes never pull to one side.

As a clever compromise, the ride is just about perfect, giving a suggestion of the taut suspension of a competition car yet soaking up bad bumps most satisfactorily. Again, the seats afford the firm location that is so necessary for fast cornering while being soft enough to give a feeling of luxury. There is a welcome absence of that pitching movement to which some mid-engined cars are addicted. Certainly, this Ferrari would be a most comfortable and untiring car for long journeys.

The machine is by no means silent but it is perhaps the quietest Ferrari yet. Such sounds as there are would be missed by the enthusiast if they were not present, but they

Road test

Easy entry to the cockpit is provided by the wide doors. Interior is similar to the 246.

Above: the boot is sensibly shaped; Below: the transverse alloy V8 is a tight fit.

are never obtrusive. The traditional Ferrari exhaust note is absent, for only six or twelve cylinders can make that musical sound, but the lower pitch of this V8 is deep and satisfying. Apart from the exhaust and the transfer gears, the machinery runs very quietly, the new camshaft drive deleting another source of noise.

The retractable headlamps rise automatically as soon as they are switched on. Though there are four of them, they were not powerful enough to allow the full performance to be exploited at night. I think that the lamps of the test car had been adjusted to point down too much towards the road, however. The heating is effective but I was at first shocked to observe that there were no eyeball ventilators on a separate fresh-air circuit. I was glad to find that the aerodynamics were such that if the driver's window were lowered to leave a gap, successful ventilation was achieved without any wind noise. The test car was fitted with refrigerated air conditioning but I had no occasion to use this.

In spite of its greater performance, the Dino is perhaps more economical than its predecessor. I would suggest 18 mpg as a likely figure, with 17 mpg for really hard driving or over 20 mpg if a police car has been spotted. There is no doubt that the reduced gearchanging at once gives improved consumption figures, the outstanding torque and flexibility of the engine encouraging the driver to make less and less use of the box. Though it is just about the ultimate in sports cars, the 308 GT4 is completely practical as a businessman's express or even a town carriage. It is delightful to drive and worthy of bearing that most famous name.

SPECIFICATION AND PERFORMANCE DATA

Car tested: Ferrari Dino 308 GT4 2+2, price £8,339.76 including car tax and VAT.
Engine: Eight cylinders 81 mm x 71 mm (2926 cc). Compression ratio 8.8 to 1, 250 bhp at 7,700 rpm. Four belt-driven overhead camshafts. Four Weber twin-choke carburetters.
Transmission: Single dry plate clutch, 5-speed gearbox with Porsche-type synchromesh and central remote control, ratios 0.952, 1.244, 1.693, 2.353 and 3.418 to 1. Final and primary drives by helical spur gears. Overall ratio of 5th gear 3.529 to 1.
Chassis: Multi-tubular steel chassis reinforced by floor pan and steel body shell. Independent front and rear suspension by unequal-length wishbones with coil spring damper units and anti-roll bars. Rack and pinion steering. Ventilated disc brakes with dual circuits and vacuum servo. Bolt-on magnesium alloy wheels fitted 205-14 tyres.
Equipment: 12-volt lighting and starting. Speedometer. Rev-counter. Oil pressure, oil temperature, water temperature, and fuel gauges. Clock, heating and demisting system. Cigar lighter. Reversing lights. Two-speed windscreen wipers and washers. Extra: Heated rear window, electric door windows, air conditioning.
Dimensions: Wheelbase 8 ft 4 in. Track (front) 4 ft 9¾ in (rear) 4 ft 9½ in. Overall length 14 ft 2 in. Width 5 ft 8¾ in. Weight 23 cwt.
Performance: Maximum speed 160 mph. Speeds in gears: Fourth 118 mph, third 86 mph, second 61 mph, first 41 mph. Standing quarter-mile 14.5 s. Acceleration: 0-30 mph 2.8 s, 0-50 mph 5.7 s, 0-60 mph 6.7 s, 0-80 mph 11.4 s, 0-100 mph 16.7 s, 0-120 mph 27.6 s.
Fuel consumption: 17 to 21 mpg

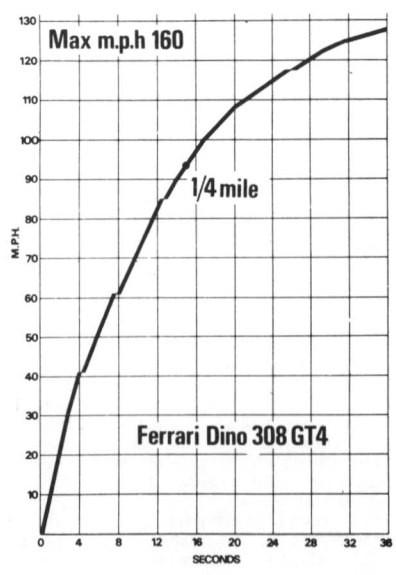

Dinos
Is newer better?

Marketing considerations required the change from pure two-seater to compromise 2+2. Was the change justified or has it killed the opportunity of a Ferrari for the masses?

DINO FERRARI was Enzo Ferrari's only son. When the young man died, it was a double tragedy for the founder of the world's leading volume manufacturer of sports cars, for he had been grooming Dino as his successor – the man to continue the autocratic tradition that made the name of Ferrari known the world over.

In memory of Dino Ferrari, the first "small" Ferrari sports racing car, a 2-litre coupé, was named after him. It appeared first at the Grand Prix of Rome in 1964 with Lorenzo Bandini at the wheel, but it earned its greatest success through giving Lodovico Scarfiotti the European Hill Climb Championship.

Like all sports-racing cars of the time, the Dino was a conventional mid-engined design with the power unit and gearbox arranged fore-and-aft, a configuration retained in Pininfarina's show-stealing effort at the Turin Show in 1966. This car, the Dino GT Berlinetta, was produced in limited numbers at Maranello – one per day at most. This was hardly the "Ferrari for the masses" and indeed, the Dino GT Berlinetta was priced very close to its larger-engined Ferrari stablemates.

It took Fiat to create the "popular" Ferrari, which came about in an odd way. Ferrari desperately needed a suitable power unit to campaign the 1967 Formula 2 programme. This called for a 1·6-litre engine of not more than six cylinders, and at least 500 examples of cars with this engine had to be produced annually. Such volumes were quite out of the question for Ferrari, but could not Fiat, Ferrari's new partner, justify using a suitable engine? They could, and they did. Thus was born the Fiat with the appropriate Ferrari-designed engine.

It was not long before Ferrari began to produce his own Dino with this engine. This car was the first of the Dino GTs as we recognize them in the 246 GT, but it had a 2-litre engine developed from the Formula 2 unit. As installed in the Ferrari, it produced 218 bhp at 9,000 rpm compared with 160 bhp when installed in the equivalent Fiat. Very few of these 206 GT Dinos, as they were known, ever came to this country and in fact the Ferrari Owners' Register only records six. In contrast to the definitive steel-bodied Dino 246 GT, the earlier car had all-aluminium bodywork.

Meanwhile, taking advantage of the maximum economical fiscal rating in Italy of 26CV, Fiat increased the capacity of the V6 engine up to 2,418 c.c. and in making the change, adopted a cast iron block with wet liners in place of aluminium. With the costly tooling for mass manufacture already installed at Maranello – at Fiat's expense –

The two pictures above tell much about the differences between the two cars. Bertone's 308 Dino makes use of fashionably thin pillars and a windscreen top that is level with the roof edge. In contrast, Pininfarina favours curves, bulges and sculpture but then he did not have to worry about 5 mph bumpers, did he?

There is much similarity in the interiors of the two Dinos. Both have leather seats and trimming but the 246 has suede-like material on the facia top. The Dino 308 (right) has the gearchange taken to the position for right hand drive although there is no problem in the position of the gearchange in the earlier car. The seating position relative to the two car's centre line is the same and thus, one sits further from the doors in the wider 308

and the necessary extensions to the factory carried out, the stage was finally set for the first "People's Ferrari" to be produced and at the Turin Show in 1969, the Dino 246 GT was announced.

This car was an eminently practical super sports car – at a reasonable price. When it first appeared in this country in October 1970, it cost £5,598.56, comparable with the top Porsche models of the day and around half the price of any other new Ferrari on offer. Above all, this car proved once and for all that given sufficient thought, it was possible to produce a *practical* mid-engined design. To accomplish this, Pininfarina departed from convention by specifying a glass rear quarter panel and using a cantilever from the rear wing top, leaning forward to support the roof. To further enhance the rear vision, the car was given an upright rear window with swept-back corners which finished level with the rear quarter windows.

In the original Dino Berlinetta, scant attention was paid to luggage space. The adoption of a transverse location for the V6 engine in the Dino 246 enabled the situation to be put right for within the elegant tail, a very good boot was provided. Space inside the car, however, was still limited, to the extent that the seat backs had little room behind them and "on journey" storage was severely limited.

In terms of handling and ride, there is no doubt that the Dino 246 set classic standards. The roadholding was such that it had a "go where you point it" ability to the extent where terms like under- and oversteer seemed irrelevant in all but the most demanding road conditions.

The use of unequal length wishbones front and rear gave the chassis designers more scope than usual to keep the

Dinos compared

wheels at the right angles even in roll, and advantage was certainly taken of that freedom. Steering was naturally by rack and pinion with a carefully chosen rack height. No bump-steer is evident and one can feel what the front wheels are doing without vicious kickback. The wide track and low centre of gravity are two significant advantages in reducing roll in cornering, thus allowing fairly soft springs. In the Dino 246's case, this degree of softness is allied to firm damping, giving a very comfortable ride regardless of conditions.

But then one comes to the question of looks... there is no doubt that the Dino 246 GT represents an absolute pinnacle of the specialist car designer's art. All the lines and shapes flow together beautifully, yet there is no need for aerodynamic appendages anywhere on the car to make it run rock steady – even in a side wind.

Evolution of the 308

Despite the commercial success of the Dino 246 and its timeless elegance, there was still room for improvement, especially through a widening of appeal. After all, the 246 is a most selfish two-seater and it was clear from the outset that token seating for two adults, or adequate seating for two children, would dramatically improve the Dino's appeal. Since the introduction of the Dino 246, most of the competition had brought in 2+2 models with comparable performance, thanks to engines a little bigger than the Dino's 2·4 litres. Freed of the constraints of that original V6, Ferrari engineers were able to consider a much greater love of theirs, a V8.

Taking advantage of research instituted at Fiat by Aurelio Lampredi, Ferrari adopted belt-drive for the four camshafts to cure one of the few shortcomings of the Dino 246 – engine noise. Since it was envisaged that the new Dino would have a greater payload than its predecessor, a well-known bore size of 71mm was dusted down and used to give a cubic capacity of 3 litres – adequate to give the required 250 bhp (DIN). It was of course important that the engine should be suitable for building with the available production machinery.

Having proved that the 246 Dino chassis was a good one, there was clearly little point in changing it unnecessarily, so the new car inherited the full wishbone suspension and rack-and-pinion steering although on a chassis with an extra 8in. in the wheelbase and a wider track front and rear. However, the overall length was kept down to 14ft 1in., only 3in. greater than the Dino. The credit for this design goes to Bertone, not Pininfarina, Ferrari choosing the former presumably because he had more experience in the exotic four-seater field.

The design is a marked contrast to the earlier one, and considerably more angular in its line. This was inevitably so to give adequate headroom for four passengers, and the relatively larger cabin dictates the cut-off tail which has a snub nose to match it. While some might argue that the 308 is less attractive to look at than the 246, all things are relative and there is no denying the admiring looks of bystanders whenever the car passes by.

Double overhead camshafts per bank – chain-driven on the 246 (above) and belt-driven on the 308. Accessibility to either installation is poor and a number of jobs are best carried out from below on both cars. Note on the 246 how the rear window is swept round to mate with the rear quarter panel to give good three-quarter rear vision – for a mid-engined design

Dino comparisons on the road

While the Dino 246 GT seems to cry out to be driven all the time, the 308 is altogether less demanding. From the sweet smooth-revving engine to the slick gearchange with its well-chosen ratios, the Dino 246 exudes liveliness and urgency, and although it is possible to make it pull from low speeds in the higher gears, the satisfaction of changing down to the "right" one is never absent. By contrast, the greater torque of the V8 in the Dino 308 permits a more leisurely approach when the driver feels lazy although, again, the ratios are admirably chosen.

In terms of handling, I would have to give first prize to the earlier car which has as much feel with less kickback and whose steering weight alters little as lock is applied. The 308 Dino is disappointing in this respect as a great deal of effort is needed to apply much lock, especially at low speeds.

Like most mid-engined cars, both Dinos understeer when power is applied smoothly. Injudicious application of too much power at low speeds can kick out the tail of either car but usually, you are coping with "running wide" at the front. Both cars tuck in quickly and controllably if the accelerator is released in mid-corner but this makes for ragged progress when driving fast on a strange road. Neither car is as fast on a strange road as, say, a Lotus Elan Sprint or Elan +2S 130. But here we are talking of ultimates; there are very few cars indeed that match the road-holding, handling and ride of either Dino.

Although, in theory, the gear linkage of the two Dinos is similar, in practice the earlier car has a better change with less baulking and also less tendency to miss the right slot when changing up from 1st to 2nd.

In terms of comfort, controls and seating position, only the crisper appearance of the facia and up-to-the-minute shape of the seats distinguish the 308 from the 246. Of course, on the current car you can have air-conditioning but the fresh air ventilation of the two cars is much the same.

The 308 GT4 that I tried had the latest-specification induction silencing and this does much to keep the noise level in the new car lower than in 246 Dino. Both cars sound "right" but it is the 246 Dino that makes the noise an enthusiast would demand.

Which would I choose?

Undoubtedly the Dino 246 is favourite for its cool good looks. The Dino 308 is nowhere near enough of a four-seater, and therefore, to my mind, its looks are unnecessarily saloon-like. In terms of pure performance, the 246 would again take the prize for its remarkable smoothness and apparent absence of any limit to the way it revs.

Where roadholding, handling and ride are concerned, there is so little to choose between the two. But my choice again, is the earlier car, for the sake of its lower incidence of kickback and more progressive feel when increasing lock and cornering force.

How nice it would be to make such a choice, though; both of them are streets ahead of the opposition, and you'll be the envy of everyone except a fellow owner whichever you choose.... □

PHOTOGRAPHY: DOUG MESNEY

CAR and DRIVER

The new Ferrari Dino and Maserati Merak will let Levittown cruise the Hamptons on not much more than a Carrera budget.

FANTASY CARS FOR EVERYMAN

BY PATRICK BEDARD

JUNE 1975

We have here a rare opportunity for a prole to rise above his station.

Consider this: For at least a year now, cutting back has been all the rage among the Beautiful People. Tightening the belt has been rumored to feel good, sending a quiver of relevance right through the old Puritan Ethic. Even in the rarefied atmosphere of upper Park Avenue in New York City, thermostats were cranked back all winter. The Rockefellers have been hosting intimate little candlelight dinner parties for 10 or 20 instead of the old horde. The pages of W are sprinkled with money-saving tips like how to squeeze more servings out of a tureen of ptarmigan bisque and travel suggestions like weekend hideaways on the Yucatan coast "for less than you ever imagined."

There is no doubt about it; society's darlings are backing off, part-throttling through the summer season. And it's the perfect opening for the upward-mobile prole to stand on the gas, pull up alongside and bask in the glory. All it takes is a little strategic effrontery. If you look right, nobody is going to check your credentials. Slide into McDonald's in your exotic mid-engine sidewinder and who is going to know that you aren't Alain Delon or some oil baron engaged in a little fashionable belt tightening? Just act like you've never been in McDonald's before—ask them, for instance, to hold the bordelaise on your Big Mac.

Of course, none of this is going to work if you show up in a Chevy or an MG. Without the grand entrance, you're just another window-washer, never mind that everything from collar to cuff is by Nino Cerutti. No, the first step toward convincing other proles that you're not one of them has to be a double-throwdown car, preferably something from one of the better houses of Italian automotive *couture*. And to borrow a breathless phrase from W, you can have one for less than you ever imagined. May we show you something in a Maserati? Or possibly a Ferrari?

That's right. A select grouping of these famous-maker Italian GTs is now available at unprecedented low, low prices. And the labels haven't even been ripped out. Well okay, the Ferrari doesn't really say Ferrari on it anywhere. It's a Dino—to get right down to the numbers, a 308 GT4—but those who know about these things won't have to be told that it's built at Maranello right in the same factory as the 40-grand Boxer Berlinetta. Besides, the Dino was named after *El Commandatore*'s son, so it's not what you'd call an A&P store brand. And the God's truth of the matter is that it's the only Ferrari product to be imported into the U.S. this season. So when you consider that these little sweeties hold their value like uncut diamonds and bankers no longer flinch at 48-month paper, the Dino is a hell of a buy at only $24,665.

Wait a minute—don't go away. If you want to get into the game at a lower ante, there is always the Maserati Merak (air conditioning, AM/FM, leather, ready to roll) for an unbelievable $19,989. That's right, almost a Porsche price—but the extra 5-grand gets you a genuine middle-motor *Maserati* and Giugiaro styling complete with flying buttresses. Porsches may be everywhere, but there's no way you'll meet yourself rolling down the freeway in Cleveland with this piece. And that, after all, is what separates those who have arrived from the multitudes who are still slogging along in the trenches. Maybe once a year you'll meet some parking-lot attendant who will recognize the Merak as Maserati's stripper model and try to get the drop on you with his knowledge of the $28,000 Bora and the $32,000 Khamsin. He might even know how the Merak achieved its price-leader status (it borrows certain innards—V-6 engine, five-speed gearbox, hydraulic braking system and instrument panel—from the Citroën SM). But so what? The SM is still pretty lofty territory; it's not like the Merak was some sort of fiberglass replica draped over a Volkswagen chassis. So don't worry about it.

Social-status-wise, both of these machines are gilt-edge securities, so we're not going to get too *Consumer Reports* about this test. A check-rated Best Buy doesn't really tell the story in the mid-engine megabuck class. Who cares if the frequency-of-repair record is poorer than the average of all models tested and ease-of-entry-and-exit is Unacceptable? Both of these cars cut a swath and do it at a fair price. They couldn't be any cheaper or everybody would have one. Twenty-grand is rock bottom for exclusivity these days.

There is no question as to the impact of these two machines: You can motor onto the scene in either one of them without even shifting out of second gear. Still, they are not exactly interchangeable. And if you're going to bust out of your rut, you might as well do it in the one that puts a smile on your face.

If intense machinery and barely tamed grand-touring racers for the street are your pleasure, the Dino is the tip. The factory has not yet raced this model, but with the current shift toward small-bore GT cars in international competition (particularly at Le-Mans), there is a good chance it will. The Dino sounds like it's ready, despite its emission-controlled state. The engine seems to scream even at idle, is cleared for operation up to a 7700-rpm redline and sings loud enough at freeway speeds to make you forget about everything but driving. Granted, it doesn't produce the exhilarating tones of the magical Ferrari V-12s, and the belt drives for the four overhead cams eliminate the chain noise customarily associated with Ferraris. But with four Webers tuned like pipe organs upstream of the intake ports, it still provides the closest imitation of the traditional Ferrari wail you could ever hope for from a V-8.

The arrangement of the cockpit reinforces the suggestion of a racer. The beltline is low, glass area is generous and the nose tapers down to give a close-up view of the pavement slipping under the car. All of this adds a sense of immediacy to driving, and the feeling is reinforced by the controls. The tools of driving cannot be ignored in this car: The round instruments on the dash are huge, complete and black-rimmed to contrast against the brushed aluminum panel; the low, leather-bound steering wheel nestles in your lap; and the shift lever sprouts up out of the tunnel so close to your leg that you invariably punch your thigh when you pull into first gear. The Dino puts everything you need right where you need it. You are encouraged to grab hold and run.

The Maserati Merak, on the other hand, is more like a stunning piece of architecture. When you have the two cars side by side, witnesses don't pay much attention to the Dino. But they crowd around the voluptuous Merak like Las Vegas chorus-girl watchers, digging each other in the ribs and making lascivious remarks. ("How'd ya like to have *that* for a night?" "I'll bet that baby gets it on.") Nobody can say that Giorgetto Giugiaro doesn't have a way with fenders. The Merak is plump and curvaceous: The grille opening sweeps into a set of lips, and there is a trace of hips in the bulging sheetmetal over the rear wheels. The message is not wasted on bystanders—they want to crawl inside and get crazy. When you have to score a fast impression, the Merak is the way to go. Nothing can touch it for those summer-weekend guest shots in the fashionable Long Island Hamptons.

To reap full benefit, you'll of course have to drive it, and that is something of a test of your commitment to upward mobility. The Merak comes up around your ears like the upturned collar of a pea coat, chopping off a good bit of the view. And then there are the bucket seats—exquisitely sculptured and covered with rolls of fine leathers but without much lumbar support and with an overall shape that makes you slink down like a surly teenager at the dinner table. Thoughtfully, 58 pounds of clutch effort have been provided to push you back in place—which ought to be enough to handle the torque of at least a supercharged Allison. And this is nicely balanced to the Citroën-designed power brake system that will lock the wheels up tight with the unaided weight of a

Gucci loafer on the pedal. But remember, nobody said this social-climbing charade was going to be easy.

Once you get used to the controls, you still have to cope with the Merak's bewildering visuals. The instrument panel with its oval dials is straight out of the Citroën SM (using the Bora panel would raise the price by $1000, according to Bob Grossman, the importer). These are unfortunately angled such that they reflect up into the swept-back windshield; they also glare on sunny days. Nightfall unhappily trades one distraction for another. The rear window is tilted just exactly right to pick up the taillights from cars ahead and beam them into the rearview mirror. Even Beautiful People tend to lose their composure when they see both headlights and taillights bearing down on them from behind.

While the personalities of these two exoticars are poles apart, they are presented in very similar packages. Both of them blanket roughly the same amount of pavement as a Mustang II. The Dino is slightly longer and narrower, most of the extra length due to its 1975 federal bumpers. This Merak (and all of the others in the country at the moment) is a 1974 model; the '75s are expected on the usual Italian timetable—that is to say, when they get here. But since there is a good supply of '74s, there is little concern over newer examples.

Both these coupes claim 2+2 passenger capacity. The Merak, with its engine cover bulging out into the middle of the rear seat, would be second-class transport even for pygmies. The transverse-engine Dino offers a bit more, if only because it has a full-width back seat, but the available space compares with that of a Porsche 911 and is best suited to backgammon sets and picnic baskets.

The greatest mechanical difference between the two cars lies under the engine lid. While the Dino's vitality radiates from a newly designed, light-alloy, 3.0-liter V-8, the Merak makes do with a 90° V-6 borrowed from the Citroën SM, also displacing three liters. It's a hard lump to love. The only reason any self-respecting engineer would stoop to this design is to save money—which is precisely the reason this engine wound up in the Merak. It is, in fact, a V-8 crippled by the amputation of two cylinders, and you feel their absence with every revolution of the crank in the form of an unequal firing sequence. At idle, it feels like two plugs are fouled, but this melts away under load to a feline purr—as though a 400-pound tomcat were lounging in the trunk. Eventually this special sound becomes one of the most endearing sensations of the car. Still, for those who find that the purr doesn't adequately offset the limping idle, the same car is available (albeit elaborately revised inside and out) with a V-8. It's called the Bora and that will be an extra $8000, please.

But even with its bargain-basement motor, the Merak can still put its belly to the ground and run when the game plan calls for the last guy to Vegas to buy the drinks. Top speed is in the 130-mph range, fast enough to land you in jail unless you really *do* know the governor. The Dino is a bit quicker, maybe as much as 10 mph if you can find room to let it all out, but with traffic and the cops being what they are, guts not horsepower will decide the race. In the short hauls, there is very little to choose between the two. The Dino will cover the first quarter mile in 15.4 seconds, shouldering past the Merak by a scant 0.2 seconds. But both of them are dog meat for Carreras and big-engine Corvettes—so a word to the wise should be sufficient here. Besides, there is no reason to get drawn into a run-off against the commoners anyway. Just smile, keep a steady foot and let that sleek Italian sheetmetal do the talking.

This is not to suggest that the Dino and the Merak are merely pretty faces. Both of them have an appetite for pavement: straight, bent, narrow, bumpy or dealer's choice. The Dino has inherited a fair share of the legendary prancing horse. When you give it its head, the engine sings, the suspension lopes over undulating roads and the tight steering lets *you* pick the path. It's damn close to a racer and it thrives on forceful inputs: The clutch wants a strong leg, and the stiff shift lever has to be muscled from one slot to the next. And your efforts are rewarded with accurate response. It's not until you approach the limits of tire adhesion that you discover the compromises built in for the protection of semi-skilled street drivers. A healthy dose of understeer goes a long way toward

specifications overleaf

MASERATI MERAK

Importer: The Grossman Motor Car Corporation
Route 59
West Nyack, New York 10904

Vehicle type: mid-engine, rear-wheel-drive, 2+2-passenger coupe

Price as tested: $19,989
(Manufacturer's suggested retail price, including all options listed below, dealer preparation and delivery charges, does not include state and local taxes, license or freight charges)

Options on test car: none

ENGINE
Type: V-6, water-cooled, aluminum block and heads, 4 main bearings
Bore x stroke .3.60x2.95 in, 91.6x75.0 mm
Displacement . . . . 180.9 cu in, 2965 cc
Compression ratio . . . . 8.8 to one
Carburetion . . . . 3x2-bbl Weber
Valve gear . . . . chain-drive double overhead cams
Power (SAE net) . . . . 190 bhp @ 4000 rpm
Torque (SAE net) . . . . 185 lbs-ft @ 3000 rpm
Specific power output . . . . 1.05 bhp/cu in, 64.0 bhp/liter
Max. recommended engine speed . . . . 6500 rpm

DRIVE TRAIN
Transmission . . . . 5-speed, all-synchro
Final drive ratio . . . . 4.85 to one

Gear	Ratio	Mph/1000 rpm	Max. test speed
I	2.92	5.4	35 mph (6500 rpm)
II	1.94	8.2	53 mph (6500 rpm)
III	1.32	12.0	78 mph (6500 rpm)
IV	0.94	17.0	110 mph (6500 rpm)
V	0.73	22.0	115 mph (5200 rpm)

DIMENSIONS AND CAPACITIES
Wheelbase . . . . 102.3 in
Track, F/R . . . . 58.0/56.9 in
Length . . . . 170.5 in
Width . . . . 69.5 in
Height . . . . 44.7 in
Curb weight . . . . 3200 lbs
Weight distribution, F/R . . . . 42.0/58.0 %
Fuel capacity . . . . 23.0 gal
Oil capacity . . . . 9.5 qts

SUSPENSION
F: . . . ind., unequal-length control arms, coil springs, anti-sway bar
R: . . . ind., unequal-length control arms, coil springs, anti-sway bar

STEERING
Type . . . . rack and pinion
Turns lock-to-lock 3.5
Turning circle curb-to-curb . . . . 40.9 ft

BRAKES
F: . . . . 11.0-in dia solid disc, power assisted
R: . . . . 11.8-in dia inboard solid disc, power assist

WHEELS AND TIRES
Wheel size 15x7.5-in
Wheel type . . . . Compagnolo cast alloy, 4-bolt
Tire make and size . . . . F: Michelin 185/70VR-15 XWX
R: Michelin 205/70VR-15 XWX
Tire type . . . . radial ply, steel belt
Test inflation pressures, F/R . . . . 32/32 psi
Tire load rating . . . . F: 1265 lbs per tire @ 36 psi
R: 1490 lbs per tire @ 36 psi

PERFORMANCE
Zero to   Seconds
30 mph 2.3
40 mph 3.8
50 mph 5.4
60 mph 7.4
70 mph 9.7
80 mph 12.4
90 mph 15.6
100 mph 20.3
Standing ¼-mile . . . . 15.6 sec @ 90.0 mph
Top speed (estimated) . . . . 130 mph
70–0 mph . . . . 180 ft (0.90 G)
Fuel mileage . . . . 11–16-mpg on premium fuel

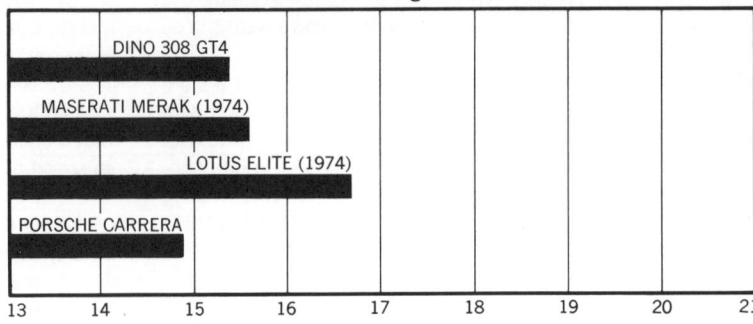

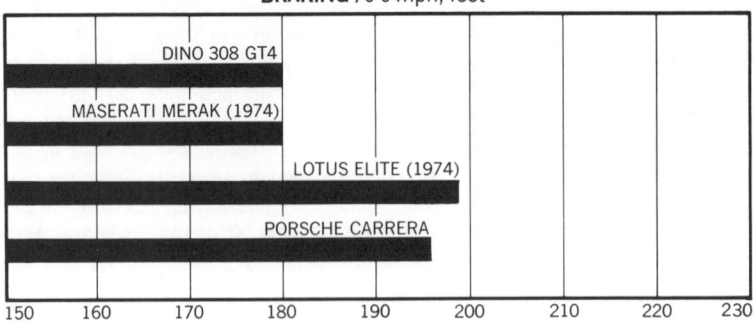

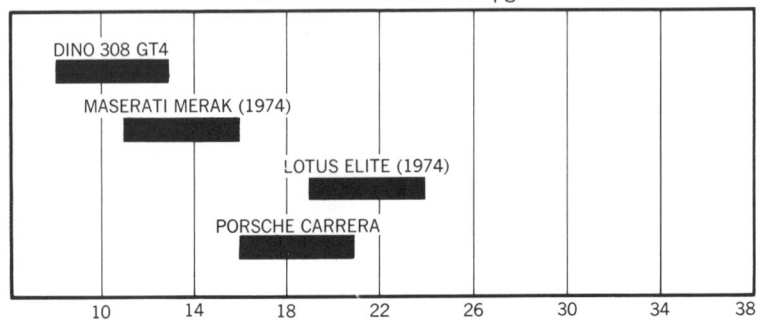

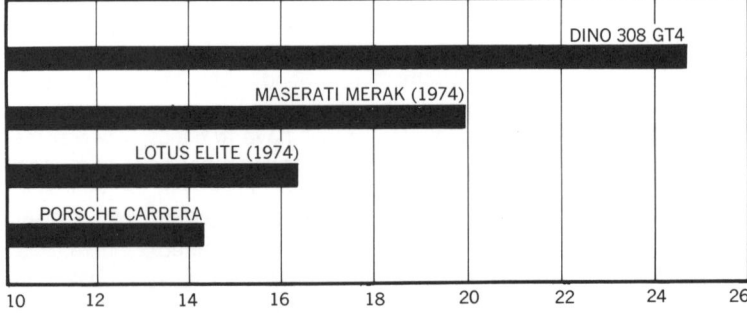

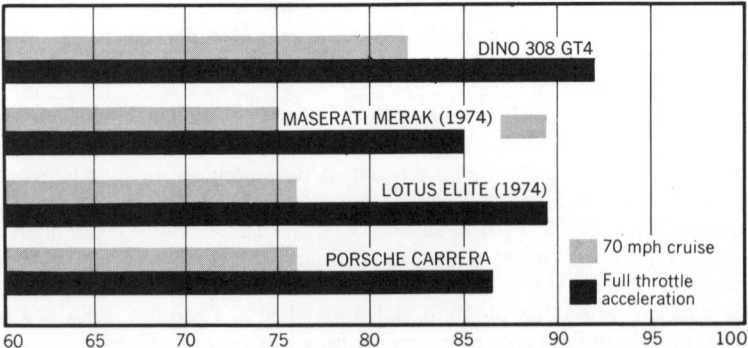

FERRARI DINO 308 GT4

Importer: Chinetti-Garthwaite Imports Co.
1100 West Swedesford Road
P.O. Box 455
Paoli, Pennsylvania 19301

Vehicle type: transverse mid-engine, rear-wheel-drive, 2+2-passenger coupe

Price as tested: $24,665
(Manufacturer's suggested retail price, including all options listed below, dealer preparation and delivery charges, does not include state and local taxes, license or freight charges)

Options on test car: base car, $22,550; air conditioning, $795; power windows, $245; antenna and speakers, $130; heater rear glass, $65; tinted glass, $90; leather interior, $640; dealer preparation, $150

ENGINE
Type: V-8, water-cooled, aluminum block and heads, 5 main bearings
Bore x stroke .... 3.19x2.79 in, 81.0x71.0 mm
Displacement .... 178.6 cu in, 2927 cc
Compression ratio .... 8.1 to one
Carburetion .... 4x2-bbl 40DCNF Weber
Valve gear .... belt-driven double overhead cam
Power (SAE net) .... 240 bhp @ 6600 rpm
Torque (SAE net) .... 195.2 lbs-ft @ 5000 rpm
Specific power output .... 1.35 bhp/cu in, 82.0 bhp/liter
Max. recommended engine speed .... 7700 rpm

DRIVE TRAIN
Transmission .... 5-speed, all-synchro
Final drive ratio .... 3.71 to one

Gear	Ratio	Mph/1000 rpm	Max. test speed
I	3.58	5.6	43 mph (7700 rpm)
II	2.35	8.4	65 mph (7700 rpm)
III	1.69	11.9	91 mph (7700 rpm)
IV	1.24	16.0	115 mph (7200 rpm)
V	0.95	21.0	120 mph (5700 rpm)

DIMENSIONS AND CAPACITIES
Wheelbase .... 100.3 in
Track, F/R .... 57.9/57.9 in
Length .... 176.7 in
Width .... 67.3 in
Height .... 47.6 in
Curb weight .... 3270 lbs
Weight distribution, F/R .... 41.9/58.1 %
Fuel capacity .... 19.8 gal
Oil capacity .... 9.3 qts

SUSPENSION
F: ...ind., unequal length control arms, coil springs, anti-sway bar
R: ...ind., unequal length control arms, coil springs, anti-sway bar

STEERING
Type .... rack and pinion
Turns lock-to-lock 3.3
Turning circle curb-to-curb .... 39.3 ft

BRAKES
F: .... 10.8-in dia disc, power assist
R: .... 11.0-in dia disc, power assist

WHEELS AND TIRES
Wheel size 14x6.5-in
Wheel type .... cast alloy, 5-bolt
Tire make and size .... Michelin XWX 205/70 VR14
Tire type .... radial ply, steel belt
Test inflation pressures, F/R .... 30/34 psi
Tire load rating .... 1490 lbs per tire @ 36 psi

PERFORMANCE
Zero to   Seconds
  30 mph 2.4
  40 mph 3.7
  50 mph 5.2
  60 mph 7.0
  70 mph 9.2
  80 mph 11.8
  90 mph 15.2
  100 mph 20.0
Standing ¼-mile .... 15.4 sec @ 90.6 mph
Top speed (estimated) .... 140 mph
70--0 mph .... 180 ft (0.90 G)
Fuel mileage .... 8–13-mpg on premium fuel

FANTASY CARS

damping any tendency a rear-heavy mid-engine car may have to spin. So you're not likely to come unstuck in the Dino and back it into the bushes somewhere.

By contrast, the Merak is somewhat remote, almost obtuse, in casual driving: The clutch is *so* heavy, the brake is *so* light and there is a certain elusive quality to the steering that keeps you from being close friends. The shifter is the only control that really cooperates. But when you hurry, the Merak collects itself up and shows a truly fine balance. It does not understeer excessively yet offers plenty of warning before it reaches its limits. Like other Maseratis, particularly the Bora and the old Ghibli, the Merak is happy in broad drift angles and you can throw it around without it getting unruly. When pushed, it's a far more graceful car than you would ever expect from normal driving.

Enough of this dissecting. Exoticars are not appliances; laying bare their

Exoticars are not appliances. Get in one, fire up the engine and then tell us how much you care about the oil-change intervals.

frame tubes and warranties with equal dispassion only confuses the issue. Do they quicken the pulse?—that's the point. Fit yourself into the Dino, fire up the sidewinder V-8 and then tell us how much you care about oil-change intervals. Picture yourself in the Merak, trolling through Southampton or Sausalito on a warm summer Saturday. There you are in the major leagues with more available options than you have hotel keys. It just wouldn't be the same in a Toyota—or five Toyotas if you're trying to keep the prices equal. Low-calorie cars don't cut it on gold coasts.

If you act now while frugality is chic, these economy models from Ferrari and Maserati will put you right up there with all those Beautiful People who call the fashion shots. Wait till the Dow-Jones punches through the 800 level and you'll need a Ferrari Boxer Berlinetta at twice the price. So if you are ever going to make your break, now is the time. ●

FERRARI 308 GTB

It took Ferrari a while to replace the Dino 246, but it was worth the wait

BY PAUL FRÈRE

Dino 246 fans, cheer up! There's a worthy two-place successor to your favorite car. And it's even better, faster, quieter and more comfortable. It also provides you with the satisfaction of officially owning a Ferrari, not the next best thing wearing a Dino label. Sergio Pininfarina was a Dino 246 fan himself and he styled the 308 GTB along the same concept, while taking advantage of the latest technological advances and his wind tunnel. In fact, it is the first Ferrari to benefit from the fullsize tunnel right from the prototype stage and it shows in the maximum speed of 154 mph I timed on an Italian *autostrada*. This is 3 mph higher than the speed I timed some 18 months ago for the Bertone-bodied Dino 308 GT4 using practically an identical power unit. It also shows up in the quite remarkable high-speed stability for which the car was notable during the maximum speed runs and in the motorway bends taken at maximum speed.

It is also the first production Ferrari featuring a fiberglass body. It's so well made you'd never spot it as fiberglass, but you get the benefit of lower weight and the non-corroding material. Though they designed it, Pininfarina doesn't make the body. It is fabricated by Scaglietti who started making racing bodies for Ferrari more than 20 years ago and finally sold out to the Prancing Horse. They made the body for the Dino 246 GT and currently produce the 365 Berlinetta Boxer in addition to the new 308. Their workmanship is as good as any and there was little to grumble about on the car I tried, which was one of the very first to come off the production line.

Though the pigskin-covered interior reminds one of the earlier Dino, modern thinking is evident in the way the padded top of the dash structure is continued into the doors to form armrests for both driver and passenger. They are very comfortable yet they don't get in the way of the driver's arms when he is hard at work. Large shelves are carried by the doors and there is just enough room behind the seats, when in a normal position, to stow a jacket or a none-too-thick wallet. The switches for the electrically-operated side windows are recessed in the armrests and the test car was fitted with the optional air conditioning which very cool weather prevented me from assessing.

I was less enthusiastic about the legibility of the instruments and the location of some of them. The clock, the only instrument of direct interest to the passenger, is tucked away on the driver's left side next to the oil temperature gauge. Both are barely visible, even to the driver. All gauge pointers are of the same color as the dials, with only a white dot at their extremity, making for difficult reading.

Mechanically, the Ferrari 308 GTB is similar to the Dino

PHOTOS BY PETE COLTRIN

accept heavier bumpers mounted on energy-absorbing struts. They will protrude slightly more from the body but not to the extent of spoiling the car's clean lines. The power unit, of course, has passed the U.S. emission tests in its American Dino 308 GT4 version so there should be no problem on this score.

The smaller Ferrari's performance is utterly deceptive, with smoothness and absence of fuss from the engine compartment that completely belie the car's performance. There is no dramatic unleashing of power as the engine gets "on the cam," just a smooth flow from, say, 3000 rpm to the 7700-rpm redline. The tachometer is certainly not just a styling item, for there is no sign of the engine running short of breath and no change in the engine note to tell you to change up. Mind you, the car isn't too quiet (that's not what you buy a Ferrari for anyway) but the engine noise never becomes really obtrusive. Furthermore, driving the 308 in heavy traffic is no problem at all and the engine will actually pull away in 5th from 1000 rpm, though a light throttle foot is needed.

Thanks to the use of an over-center spring, the pressure required to depress the clutch pedal is not excessive, but during the acceleration tests I had some trouble with the mechanism taking its job too seriously and keeping the clutch pedal right down—not a pleasant thing to happen with a highly tuned and highly expensive engine! Luckily, this happened only when the clutch was hot from a series of standing-start runs and could not be reproduced in normal driving or at rest. Otherwise, we never missed a gear change and the shifting mechanism with the typical Ferrari visible gate is beautifully smooth and precise in operation even though the car was new. At the lower speeds gear whine, probably from the primary drive, is quite noticeable in any gear but it disappears when the speed rises and wind, engine and road noise increase, though they never reach a tiring level and are definitely lower than in the Dino 246 GT.

I was lucky to be able to take advantage of Ferrari's own Misano test track, just outside the Maranello factory. It is

308 GT4. It uses virtually the same 3-liter, 90-degree four-camshaft V-8 and an identical 5-speed transmission. The main difference is that the new model has a dry sump lubrication system which should keep the oil temperature at a reasonable level even in the hottest climate. Both the engine and transmission are, of course, transversely mounted ahead of the rear wheels and the suspension, by transverse wishbones and coil spring-damper units all around, is basically the same as the Dino 308 though the spring rates and the damping are different. As in other Ferraris, the car's structure is based on a tubular steel frame reinforced by the body structure and some sheet metal elements including a riveted aluminum floor and a welded sheet-steel rear bulkhead. The wheelbase is 92.2 inches, 8 in. shorter than that of the Bertone 308.

For the time being, the 308 GTB is made to European specifications only, but it was designed with due attention to the American market and the front and rear structures will

figure-eight shaped to provide an equal number of right-hand and left-hand bends (five each) ranging from hairpin to very fast, including a series of S-bends and a curved braking zone. This is where all Ferrari racing cars are developed. The circuit was very hard on the test car's brakes, which were undisturbed by 10 consecutive high-speed laps, and the following performance tests requiring very hard braking at either end of the circuit's main straight. The general handling characteristic of the Michelin XWX-equipped car is understeer. It is definitely more pronounced than in a Boxer with which I also did several laps on the same occasion, but whereas the Boxer hardly changes attitude at all if the accelerator is lifted in a corner (in my opinion this is not the ideal behavior), the 308 reacts to the maneuver by gently closing its line. Coming out of the bend, the tail can be forced around under power in an easily controlled drift allowing full throttle quite early. Fast bends are really the 308 GTB's element, however, and they can be

FERRARI 308 GTB SPECIFICATIONS		
PRICE		
List price		est $25,000
GENERAL		
Curb weight, lb		approx 2650
Weight distribution (with driver), front/rear, %		43/57
Wheelbase, in.		92.2
Track, front/rear		57.9/57.5
Length		166.0
Width		67.0
Height		44.1
CHASSIS & BODY		
Body/frame		tubular steel chassis/fiberglass panels
Brake system		vented discs front and rear, vacuum assisted
Wheels		cast alloy, 14 x 6½J
Tires		Michelin XWX, 205/70VR-14
Steering type		rack & pinion
Turns, lock-to-lock		3.3
Suspension, front & rear: unequal-length A-arms, coil springs, tube shocks, anti-roll bar		
ENGINE & DRIVETRAIN		
Type		dohc V-8
Bore x stroke, mm		81.0 x 71.0
Displacement, cc/cu in.		2926/179
Compression ratio		8.8:1
Bhp @ rpm, net		243 @ 7700
Torque @ rpm, lb-ft		na
Fuel requirement		premium, 96-oct
Transmission		5-sp manual

Gear ratios: (0.89)	3.30:1
4th (1.20)	4.45:1
3rd (1.61)	5.97:1
2nd (2.23)	8.27:1
1st (3.23)	11.98:1
Final drive ratio	3.71:1

CALCULATED DATA

Lb/bhp (test weight)	approx 11.5
Mph/1000 rpm (5th gear)	22.2
Engine revs/mi (60 mph)	2700

ROAD TEST RESULTS

ACCELERATION

Time to distance, sec:	
0–1320 ft (¼ mi)	14.6
Speed at end of ¼ mi, mph	96.0
Time to speed, sec:	
0–30 mph	2.6
0–50 mph	5.0
0–60 mph	6.4
0–80 mph	10.2
0–100 mph	15.9

SPEEDS IN GEARS

5th gear (6900 rpm)	154
4th (7000)	116
3rd (7000)	86
2nd (7000)	63
1st (7000)	43

FUEL ECONOMY

Normal driving, mpg ... 12.5

taken in a full four-wheel drift with a feeling of utter stability. The side forces generated are obviously high, even by sports car standards, and are of the same order as the Dino 246 GT.

The 308's main asset, as far as road behavior is concerned, is the easy way it can be controlled. The reasonably geared (3⅓ turns lock-to-lock) and commendably light rack-and-pinion steering also plays an important part. The excellent handling was confirmed when the car was taken out on the *autostrada* and to winding mountain roads in the Appenine. At speed it feels dead stable and its behavior on 140-mph bends suggests that the front spoiler does its job of keeping the front wheels firmly down on the tarmac, while the mountain roads indicated that even bad bumps in corners do not throw the Ferrari seriously out of line, even though they cause some (though not excessive) steering kick-back. What these roads also revealed is how comfortable the suspension really is. Not only is it quite reasonably soft and by no means fiercely damped, but its travel is quite surprisingly long for such a low-built car. This is one of the biggest advances made, compared with the 246 GT. It is this sort of advancement that makes the 308 GTB the most charming car to come from Ferrari. Look for the American versions to begin arriving late this year.

'Signor, you go more slow, eh?'

THE CARABINIERI WERE WAITING AT THE TOP OF THE PASS. They were hidden in a farm entrance around a tight right-hand bend, and I didn't see them until it was much too late. Damn it, where had the radar trap been? How the hell was I going to worm out of this? Their ears alone would have told them everything. They'd have heard the wailing of that superlative little V8 rising and rising and rising as it wound through four gears to culminate in a bellow as it touched 7500rpm. And then there'd have been the pause and the smack as the next gear took up and it was all repeated, again and again. Among the wha-room . . . wha-room . . . wha-room . . . of the downward changes there'd have been the occassional squeal of a punished brake as the car was hauled down hard, Michelins scrabbling, for one of the tighter hairpins. For close to seven miles, by my reckoning, they must have been treated to this; and they must have been in no doubt that the red Ferrari now trickling to a stop at their command had been going very, very fast. It is possible that once or twice, had they perhaps needed visual proof or indeed sought only to heighten the experience, they'd have been able to catch sight of it as it devoured the mountain beneath them, racing ever faster into their arms, a red streak against the lush green of early summer.

I got out. The younger policeman took charge: 'Documenti, signor?' I gave him the wad of papers Ferrari had provided for me, hauled out my licence and told him what he already knew: that this little red devil of a Dino 308GTB was out on test from the factory now more than 100 miles back through the mountains. He studied the papers for a long time, going slowly over each document. Then he replaced them carefully in their leather folder and handed them back. I made to get back in the car, but he waved me back with his ping-pong bat, and it was then that I thought I was in real trouble.

But no; he just wanted to peer inside, and then to walk slowly around it, admiring it from this angle and that. He said only one word: 'magnifico,' held out his hand and invited me to retake my place behind the little leather steering wheel. When the V8, after that high-pitched wheeze from the starter that so characterises Ferraris, burst into life he summoned for me to blip it several times, and he shook his head slowly with appreciation and wonder and respect. 'Bon giorno, signor,' he said. 'Arrividerci' And then, with a grin, in the best English he could summon: 'Signor, you go more slow, eh?' He stepped back and waved me forward with a happy grin, and he raised his ping-pong stick in salute as I brought up a few revs and released the clutch. I will never know if they had taken a reading of my speed – and at times, even up that steep snake of a road, the car had been flat in fourth at 112mph and a few more yards of road would have required fifth – or if they were merely there to stop everybody because bandits had raided a bank in the area or if they were simply checking for stolen cars. I do know that the Carabinieri, respected rather than feared for their integrity and efficiency, looked at and listened to the newest Ferrari with pride as well as admiration. Say what you like; a Ferrari is still a Ferrari.

In fact, the Dino 308GTB does not sound especially good from the outside. The 308GT4 2+2 from which it is derived mechanically is not noted for its musical ability, but it is rather more pleasing than its new sister which has the misfortune to sound more like a Lotus than a quad cam V8. There is a reason: the exhaust pipes, in two bunches of four, are collected into two main pipes and these in turn feed into the one large silencer; the gas and the sound is then forced to emerge through only the one paltry outlet. It is a different matter from inside the cabin, for this sort of concession to the latest market requirements, the sort of sophistication that Ferrari are now building into their cars, does not extend to absolute muffling. Even if they tried to keep the music from the driver's ears, they'd stand little chance of succeeding. The modestly oversquare 3.0litre V8, sitting there transversely, is even closer to the occupants' backs than in the GT4. Top it with twin cams on each bank, feed it with four twin-choke Weber carburettors, build most of it from light alloy, let it rev to 7700rpm and tune it to provide 255horsepower and you get an engine that will not, should not, be silenced. The exhaust note might be lilly-livered, but the rest of the sounds that provide your first clue to the sheer excellence of the Ferrari 3.0litre V8 are all there, even down to the just-audible sucking at the air intakes just aft of the doors.

Whereas the 308GTB bears a strong physical resemblance to its much-loved 246 predecessor and is mechanically almost identical to the 308 GT4, it is a departure both from them and from all previous production Ferrari practice: the body is glass-fibre reinforced plastic over a tubular steel chassis. Ferrari do not feel the need to make mention of this fact in their brochure on the car; nor, I suppose, should they: it is the finest plastic coachwork I have yet seen, and most people would never suspect that it is not metal. The facts that emerge when one compares the three Dinos (discounting the

Good vision, a simple dashboard and a plain but deeply-hooded instrument binnacle (above) are sure signs of the sheer purposefulness of the Dino 308GTB. The driving position is faultless, with everything geared to clean, effortless operation and fatigue-free motoring for hundreds of fast miles

'...the feeling of sheer security imparted by the car when

original 206 and the current 208) show that the 308GTB represents a swing back to the concept and character of the beloved 246 – a small, light, fast and intensely personal two-seater. Gone is the compromise of the 2+2 but present are the lessons of greater sophistication and practicality gleaned from it. Dimensionally, in all but width (where it is broader by a mere half-inch anyway) the 308GTB is actually smaller than the Dino 246. Fractionally so, but it is smaller. Interestingly enough it is not lighter, although of course it has a significant weight advantage over the 308GT4 – 2403lb compared with 2576lb. And since both cars have precisely the same engine, the GTB has a decent edge over its 2+2 sister in its power to weight ratio: it has to pull 9.4lb/bhp compared with 10.1 (the 195bhp Dino 246's figure was 12.2).

Translate all this into on-the-road terms, into the barking of the engines, the howling of the cams, the sucking of the carburettors, the chirp of the tyres and the slap of the thin metal gearlever in its clinical metal gate and the 308GTB forges firmly away from the GT4. It covers the standing quarter mile in 14.1secs instead of 14.4; it emerges from the standing kilometre at 136.7mph after a 25.4sec run instead of 131mph and 26.2secs.

There's another hint of the way Maranello, and indeed the market place, has been thinking to be found in the gearing. The final drive and the first four ratios in the gearbox that nestles beneath the V8 are identical for the two 308s. Take them to 7000rpm, the power peak but not the absolute rev limit, in those first four gears and both will provide you with a potent 41, 59, 83 and 112mph. But in the GTB Ferrari, freed from the need to actually make the GTB achieve a higher top speed than the GT4, have raised fifth gear to provide instead for more economical cruising. The GTB's weight advantage means that the new car will still in fact accelerate quicker in top anyway. The difference in the final gearing in fifth isn't great – 21.6mph/1000rpm rather than 21.08, but at the identical top speeds of 156mph you'll have 7250rpm on the GTB's tachometer while the man in the GT4 is looking at 7400rpm.

And does the 308 GTB really do as Ferrari claim? Of course it does! After being given carte-blanche by the Carabinieri, after being waved on by another group a couple of hundred miles later as I pelted through the Futa Pass – what joy! – I rejoined the motorway from Bologna back to Modena and after watching a Daytona closing steadily on me as I sat on a relaxed 130mph and then seeing him disappear into the distance at something well in excess of 160mph with no apparent concern on the face I saw in his mirror, I opened the GTB right up. It answers like the throughbred that it is. It surges forward with spirit, the tachometer and speedometer needles climbing in unison, and the sound behind your head growing more and more intense, until there is just no more to come. In fact, the speedo was reading just over 160mph, but the tachometer said it was incorrect. At this speed, like the GT4, the 308GTB is outstandingly stable. It moves not an inch from the path you have selected, goes unmolested by crosswinds and sits down on the road with a security you can feel through your backside.
It's a *good* feeling.

So too is the feeling of sheer security imparted by the car when it is flat-out through a bend or being thundered through a dip, as it was for mile after mile on the way from Maranello south through the mountains, down the other side towards Florence and then back again along that magnificent driving road called the Passo della Futa. Like the 308GT4, the new Dino is at once a very easy car to drive; it

puts you at ease. There is the same taut but undemanding drive train, with a clutch that has plenty of firmness to its springing but no too much weight and a gearshift that moves with the sort of chunky, metal-to-metal deliberation that one has come to expect. There is the same lack of fuss from the 3.0litre V8, the same instant response and the same magnificent flexibility. There is no flab about any of the functions with which you must deal in the Dino; you just get into it, feel so very much at home, and get on with the job.

The driving position is low, but quite 'normal'. There's the usual Italian rake to the wheel, but the positioning is fine and the distance from the seat to the rim and to the pedals makes for immediate comfort. The instruments are arranged differently from the 308GT4: there is a rather tall binnacle jutting up from the fascia, and deeply hooded. Big, simply marked speedo and tachometer reside there, along with oil pressure, water temperature and fuel gauges. A clock and the oil temperature gauge are tucked away on the lower extremity of the dash where you won't see either in a hurry. Visual access to the main instruments is excellent, but Ferrari spoiled it by failing to eliminate reflections on their faces. Three stalks operate the minor controls, and toggles for the auxiliaries, the air conditioning knobs and the heater/ventilator slides are grouped on the little central console just behind the gear lever. The cabin is very much like the Boxer's, but perhaps a bit neater and more elegantly simple. There is an unusual appeal about the way the long, thin armrests on the doors mate up at their forward ends with the edges of the dashboard and although the cabin isn't very big there is an air of sufficient spaciousness to it thanks to the use of pleasant trim materials. The vision creates no problems, and combined with the other well-considered aspects of the GTB's passenger cell makes it such an accommodating vehicle.

The seats – very plain leather-covered affairs – prove to be comfortable and to have sufficient grip as you begin driving, in this instance a few familiarisation laps around the Ferrari test track over the way from the factory. Nicest of all, as you begin, is the steering. Light and smooth and pleasantly geared; the sort of steering that lets you work the wheel through the hands with a soft, almost caressing motion, taking the messages through the skin of the palm and the very tips of the fingers and thumbs. And yet, when you encounter a bump there is enough feedback to remind you – if that should be necessary – just how directly the rim in your hands is linked to the front wheels.

The Dino comes so smoothly, so effortlessly into the bends. It has more of a softness about it than the 308GT4, a gentleness; even more poise. There is a manifest feeling of sweetness, and you tend to trust it implicitly from the outset for you can sense how outstandingly well-balanced it is. You begin to go faster, unable to resist that track with its corners from the world's Grand Prix circuits. There is a gentle pushing out at the nose into the very tight ones. A gentle lifting of the throttle stops the pushing and, with the steering still so light in the hands and the car feeling so beautifully poised, it adheres to the line you wish it to follow, and you can then really push open the long-travel throttle and let the V8 propel you forward at a most pleasing rate. With such an introduction I was ready for the mountains ahead, and something close to 300 miles. Except for the few miles on the autostrada that took us between the two passes that we chose to negotiate and then back to Maranello when we had left the hills behind, except for them we were in series after series of bends; some hairpin-tight, others long and open.

is flat out...'

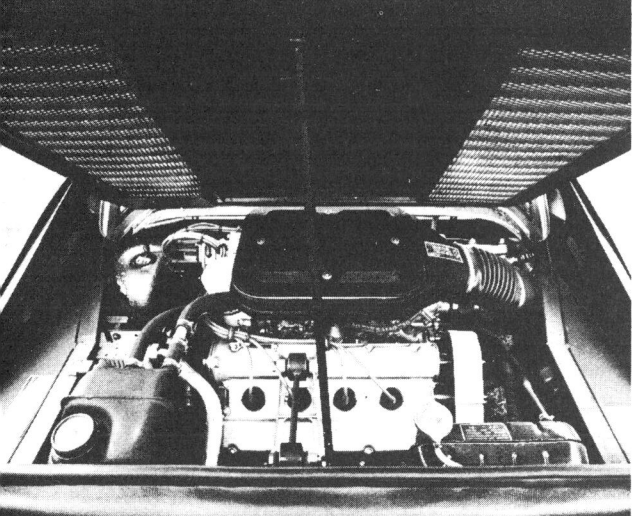

Most of the time I drove at a fast, steady pace so that Richard, my passenger, might not have to endure too much by way of g-force. Even so, it was a quick enough pace for him to remark how exceptional, how extraordinary it seemed to be able to cover so many miles so quickly and easily and so obviously securely. For this little Dino is like that; one feels so relaxed in it. There seems to be plenty of time to position it correctly, to brake it, to steer it, to sweep through the bends. One remains quite unflustered, even when you decide to use even more of the performance; to take it to 7700rpm out of the bends, to come in hard and bloody fast under brakes, to approach the limits of the roadholding, to see if the balance can be upset. It can't. The Dino merely goes faster, swinging from one lock to the other with lightning response – and yet with gentleness – and outstanding precision. It's poise is never lost, its stability never upset. It feels so damned *safe* all the time. I would venture to suggest, with nothing more than what may be an unreliable memory as the measuring stick, that it has decidely more roadholding than the 308GT4. And good as that car is, this is one notably better because it has such magnificent balance, and when it is going to go it gives you the warning, loud and clear, that the 2+2 doesn't; and it easy to catch. This one is a jewel, an absolute honey; a fun car par excellence. A sports car. Ferrari's prescription for post-crisis performance car motoring is indeed delicious; their diagnosis, it would appear, having been absolutely spot-on. Their approach to furthering the appeal of the cars, if this one is typical, is magnificently thorough: if the 308GTB has faults it is that the zip-on cover the luggage compartment has is perhaps a bit fiddly and that there is rather too much windnoise around the trailing edges of the doors above 120mph. To criticise beyond that, it seems to me, is to totally fail to appreciate the sheer accomplishment of the thing; the profound beauty of its character and indeed its ability. It leaves me with only one agonising question: How to choose between it and the Lamborghini Urraco?

The Dino rolls a little when corning very hard (top) but its balance cannot be upset. It is outstandingly fast through the bends, and gives the driver a sense of profound safety. The exhaust note might be half-baked but all the right sounds reach the cabin from the magnificent 3.0litre V8 (above left) mounted transversely behind the seats. There is a visual harmony about the 308GTB not present in its 2+2 sister and the finish of the glass fibre body is exceptional. Fuel consumption is 13mpg flat-out, upwards of 15mpg when driving a little less enthusiastically

FERRARI 308 GTB

Not a grand tourer in the old style, reports Alessandro Stefanini; but this compact two-seater, powered by a mid-mounted twin-cam V8, is a car of intoxicating excitement...

WITH A model designation like this — 308 GTB — the latest Ferrari has a lot to live up to. It evokes memories of the great 275 and 365 touring saloons built between 1964 and 1972, although it is technically far closer to the more recent Dino series. Its striking shape derives from the Dino 246 GT, in fact, and its beautiful 90-degree V8 motor is transversely mounted behind the driver, who is protected by a fireproof bulkhead.

The graceful yet aggressive lines were created by Pininfarina, who has been responsible for Ferrari bodies in recent years, working with the aid of a wind tunnel large enough to take full-size mock-ups. His use of contours extending well beyond both axles makes the car appear relatively long, but when it is parked beside another vehicle, you realise how compact it really is.

In this model, Ferrari has fully exploited the accurate use of glass reinforced plastics and it requires an expert appraisal to spot the slight differences in detail finish, in comparison with a metal body. The car is devoid of frills or embellishments of any kind — there is no chrome plating and all trim, even in the window openings, is finished in anti-reflecting materials.

For structural rigidity, the tubular steel frame is reinforced with steel sheets which also provide the necessary fireproof barrier between cockpit and engine. With an overall length of only 2 340 mm the car can accommodate only two people but its interior will also accept one small suitcase and a well-folded coat. So it's not a grand tourer in the old style. But it is a car of intoxicating excitement, as I found when I drove it under varied conditions which included Ferrari's Fiorano test track, ordinary country roads and, for good measure, thick city traffic...

Both engine and suspension are derived from Ferrari's racing experience. The 2 926 cm^3 V8 is based on the unit which won the 1964 Formula 1 Championship for Ferrari, with twin overhead camshafts per bank driven by toothed belts and an array of four twin-choke Weber carburettors.

The dry sump is drained by twin pumps and the engine's massive output of up to 190 kW is fed to the rear wheels through a cushioned-centre clutch and a five-speed all-synchro gearbox.

A limited slip diff is used and ventilated disc brakes are fitted to each independently sprung hub, with spring-damper struts working with unequal length wishbones all round.

I took over the car at Fiorano test track, which had been made available to me during a pause in the practice sessions held in preparation for the South African Grand Prix. This had been arranged to give me a "safe" introduction to the 308 GTB, before taking it onto public roads.

I feel immediately at home in the driving seat, which is easy to slip into and provides an ideal position for car control. The adjustable rake supports the back perfectly and the seat grips you firmly, preventing any possibility of sliding sideways on fast bends. Even with the car facing into the sun, there is no reflected light to confuse the eyes and the steering wheel and pedals are placed so that you need not shift your position to drive really fast or to relax. You can reach every control without leaning forwards and the dashboard is comprehensively equipped, laid out so that the wheel does not impede your view of either instruments or warning lights.

Lights, indicators and wipers are operated by stalk switches behind the wheel while those for the air-conditioning, heating and emergency flasher system are on the central console behind the remote gear lever.

At the twist of the key, the engine starts at once. At this stage, you hear it as a soft roar which does not disturb you but the decibels rise steeply with the throttle action, reminding you that this is no luxury saloon but a full-blooded sports model. The pedals are surprisingly light, though the clutch is unusually "deep", and the gears engage precisely and easily except for 2nd, which requires a little extra effort — a sign that the car is brand new and has still to loosen up.

Frontal visibility is excellent and distortion-free, but to the rear quarters your view is restricted by the struts connecting to the engine housing. I move off with a slight squirt of acceleration but without much clutch slippage, noting that the motor is obviously more flexible than some sports units, and move through the gears to lap the circuit slowly at first, learning the corners and becoming accus-

I took over the car at Fiorano ...

tomed to the car. After three laps, I note that all gauges, including engine temperature, are "normal" and thrust harder on the throttle pedal, calling on the motor to show what it is made of. It responds with a muscular pressure in the small of my back as the car gathers speed rapidly and the succession of bends seem to change character immediately, tightening up and calling for far more activity with the gear-lever and brakes.

The engine is now a muted scream behind me, incredibly satisfying to my wilder side, and I enjoy the stable braking into each corner, which calls for less pedal effort than I expected, followed by a rush of hefty acceleration out into the next straight. As this is the only car on the track, I can use its full width with no worries about other traffic and I am soon circulating at over 150 km/h, despite the restricted nature of the circuit. I think I am most impressed by the stopping power, which will haul the car down to a mere fraction of its speed without calling for any steering correction.

All too soon, my time is up and I nose the Ferrari out onto quiet country roads, then onto a motorway, where I give it its head for the first time, seeing 245 km/h at 7 000 r/min on the two main instruments. Braking hard to a standstill, I drop the clutch at over 6 000 r/min to send the car skating away from rest in a tremendous roar and a burst of blue smoke as the tyres fight to grip the tarmac. The needles spin round the dials — the rev-counter's in a swift succession of arcing swings — and I cover the first 400 metres in 14,1 seconds and the standing kilometre in 25,4, using the Ferrari marks!

Having now got used to the car, I move onto State Road No 12 — Ferrari's routine test route for the road-going prototypes they have built over the last ten years. Here I can really explore the car's handling at ultra-high speeds, nudging it through bends on a touch of lock and a little extra pressure from the right foot. There are absolutely no problems, for the car behaves perfectly, exhibiting neutral handling and fantastic braking power. You obtain peak acceleration by keeping the revs above 4 500, which gives immediate throttle response in any gear, and I note that there is no sign of brake fade in repeated stops and that the engine temperature remains constant whatever I am doing with the car. The ride is taut but rubbery, with a sense of wide treads sucking at the road surface during each change of direction, and this quickly builds confidence and invites you to exploit the engine power.

To show up the car's negative features, I take it onto city roads — narrow and full of rush-hour traffic. But again, the results are positive: the V8 proves so flexible I can burble along at 1 500-2 000 r/min on a whiff of fuel in either 3rd or 4th, taking care with the throttle but never needing to intervene with the clutch. And the oil and water temperatures stay within normal limits, as does the interior of the car, which is effectively insulated from the engine bay.

With the Ferrari works in sight, I finish with a final burst of speed to see whether the slow city driving has cluttered up the plugs; but the car responds with vivid, clean acceleration.

A check on the twin alloy fuel tanks reveals that the overall consumption has been 18,5 litres/100 km and I have only one criticism worth making, which would apply equally to other models with concealed headlamps: for driving in traffic, I am used to flashing my beams occasionally to signal other drivers — and in the 308 that just can't be done. But if I had the cash, this certainly would not deter me!

The cut-off tail has just the hint of a spoiler (left), instrumentation is comprehensive and the cockpit, impressive . . .

road test

The Ferrari 308 GTB possesses outstanding high-speed stability in spite of a short wheel base.

Ferrari 308 GTB: bravura!

When Ferrari replaced the much-loved Dino 246 with the 308 GT4 2+2, there were murmurs that much of the very special character of the original car had been lost. The new V8 engine was a vast improvement on the V6, which was very short of torque in the lower-middle ranges, but Bertone's practical body was less sporting than its very attractive predecessor.

Now, Ferrari have produced an additional model which has all the glamour of the Dino 246 plus the extra horsepower and, above all, the massive torque of the V8. The engine is a dry-sump version of the unit from the 2+2 and the transmission arrangements are also similar. However, the new 2-seater is much shorter in the wheelbase, by a full 8in, fractionally narrower, and no less than 3in lower in the roof line. All this adds up to a useful reduction in frontal area and a very substantial decrease in weight.

Better still, the body has been designed in Pininfarina's new wind tunnel. The resulting shape has given the car outstanding high-speed stability, in spite of its short wheelbase, without unsightly aerodynamic aids attached as an afterthought.

The 308 GTB is beautiful, largely because it is truly functional, and I am glad that the present ugly fashion of excessive angularity has not been followed. A really effective aerodynamic shape must depend largely on a subtle blending of curves, so stylists should forget their flat panels and uncompromising angles.

This body also breaks new ground by being Pininfarina's first glassfibre production. It is built on a steel space frame of round and square tubes and is entirely immune from the effects of corrosion, even when salt is spread on icy roads. The actual construction is in the hands of Scaglietti, so the arrangement is really parallel to the building of the Camargue by Mulliner Park Ward, the design again being the work of Pininfarina. The finish is superb and I defy anybody to distinguish it from the best metal-panelled coachwork, while the interior trim is on an altogether higher plane than that of the Dino 246.

The chassis is carried on wishbones all round, both ends having anti-roll bars, while all the brakes have large, ventilated discs. Though this is a mid-engined car, the power unit is not in the fore and aft position of a grand prix machine, but has the East-West location that Sir Alec Issigonis pioneered. The design also follows Sir Alec in having a cascade of pinions for the primary drive, the urge going to the limited-slip differential through a pair of spur gears.

The space saved by this compact arrangement of the main components permits the short-wheelbase car to carry a luxurious 2-seater body, with some luggage capacity. There is quite a useful boot behind the engine and a bit more room above the spare wheel in the front bonnet, just behind the radiator and electric fan. One would not

Glassfibre body is comparable to the best metal-pannelled coachwork.   *Space-conscious east-west V8 engine location.*

road test
by John Bolster

Interior is luxurious and the cabin noise level is commendably low..

advise the use of either of these positions for the carriage of ice cream, however.

Speaking of ice, the test car was furnished with the optional air conditioning; this equipment now being fitted to the majority of luxurious or exotic cars. Uncomfortable sports cars are things of the past; if a man is a connoisseur of fine automobiles, he is likely to appreciate civilised living. Electric windows and tinted glass are other attributes which he takes for granted.

The doors fit perfectly and close without slamming. The driving position is not noticeably Italian and the all-round view is unobstructed. The rear quarters looked as though they might get in the way, but this is not so, and the view in the driving mirror is exceptionally wide and clear. The engine starts at once without the choke in summer at any rate, and while it warms up it idles evenly, without any tendency to stall.

A great deal of work has gone into silencing the power unit, a very elaborate exhaust system having been evolved, as well as new air inlet arrangements. Mid-engined cars have a bad reputation for high interior noise levels, and so it is necessary for such vehicles to be beyond reproach in this respect, before a suspicious customer will accept them. The four belt-driven camshafts are remarkably quiet and so is the gearbox, the most obtrusive sound coming from the primary driving gears, which is by no means objectionable. The exhaust note is a deep, masculine tone, far removed from the traditional scream of racing Ferrari's. Though the engine is not silent, there is no need to keep the revs down while following bulletins on the radio.

A remarkable feature of the 308 GTB is the comfort of its ride. Let us forget this stupid tradition that a car must feel as hard as a board to handle well. The Ferrari does not have those very low profile tyres, which lack resilience, and its ride is rather what one would expect if Rolls-Royce made a 2-seater. There is never the slightest sign of pitching and the angle of roll is moderate during hard cornering, though just enough to add to the feel.

The steering is a delight, transmitting all the messages from the road. There is a little kick-back on occasion, which I like, with no sponginess or damping to stop the feel from getting through to the driver's fingers. The car can be taken very fast indeed through the quicker curves, settling down to a moderate angle of drift and holding its line beautifully. On the slower corners, there is more understeer and it's best not to put on full power too early, or you may have to lift off sharpish. However, one can seldom drive as hard as that on the public road. No car could be less twitchy and the stability at maximum speed is outstanding, which is rare among mid-engined sports cars.

One never seems to think about the brakes, for they always behave as expected, while the front wheels do not tend to lock, as they do on some other mid-engined cars and even on some previous Ferraris. The clutch pedal is not light and must be pushed well down, but it does not stick down when the clutch gets hot, as did that of the 2+2 during my testing of performance figures, so there has evidently been some unseen modification.

Nothing could be easier than taking performance figures with this car, for the engine torque matches the traction perfectly and wheelspin can be controlled to a nicety. The gear gate is not of the more popular 5-speed pattern, with fifth out on a limb to the right, but has first left and back, which means that the first-to-second change must be slowed by the wiggle in the middle. I was therefore amazed to do a 0-60mph repeatedly in 5.8s. so evidently the wiggle doesn't matter unduly. There are arguments in favour of both gate configurations, but although I prefer what is usually called the Alfa pattern, I don't really feel strongly either way.

The gear ratios are well chosen and if you charge into fifth at 120mph or so, the acceleration continues surprisingly vividly past 130mph. At 140mph, the speed is still building up slowly, but it goes on going on; like some other very slippery car, such as the Citroën SM, one can never be sure that the ultimate maximum has been reached. I still had about 300 revs in hand, so if anybody can find a long enough straight, he will probably beat my 156mph maximum.

Such speeds scarcely concern everyday motoring. The fuel gauge seems to go down pretty quickly at these velocities, but a reasonably law-abiding driver should get 20mpg. The car will crawl in traffic for hours without a hiccup, which is a contrast indeed to some earlier Ferraris. The engine has a lot of torque and is flexible on the higher gears, but being a thoroughbred unit, it shows its appreciation of sensitive handling.

The 308 GTB is a civilised car that anybody could drive. Its pop-up headlamps are quite powerful, though not up to 150mph motoring, of course! There is a little road noise on certain surfaces, but it is well subdued and wind noise is absent, except at very high speeds.

From the above, this car may seem like a paragon of all the virtues, so now I am going to find fault! The hand brake has to be jerked on jolly hard to park safely on a gradient, so please, Commendatore, can we have a bit more leverage? Then there's that magnificent pigskin covering over the instrument binnacle; it's lovely, and it looks a million dollars, but it's got a very shiny surface. In brilliant sunshine, particularly driving into the westering sun, that shiny covering reflects in the screen and spoils the pleasure of driving. The screen is at just the right angle and there is no reflection trouble at night, but in bright sunlight it's a different story.

There are exotic cars, such as the Stratos, which are marvellous toys for rich men but could not conceivably be regarded as everyday transport. The Ferrari 308 GTB could perfectly well be a man's only car, if he needed no more than two seats. Though its appearance is just about the ultimate in functional aerodynamics, there is nothing vulgar or boy-racer about it and it has the dignity of perfect proportions, so it would never look out of place.

Here is a warning! If you really ought not to spend £10,000, don't try this car. Once you have driven it, you will have to buy it, whether you can afford it or not.

SPECIFICATION & PERFORMANCE DATA

Car tested: Ferrari 308 GTB 2-seater coupé, price £10,501 including car tax and VAT.
Engine: Eight cylinders 81 x 71mm (2927cc). Compression ratio 8.8 to 1. 250bhp DIN at 7,700rpm. Twin belt-driven overhead camshafts. Weber quadruple carburettor. Dry sump lubrication.
Transmission: Single dry plate clutch. 5-speed synchromesh gearbox with central remote control, ratios 0.95, 1.24, 1.69, 2.35, and 3.42 to 1. Primary drive, with idler pinion, and final drive, all by helical-toothed spur gears, overall reduction 3.71 to 1. Limited-slip differential.
Chassis: Glassfibre body with built-in multi-tubular steel frame. Independent suspension of all four wheels by upper and lower wishbones with coil springs and telescopic dampers, plus anti-roll bars both ends. Rack and pinion steering. Servo-assisted 10¼in ventilated disc brakes on all four wheels. Bolt-on light-alloy wheels fitted 205/70 VR 14 tyres.
Equipment: 12-volt lighting and starting. Speedometer. Rev-counter. Oil pressure, oil temperature, water temperature, and fuel gauges. Clock. Heating, demisting and ventilation system. Electric windows. Flashing direction indicators with hazard warning. Reversing lights. 2-speed windscreen wipers and washers. Extras on test car: refrigerated air conditioning and radio with electric aerial.
Dimensions: Wheelbase 7ft 8in. Track left 9in. Overall length 13ft 10in. Width 5ft 8in. Weight 21¼ cwt.
Performance: Maximum speed 156mph. Speeds in gears: fourth 124mph, third 91mph, second 65mph, first 45mph. Standing quarter mile 14.5s. Acceleration: 0-30mph 2.1s, 0-50mph 4.8s, 0-60mph 5.8s, 0-80mph 10.1s, 0-100mph 15.8s, 0-120mph 24.2s.
Fuel Consumption: 15 to 25mpg.

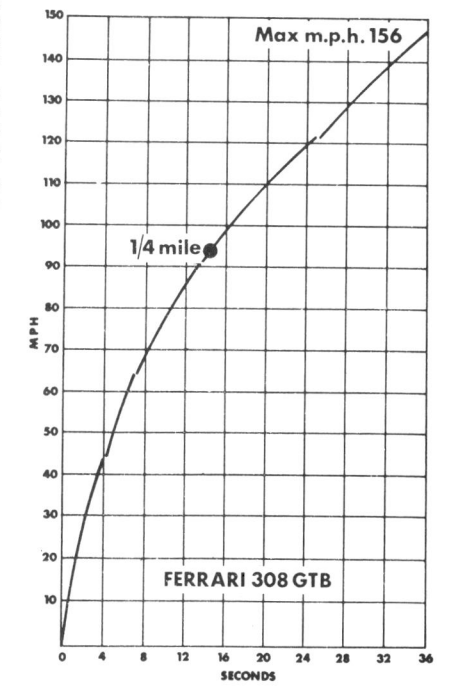

The Ferrari 308 GTB

The Dino is dead, long live Ferrari

THERE ARE some games you just cannot win and it would seem that my game with the Ferrari 308 GTB is one of them. This particular model got under way in production early this year and was notable for a number of things, the body is in glassfibre, the shape has regained the faith of Ferrari enthusiasts and the name Dino has been dropped. Back in February a small group of us were taken to the Paul Ricard circuit by Maranello Concessionaires of Egham, Surrey, for a number of reasons, but principally to see the first handful of production 308 GTB coupés, to ride in them and drive them. An added bonus was the fact that Lauda and Regazzoni were there testing the 1976 Grand Prix Ferraris, with and without the de Dion rear suspension. Also, there was a Ferrari Berlinetta Boxer there for our use as well. Needless to say such a trip was accepted without second thoughts, but through no fault of Maranello Concessionaires it turned out to be a disaster.

It started well enough when Niki Lauda took me for a couple of laps in the 4.4-litre flat-12-cylinder Berlinetta Boxer, a car that was so impressive that it left very little impression. When you look at a Boxer Ferrari, at the overall size and shape, and think about that incredible flat-12-cylinder engine in unit with the final drive, the size of the rear tyres, the disc brakes, the racing-like driving position and seats, the view through the huge sloping windscreen, you can't help gulping and saying "Cor!" Put the number one Ferrari driver behind the wheel and have him tell you that this particular car is mounted on an experimental tyre that he is interested in finding out about, and the result has got to be memorable. Memorable it was, but not unforgettable. The Boxer is almost too good, so that it did everything expected of it in a most dignified and perfect way. We conversed easily at *very* high speed down the long back straight, the acceleration out of corners was smooth and progressive with the big back tyres transmitting all the power to the road without fuss, it cornered in a very flat stance, responding to every whim of the driver without any drama, it slid about on all four tyres as Lauda deliberately went over the break-away point, in a very smooth and refined manner; it rode over the bevel kerbs without fuss as he "straight-lined" ess-bends and when we stopped he summed it all up by saying "This is a good car." To me it was almost too good, for even with Lauda throwing it about it was devoid of excitement, so that a lesser mortal at the wheel would never approach the threshold of excitement. Satisfaction, yes indeed. I got out thinking "That is exactly how you would expect Ferrari's latest masterpiece to be." It was so impressive that I was left without an impression, if you can follow me.

As I said earlier, the real reason for the trip was to try the new glassfibre-bodied Ferrari 308 GTB and the set of new cars that the French importer had lined up looked very good. The vast crowd of people to be accommodated looked very bad, but a drive on the next morning was promised. Without warning I keeled over and became horizontal with a severe attack of 'flu and next day could only remain vertical long enough to realise that the possibility of actually driving a 308 GTB was pretty remote. A lap in the passenger seat with a known or unknown French rally driver was about all that my colleagues could hope for. Surfacing every now and then from my fever I found them getting angrier and angrier as more and more "unknowns" had joy-rides in the 308 GTB and when I was finally carried away for the return trip home I enquired how they had got on with their test-drives. A stony silence greeted me and I realised that I hadn't missed much by being struck down by the 'flu bug. The UK Ferrari man tried to console us by saying that they were expecting the first of the right-hand-drive 308 GTBs into England pretty soon and he would make sure we were looked after better than the French had looked after us.

True to their word Maranello Concessionaires got the first consignment of right-hand-drive cars in, and a car was offered to try. Not for a few laps round Silverstone, or a 30-minute drive down the A30, but to take away and use. It was a glorious sunny afternoon when I collected the silver 308 GTB from Mark Konig at the new Ferrari Centre at Thorpe, a mile or two from Egham, and it did not take more than a mile or two to know that I was going to enjoy it. I managed to acclimatise myself to it that evening and looked forward to an interesting B to C journey from Hampshire to darkest Wales on the following day, to visit the lair of The Bod. As I say, there are some games you just cannot win, and the Ferrari 308 GTB is one of them. I used the car for a round trip to Wales, the West Midlands, across to the east and back down south and suffered heavy rain, poor visibility, unbelievable traffic congestion, darkness, and everything else except snow. If it was not raining, it was dark, if it was not dark, there was solid traffic, if there was not solid traffic it was raining, and so it went on. I tested the two-speed wipers very thoroughly, also the tractability and slow-speed running, and the all-round visibility and the lights and they all passed with flying colours. On the morning I took the car back to Thorpe the sun came out and summer started all over again!

In spite of all this the Ferrari 308 GTB left some very interesting impressions, but before considering them a little Ferrari history will not go amiss. For years Ferrari road cars have had glorious V12-cylinder engines mounted ahead of the driver, some with two overhead camshafts, some with four overhead camshafts, but all making the lovely sounds that only a V12 Ferrari seemed capable of doing. Then in 1967 there appeared a beautiful little car called the Dino 206 GT, a toy car compared with a Daytona Ferrari, but none-the-less a nice toy. In 1969 the Dino 246 GT went into serious production and its 2.4-litre V6 engine mounted transversely over the rear axle was pure Ferrari in the way it hummed and sang to itself at 7,500 r.p.m. It was a beautiful looking little 2-seater coupé and though one was prepared to accept that it was not a *real* Ferrari, compared to the 4.4-litre V12 cars, it was truly ". . . a son of Ferrari . . ."

In 1973 the Ferrari BB was introduced, of similar general lines to the Dino 246 GT, with its flat-12-cylinder engine mounted amid-

38

REDRESSING THE BALANCE.—The 308 GTB is a masterpiece of mould-making and finish, though silver paint did not do justice to its lines, left. "While at the controls there is a lot to be desired . . . Instrumentation . . . 'styled' by someone in Turin more used to 850 c.c. Fiats than Ferraris" . . . a "riot of stylistic art", above and right. Below, in the nose lies a Michelin "get-you-home-tyre" and in the tail a leaky "boot" and a "smooth and effortless producer of power".

ships, though the whole thing seemed unreal and the feeling was "mmm! yes, but . . ." In no time at all the BB, or Berlinetta Boxer, was in full production and over 50 had been sold in Great Britain alone, and that was a year ago. This "ultimate" Ferrari has now been enlarged to 5-litres and is called the 512 BB, an illustrious name indeed. The mid-engined concept from Ferrari was now serious. The Dino series had been an offshoot or probe, and the 246 GT has become a classic, mostly because of what followed. With little warning the beautiful little Dino 246 GT, styled by Pininfarina, disappeared and was replaced by an angular looking car called the Dino 308 GT/4 whose only similarity to the previous Dino lay in the name. This was styled by Bertone and exuded nothing compared to the Boxer and the 246 GT. It had a four-cam V8 engine mounted transversely amidships, integral with a 5-speed gearbox, and was a 3-litre, though a 2-litre was also offered as the Dino 208 GT4, but it seemed to be still-born. The Dino enthusiasts wept when they saw the new car, but the *real* Ferrari people were a bit more tolerant as they felt a 3-litre was a bit more like a proper Ferrari, but the V8 layout left them puzzled. At least the V6 Dino 246 GT was "half a proper Ferrari," but a V8 was something from Detroit or St. Agate, for Lamborghini was making the 3-litre V8 Uracco, with the engine mounted transversely, amidships atop a 5-speed gearbox!

What was much more worrying to the pure-Ferrari enthusiast was the fact that the front-engined V12-cylinder, four-camshaft, 4.4-litre 365 GT was the only "proper" Ferrari offered for sale, and it was more of a "limousine" than a "sports car" and even had power-steering. Worse was to follow, for in 400 GT guise, it has now been given the benefit of automatic transmission. If you view the trend in Ferrari production cars dispassionately, with the introduction of the Dino 246 GT, the Boxer, the 308 GT4, and the latest offering, the 308 GTB, then the end of the road for the traditional V12 Ferrari with front-mounted engine is in sight. The Bertone-styled 308 GT4 was not acclaimed as a roaring success and Pininfarina were soon at work making amends. In the 308 GTB they have surely redressed the balance, and what is more it has been done in glassfibre and is a masterpiece in the art of mould-making and finish. Underneath it is still the 3-litre V8, four-camshaft, transverse mid-engine and 5-speed gearbox, with all-round independent suspension, ventilated disc brakes and ride and handling that is everything you would expect. The name Dino has been dropped completely and this latest offering is unashamedly the Ferrari 308 GTB, little brother to the fabulous Ferrari Berlinetta Boxer. All the badges and name-plates now depict the Prancing Horse or, say Ferrari, the rectangular Dino badge has gone.

Throughout the motor car manufacturing world there have been instances of a firm making a long-lived and successful product and then the time has come when somebody within the organisation has to make an irrevocable decision. I am thinking of the day when Porsche said "Stop" to the production of the 356 series, and the 911 took over; when the last E-type Jaguar was made; when the DS-series Citroëns ceased; when Lotus made the last Elan; even when BMC made the last Morris 1000. The day has got to come when Ferrari makes the last front-engined V12 Ferrari, but it will have to be a courageous man who makes the decision. Enzo himself, surely. I say this after spending a few days with the 308 GTB, even in appalling conditions, and reflecting on experiences in the Dino 246 GT and the Boxer. Ferrari makes modern cars, using all the knowledge and technology that stems from his racing department. The last V12 racing Ferrari was so long ago, I don't really recall it. The Ferrari 308 GTB is the Maranello expression of a modern car for today. It is functional, efficient, and excellent, but exudes little in character other than being absolutely right. Not far from Maranello the Lamborghini Uracco P300 is being built, and not long ago I said of that car that it was a proper car. The 308 GTB Ferrari is virtually identical and in the dark I would defy anyone to tell the difference when driving it. The only subtle difference is that the Ferrari exudes

39

a certain "feel" about the way things work that suggest the obvious fact that Ferrari has been building production cars that much longer than Lamborghini.

A 3-litre V8 engine, with toothed belts driving the four overhead camshafts and feeding from four double-choke Weber carburetters, is not necessarily a Ferrari engine, nor a Lamborghini for that matter. It is a very smooth and effortless producer of power, whether you want it at 1,500 r.p.m. or 7,500 r.pm. and when a friend looked under the engine lid and said "Good grief, a V8 Fiat . . ." he wasn't far wrong. With a bore and stroke of 81 x 71 mm. the 90-degree V8 has a capacity of 2,926 c.c. and at 7,600 r.p.m. it produces a claimed 255 b.h.p. using an 8.8 to 1 compression ratio. As the power comes on and the revs rise the engine gives a hard purposeful growl, but at low r.p.m. and no load it bleats from its exhaust like a sheep. A top speed of 158 m.p.h. is claimed and a weekly magazine noted for its serious approach to facts and figures said it would do 154 m.p.h. (4 m.p.h. short of its claimed maximum, they complained!). Personally I never saw more than 130 m.p.h., for it even poured with rain when we took the car to a test-track, and 150 m.p.h. or more always seems pretty academic to me anyway. What was usual, was 100-110 m.p.h. in fourth gear, with no strain at all, and then into fifth gear and onwards. The steering was beautiful, with very positive feel, and the natural tendency of the car was to understeer, even to final front end breakaway, but an over-steering "tail-out" stance could be invoked on a skid-pan and increasing the speed made the tail go on out until a spin was provoked. Even on the streaming wet test-track the fat Michelin tyres hung on incredibly well, sending out a piercing shriek before reaching the limit of adhesion. They are 205/70 VR14 XWX and look fat enough to have been Grand Prix tyres in 1966/7 (there is a purpose in racing). In the nose of the Ferrari 308 GTB lies a Michelin "get-you-home" tyre, looking like a motorcycle tyre, this small section allowing more space in the nose for coats or boots or suchlike. A punctured 205/70 will just squash into the space taken by the "get-you-home" tyre, but oddments would have to be removed. The space-saving tyre mainly aids the drooping-snoot line of the front of the car.

For all normal (and most abnormal) road speeds and cornering the Ferrari rides incredibly flat, with no apparent roll or pitch, but below 50 m.p.h. there is a lot of thumping and banging from the suspension. Once on its way beyond 50 m.p.h. everything smooths out, the spring rates are dead right, as are the shock-absorbers and you really feel that all four wheels are doing their job with complete confidence and the messages sent back through the steering and the seats are to say "all is well, press on". If the harshness at low speeds was eliminated, either by the introduction of rubber, or different spring and shock-absorber settings, the wonderful feel at over 100 m.p.h. would be lost, and the Ferrari 308 GTB *is* supposed to be a fast car, even though "they" are trying to stamp out good cars with legislation.

As a car to live with I found this 308 GTB lacking on many detail points. It was not water-proof (and it certainly got the full test-treatment). for drips came in at the top of the driver's door, but much worse, the boot at the rear filled with water in its corners. Behind the engine compartment is the full-width luggage boot in the extremity of the tail and at each side are deep wells, which were very useful until it rained, when the bottoms became soggy. Another tiresome design fault is that the rear lid over the engine and the luggage compartment is in one, so even to remove a briefcase you have to lift up the whole tail lid. Even worse was the fact that water lies on this flat lid and when you lift it up, hinged at the front, it dumps half a gallon of water all over the engine and the ignition distributor, which is on the left side of the front cylinder head. The impressive thing was that the electrical system did not object to being doused in water.

Inside the car, while at the controls, there is a lot to be desired, especially the instrumentation which would appear to have been "styled" by someone in Turin more used to 850 c.c. Fiats than fast Ferraris. Can you imagine designing instruments with black dials and dark green needles? The tips of the needles are white, which is fine until it is in line with the white markings on the dial face! As for a rev-counter whose figures go from white, to orange, to dark red on a black background as you go up the rev-range, and the instrument is buried deep down in a well with the important 6,500 to 8,500 portion in the shadow . . .! Below the instrument panel and to the right were two extra dials, one for oil temperature and the other a clock. You could read them very well when the door was open! As the speedometer read 10 m.p.h. when you were stationary there did not seem much point in worrying too much about the whole Turin-styling exercise spread out in front of you. However, full marks for the high-beam indicator light that was discreet and sensible like a 1939 BMW 327 coupé. The final touch to this riot of stylistic art on a very fast car, was the binnacle over the main instruments that reflected the sunshine (what little there was) up onto the windscreen to form a nice fuzzy glare right in front of the driver's face. However, the real prize went to the stalk on the left of the steering column that controlled the lights. Rotating the end put on the side and tail lights and moving the lever down a notch operated electric motors that wound up the hidden headlamps very smartly and had them on "dipped". Moving the lever down another notch put the headlamps on to full beam and excellent "full" beams they were. Flick the lever up a notch and you were back on "dipped" which was fine, except that I have a simple brain that thinks a lever control moving "upwards" should put things "up" and a lever moving downwards should put things "down". Not on a Ferrari. In the heat of the moment, such as overtaking in a hurry, flicking the lever inadvertently the wrong way not only does not give you full beam, but takes away the "dipped" beam and the headlamps fold smartly out of sight. Presumably, if Ferrari owners live long enough they will get used to this very dangerous lack of "fail safe thinking".

When I heard that the Ferrari 400 GT was going to have automatic transmission I began to wonder if it would not be a bad idea for the mid-engined 308 GTB, for the gearchange on the 5-speed box leaves a lot to be desired. The ratios are marvellous, and the clear and open gate are fine, but the movement is heavy and sticky, especially at low revs. When you have got it all wound on in third gear, the lever "wangs" across into fourth all right, and 7,500 in fourth and a haul back into fifth is fine. But unless you are doing that, gear-changing is hard work and there is no encouragement to "play-tunes" on the lever. Anyway, the V8 engine doesn't sing to itself like the little V6 Dino engine, so you might as well let it burble away at 3,500 r.p.m. There always seemed to be ample power and torque for all normal needs.

When all is said and done you find that this latest "Ferrari for today" costs £12,000, with another £1,000 for optional extras, but for that money you have got a proper car, and a Ferrari at that. When Fiat took over the Ferrari empire, Enzo Ferrari said "that's fine, you look after production, and I can concentrate on the motor racing, which is all I have ever really enjoyed". In the Ferrari 308 GTB we have a modern, efficient, effective and first-class automobile, but somehow it does not exude the magic of Enzo Ferrari. The idiosyncrasies and character that Enzo Ferrari put into his cars seem to have been discreetly swept away by the dead-hand of Fiat.—D.S.J.

AutoTEST

Ferrari 308GTB

**Ultimate Grand Touring car for two very lucky people.
Fast, obviously; economical, surprisingly;
comfortable, thankfully; quiet, usually; infuriating, sometimes;
satisfying, most times; self-expressive, always.
Outrageously appealing — outrageously expensive but still the best Ferrari yet.**

A FERRARI is on most people's list when the disposal of a big Pools win is under discussion. The stuff of which such reputations are made is, in the sports car world anyway, a combination of high performance, superb roadholding, a cossetting interior, and the best of good looks. On each of these counts, the Ferrari Dino 308 GTB scores most of the points that are there to be won.

Of course, there are shortcomings. The price, for a start, is enough to ensure that the Pools win must be a big one. There are others. Like the mistaken choice of facia covering that allows bad reflections in the windscreen, and the fact that the beautiful lines spell out "design triumphs over engineering" to the extent that there is not room for a proper spare wheel. But let us see the car in context:

Two-seater cars in the class of the Ferrari are ultimate expressions of hedonism. Most of the time, circumstances — speed limits and sensible caution — keep the car well among the herd where something less exotic would do the job just as well. But it is when the open road beckons, when a touch of adrenalin is wanted and the spirits need to soar, that a car like this particular Ferrari is the best tonic available. We should not support that aspect of the pursuit of pleasure that seeks for flattery but there is no mistaking the admiring looks of passers-by of any age and the questioning yet furtive look inside the car to query, "who is the lucky so-and-so who owns that?"

About the Dino 308 GTB

When the Ferrari Dino 308 GT4 2+2 coupé was brought out to succeed the Dino 246 GT, not all enthusiasts approved. Admittedly, the presence of two reasonable short-journey rear seats widened the appeal of the successor, but in finding space for them, some of the unmistakable goods looks of the first Dino were inevitably lost. Morale among the disappointed Ferrari fans was boosted with rumours of a two-seater version,

AutoTEST
Ferrari 308 GTB

Body roll is well controlled at high cornering speeds. The lines are beautifully balanced from any angle

and boosted still further when it became known that the design would be executed by Pininfarina, the architect of the first much-loved Dino production mid-engined car. Sneak pictures from Italy at the beginning of last year revealed a long, low, dart-shaped body with a below-nose spoiler and just the hint of a tail-end spoiler on the engine lid. Not surprisingly, when the 308 GTB was shown at the Paris and London Shows a year ago, it was swamped with admirers.

The 308 GTB is the first production Ferrari to feature glass-fibre for the complete construction of the body. The standard of finish is superb, with little or no hint that the material used is out of the ordinary. The glass-fibre body is constructed by Scaglietti who, as well as producing the "Lightweight" light-alloy versions of earlier Ferraris also makes all the bodywork for the Ferrari formula 1 racing cars as well as the glass-fibre body sections used on other road cars in the range (the Boxer has the nose underside constructed in plastic).

Very much of two-seater design, the 308 GTB has little room even for discarded coats in the cockpit, and requires a wide central tunnel to take water cooling pipework to the front of the car and the gearchange and handbrake linkages to the back. Naturally, the overall height of the car at only 3ft 10in. could pose problems, but the very reclined seating position, wide door opening, and minor intrusion of the roof into the door opening in the vertical plane all combine to allow entry, exit, and the seating position, to be comfortable.

Suspension details of the Dino 308 GTB parallel those of the GT4 2+2, with double wishbones and coil springs front and rear as well as substantial anti-roll bars at both ends. The low body lines do not permit very much suspension travel, and the need to control roll is very important.

The familiar 3-litre V8 dohc engine is mounted transversely in the middle of the car, and the drive is taken from the left end of the engine via the clutch to a set of drop gears that reverse the direction of drive from right-to-left back to left-to-right towards the five-speed all-synchromesh gearbox. The gears sit in their own separate sump partitioned off from the engine sump. The final drive has a limited-slip differential as standard.

Both the 308 GT4 2+2 and 308 GTB Dino engines produce the same power output of 255 bhp (SAE net) at 7,700 rpm. In the case of the GTB, however, dry sump oil lubrication is employed, taking the oil capacity up from 16 to 19½ pints. The reasons for the adoption of dry sump lubrication are legion but added oil cooling capacity, a reduction in engine height, avoidance of oil surge, and greater possibilities for racing development are just a few that influenced the change. As well as the high specific output of the Ferrari's engine, it should not be forgotten that the V8 configuration, at 3-litres capacity, in a car only weighing 25.6 cwt, allows the benefits of considerable torque to be enjoyed. The maximum torque given is 210 lb. ft. at 5,000 rpm, and such a high figure means tht one is not always changing gear to sustain rapid progress.

A significant difference between the GT4 2+2 and the GTB is the wheelbase, which is 8½in. shorter in the two-seater. However, most of this shortening in the chassis is recovered by the longer nose and tail, and the overall length of the two cars is within 2¾ in. The GTB is ½ in. wider than the 2+2 and no less than 3½ in. lower. Suspension details are shared, with unequal length wishbones, coil springs, telescopic dampers all round, and anti-roll bars front and rear.

The body shape of the 308 GTB demands that it can only be a two-seater, but the long tail does permit reasonable luggage accommodation. To get at the boot, one must first raise the engine cover and, with this out of the way, a neat, zipped, pvc cover can be opened to reveal a deep, full width, carpeted area in which the jack, wheelbrace and toolkit also live. There is additional minimal luggage space beneath the front bonnet and again, the contents of the luggage area are kept clean beneath a zipped cover.

Performance and economy

Favourable weight distribution, good aerodynamic shape, excellent gearing and the unusually high power-to-weight ratio ensure that the Ferrari gives exceptional performance. The speeds that it can attain are deceptively fast since there is little or no wind and road noise. For this reason, one could be forgiven for imagining subjectively that the noisier GT4 2+2 is quicker. To dispel this thought, the performance figures for the bigger car are given in brackets in the details that follow.

Wide tyres, the limited slip differential, and inherently good traction ensure that the step-off from rest (even when searching for ultimates) is without drama. Dropping the clutch sharply with the engine revs at 4.500 rpm gives enough immediate wheelspin to give the clutch an easier time. While the revs stay approximately the same, the wheels catch up within 20 ft. or so, when all slack is taken up and the car rockets forward towards the maximum engine speed allowed of 7,700 rpm. First gear takes you to 44 mph, passing 30 mph in just 2.3 sec (2.5 sec) and 40 mph in 3.3 sec (3.6 sec). First gear is out to the left and back in the gated six-position gearchange, but there is enough strong spring centring to ensure that the change up to second gear can be as fast as the hand can move. The improved synchromesh cannot be beaten, and the only considerations that can slow the change are the need for full clutch clearance for each upward (or downward) change and the inertia of the long linkage. 2nd gear takes you onwards to 65 mph with the magic 60 mph mark coming up in 6.5 sec (6.9 sec), when it is time to snatch the gearlever hastily back into 3rd gear to rush the car onwards to 92 mph. 80 mph is reached in the amazing time of 10.8 sec (11.4 sec) and the better shape and lower weight of the 2-seater begin to be noticeable. The change to 4th gear is again a dog-leg forward and out to the right, taking care to push deliberately against the spring pressure. 100 mph comes up quickly in just 17.0 sec (18.1 sec), the ¼-mile mark having passed in 14.8 sec (14.9 sec) at 93 mph (89 mph). 4th gear is good for 124 mph, 120 mph passing in just 25.0 sec (30.3 sec) — perhaps half the time it has taken you to read this paragraph.

Acceleration in each gear reveals

This engine shot shows the air filter removed to reveal the four twin-choke Weber carburettors. The oil filter is positioned conveniently on the engine top. There are two contact breakers, one for each cylinder bank. The engine cover may be removed completely for better access

42

the useful spread of torque, as well as the absence of snatch in the driveline or holes in the torque curve. One can accelerate from as low as 10 mph in third gear, and there is no feeling that fifth gear should be used as an overdrive, since you can pull away strongly in this gear from 30 mph. As one would hope in this sort of car, the gear ratios are ideally chosen, the gaps closing up progressively the higher you go to overcome aerodynamic resistance.

When measuring the maximum speed, a long run in was needed before the car would settle on 7,050 rpm (7,150 rpm on the test car's slightly optimistic instrument), equivalent to 154 mph. This is some 4 mph less than Ferrari's claim. (It should be pointed out that our figures were found with overnight luggage and other test gear aboard which, in a car with a rear boot, can be enough to bring the nose up and to knock off the last few mph.) Such a speed is only attainable in quite still air, the slightest cross wind leading to very disconcerting weaving at speeds in excess of 130 mph.

Naturally, one pays a penalty for such performance in fuel consumption. The overall figure of 19.2 mpg is, however, good for the class, and results from a more-than-average amount of open

Above: Superb attention to detail is shown in the door pulls which are aerodynamically-shaped yet easy to use

road driving and would not be matched if much urban running was included. The best return was a laudable 25.6 mpg over 270 miles while keeping pace with returning holiday traffic in Germany, when more than 80 mph was out of the question. By contrast, the worst result included the performance testing at MIRA, which increased the consumption to little over 17 mpg. The *Autocar* economy formula gives an excellent guide to the economy potential of this particular test car, and one would expect careful owners to better 20 mpg regularly. Thus, with 16¼ gal available in the balanced, twin fuel tanks, the range between fuel stops could be as high as 300 miles; there is no excuse for running out as there is a reserve of 3.3 gal whose use is indicated by a warning lamp in the fuel gauge face. When filling the twin tanks only one filler is used, its screw cap and thoughtful anti-spill bib being hidden behind a louvred flap on the left-hand quarter panel; only the last one-third of a gallon takes extra time to get in.

Though the handbook suggests that an oil consumption of 600 miles per pint may be expected, the test car did not use a measurable amount of multigrade.

Above left: Head-on view shows the diminutive bib spoiler and extraction vents on the wing tops for brake cooling
Above right: Squat tail emphasizes the width of the 308 GTB in relation to its diminutive height. Reversing lamps are neatly built into the rear plastic-covered bumper
Left: The headlamps are driven up and down by links to an electric motor. Should the motor fail, the lamps can be cranked up by hand

Ride, handling and roadholding

The aesthetically satisfying lines of the 308 cannot promise very much wheel travel, and the suspension settings for both springs and dampers are necessarily hard. At low speeds, on less than smooth surfaces, the 308 GTB's ride is very bumpy, with potholes giving rise to much crashing and banging. However, once into the mid-range of the car's performance the ride improves greatly, and only the worst of undulations are not soaked up adequately. At very high speed the ride gives great confidence, with no tendency to pitch, and not a trace of float. Under most conditions the handling is reassuring too, though there is a strong tendency to follow camber changes, especially at the front of the car, and a strong grip on the steering wheel should be avoided.

Curved side windows improve the available door opening. Note the sun visors set into the roof panel and the downward-curving padded roll on the doors that makes a useful armrest. Rake adjustment of the seat backs is by infinitely adjustable wheels; the seat backs can be tipped forward

There is some kickback evident through the steering, and Ferrari argue that some is necessary to give the necessary feel to the steering. One can accept this — the steering gives all the right messages to the hands to tell what the front wheels are doing. Since the predominant characteristic is extra understeer with extra speed, the feeling of increased weight in the steering felt at the steering wheel rim is very pleasantly *pro rata*. Though 3.3 turns of the wheel from lock-to-lock might suggest high-geared steering, one must not lose sight of the dreadful turning circle of around 40ft. The steering ratio could well be higher still without making parking-speed manouevring impossibly heavy.

On the move, the steering is about right, the slightly low gearing helping to avoid any suddenness in the steering reaction. Provided that the air is still, high natural stability is not affected. The behaviour in crosswinds was criticised earlier.

While the handling has been praised generally, it would not be right to leave the subject without reference to throttle open/throttle closed response. While the natural tendency is to understeer, if the accelerator is released in mid-corner the resultant weight transfer does lead to an immediate tightening of the line. This is, of course, no penalty under most conditions (and can be used to advantage) but some caution is needed on wet or slippery roads. Once the natural understeer has been killed by lifting off the accelerator, immediate application of power can be used to set up a very satisfying four-wheel drift which will please the skilled and determined driver immensely. However, the relatively short wheelbase and low polar moment of inertia of the design mean that extreme angles of drift cannot be held despite the quickness and accuracy of the steering.

Michelin XWX radial ply tyres, of 70 per cent profile and 205 section, give excellent roadholding and splendid traction on wet surfaces. They are guilty of mild bump-thumping which suspension compliance does not completely isolate from the car's occupants, but

43

AutoTEST
Ferrari 308 GTB

it is not possible to make the tyres squeal. There is only one recommended tyre pressure setting for all speeds and load conditions, and owners may wish to play around with slightly lower pressures to improve the ride — having little effect on roadholding through maintaining the same differential between front and rear settings. The gains to be made by trying this are marked since the recommended settings are a compromise, leaning towards the full performance potential rather than that which road conditions generally permit.

Brakes

Ventilated disc brakes of generous diameter and thickness are used for all four wheels, and the hydraulic circuit to them is split front-and-rear. A single in-line servo serves for both circuits, the rear of which has a pressure-limiting valve. The handbrake warning light doubles as an indicator of fluid loss from either circuit. The handbrake operates a rather ineffective mechanical linkage to the rear inboard pads.

Though at low speeds there is a feeling that the braking system is over-servoed, this is only to ensure bite from the hard linings and well-cooled discs. At speed such feelings disappear, and the system gives reassuring performance through close progression in efficiency with increased pedal effort, coupled to total resistance to fade. A pedal pressure of 80 lb is needed to produce an optimum 1.00 g, while only just under half this is needed to give 0.50 g when the brakes are cold. As they warm, the pressure required reduces slightly, levelling out at 20/25 lb for 0.50 g and 60 lb for 1.00 g. The handbrake could only hold the car on a 1-in-4 incline and returned an entirely inadequate 0.22 g when used on its own as an emergency brake.

Noise

Such evident refinement as the comfort of the interior of the Ferrari and its eye-catching exterior would be spoilt by an excess of noise. Suffice it to say that the noises the GTB makes are all pleasant ones. The exhaust note is well subdued at most speeds and only full acceleration in confined surroundings could annoy anyone. The note itself is very close to that of the Big Valve Ford Twin-Cam, particularly as installed in the Lotus Elan Sprint. There is just the slightest increase in wind noise at speed and this is not enough to require the radio volume to be turned up.

Only the lowest fan speed of the optional air conditioning is acceptably quiet but higher fan speeds for the system are only required for strictly limited periods.

Equipment and accommodation

As one should surely expect at the price, the Ferrari is well-equipped. Electric windows, tinted glass, a laminated windscreen with tinted top section, heated rear window and leather upholstery are all standard. The options list is confined to air-conditioning, metallic body paint, and wide-rim wheels (carrying tyres of the same size).

Once in the seats the occupants are comfortably positioned, with generous rearward and backrest rake adjustment. Ahead of the driver, there is a very full complement of clear, round instruments, including a speedometer with inset trip and total mileometers (but no kph equivalents), rev counter, and three smaller dials for oil pressure, fuel contents, and water temperature set between them. Set in the binnacle that houses the main instruments is the warning lamp for the hazard warning flashers, as well as a matching warning lamp for generator charge. Set low down to the right of the facia are two further instruments — an oil temperature gauge and a clock.

Major function controls are operated by three fingertip stalks, with wash/wipe to the right of the steering wheel and indicators and lights controls to the left. The splendid air horns are operated by a horn-push in the centre of the Momo leather-trimmed steering wheel. The steering/ignition lock is conveniently positioned to the right of the steering column (on the facia edge).

The driver's office. All instruments are clearly visible through the leather-rimmed wheel though the two (oil temperature and the clock) to the right of the main console are in the shade of the facia. Three fingertip stalks look after the usual functions while the horn button in the steering wheel centre operates strident air horns. The controls on the central tunnel are for air-conditioning, heating and ventilation, heated rear window, hazard warning flashers and windscreen wiper speed. The two air vents beneath the radio are for conditioned air only, those on the facia top may pass a mixture of conditioned and heated air. Behind the handbrake is a useful lockable box which can accommodate a camera

Specification

ENGINE
	Rear; rear drive
Cylinders	8, 90deg V
Main bearings	5
Cooling	Water
Fan	Electric
Bore, mm (in.)	81 (3.19)
Stroke, mm (in.)	71 (2.79)
Capacity, c.c. c.c. (in³)	2,927 (178.6)
Valve gear	Dohc
Camshaft drive	Toothed belt
Compression ratio	8.8-to-1
Octane rating	98RM
Carburettor	4 x Weber 40 DCNF
Max power	255 bhp (SAE) at 7,700 rpm
Max torque	210lb. ft. at 5,000 rpm

TRANSMISSION
Type	5-speed all synchromesh
Clutch	single dry plate

Gear	Ratio	mph/1,000 rpm
Top	0.918	21.8
4th	1.244	16.1
3rd	1.693	12.0
2nd	2.353	8.4
1st	3.418	5.8

Final drive gear	Helical spur (lsd standard)
Ratio	3.71-to-1

SUSPENSION
Front—location	Unequal length wishbones
springs	Coil
dampers	Telescopic
anti-roll bar	Yes
Rear—location	Unequal length wishbones
springs	Coil
dampers	Telescopic
anti-roll bar	Yes

STEERING
Type	Rack and pinion
Power assistance	None
Wheel diameter	14 in.

BRAKES
Front	10.8 in. dia. ventilated disc
Rear	11.0 in. dia. ventilated disc
Servo	Vacuum, direct-acting

WHEELS
Type	Cast light alloy
Rim width	6½ in.
Tyres—make	Michelin XWX
— type	Radial ply tubeless
— size	205/70 VR 14 (105 R 18)

EQUIPMENT
Battery	12 volt 60 Ah
Alternator	55 amp
Headlamps	110/110 halogen
Reversing lamp	Standard
Hazard warning	Standard
Electric fuses	18 fuses, 11 relays
Screen wipers	2-speed + intermittent
Screen washer	Electric
Interior heater	Water valve
Interior trim	Leather seats, pvc headlining
Floor covering	Carpet
Jack	Screw parallelogram
Jacking points	2 each side
Windscreen	Laminated
Underbody protection	Grp body, mastic in wheel arches

MAINTENANCE
Fuel tank	16.3 Imp. galls (74 litres)
Cooling system	31.6 pints (inc. heater)
Engine sump	19.4 pints SAE 10W/50
Gearbox	7 pints SAE 85W/90
Grease	4 points
Valve clearance	Inlet 0.008-0.010 in. (cold) Exhaust 0.012-0.014 in. (cold)
Contact breaker	0.012-0.015 in. gap
Ignition timing	6 deg BTDC (static) 33 deg BTDC (stroboscopic at 5,000 rpm)
Spark plug type gap	Champion N7Y 0.024-0.028
Tyre pressures	F 28, R 34 (normal driving)

Maximum Speeds

Gear	mph	kph	rpm
Top (mean)	154	248	7,050
(best)	154	248	7,050
4th	124	200	7,700
3rd	92	148	7,700
2nd	65	104	7,700
1st	44	71	7,700

Acceleration

True mph	Time secs	Speedo mph
30	2.3	28
40	3.3	40
50	5.1	50
60	6.5	61
70	8.7	71
80	10.8	81
90	13.8	91
100	17.0	102
110	20.4	112
120	25.0	124

Standing ¼-mile: **14.8 sec**, 93 mph
Standing kilometre: **26.9 sec**, 124 mph

mph	Top	4th	3rd	2nd
10-30	—	—	5.1	3.3
20-40	—	6.4	4.1	2.7
30-50	9.5	5.5	3.5	2.6
40-60	8.4	5.0	3.6	4.8
50-70	7.8	5.1	3.6	—
60-80	7.2	5.1	3.9	—
70-90	7.7	5.5	4.4	—
80-100	8.8	6.0	—	—
90-110	9.2	6.7	—	—
100-120	10.2	8.0	—	—

Consumption

Fuel
Overall mpg: 19.2
(14.7 litres/100km)
Constant speed: No figures; system incompatible with Test equipment.

Autocar formula
Hard driving, difficult conditions 17 mpg
Average driving, average conditions 21 mpg
Gentle driving, easy conditions 25 mpg

Grade of fuel: Premium, 4-star (98 RM)
Mileage recorder: 2.5 per cent over reading

Oil
Consumption (SAE 10W/50). Negligible

Brakes

Fade (from 70 mph in neutral)
Pedal load for 0.5g stops in lb

	start/end		start/end
1	25-30	6	30-25
2	25-30	7	20-20
3	30-35	8	20-20
4	35-35	9	20-25
5	30-30	10	20-25

Response from 30 mph in neutral

Load	g	Dist.
20lb	0.25	120.0ft
40lb	0.50	60.0ft
60lb	0.75	40.0ft
80lb	1.00	30.1ft
Handbrake	0.22	137.0ft

Max. gradient 1-in-4

Clutch Pedal 60lb and 7in

Test Conditions

Wind: 0-5 mph
Temperature: 18 deg C (65 deg F)
Barometer: 29.4 in. Hg
Humidity: 20 per cent
Surface: dry asphalt and concrete
Test distance 2,400 miles

Figures taken at 9,400 miles by our own staff at the Motor Industry Research Association proving ground at Nuneaton, and on the Continent.

All Autocar test results are subject to world copyright and may not be reproduced in whole or part without the Editor's written permission.

Regular Service

Interval

Change	3,000	6,000	12,000
Engine oil	Yes	Yes	Yes
Oil filter	No	Yes	Yes
Gearbox oil	No	Check	Yes
Spark plugs	No	Check	Yes
Air cleaner	No	No	Yes
C/breaker	No	Check	Yes
Total cost	**36.13**	**54.53**	**110.07**

(Assuming labour at £4.30/hour)

Parts Cost

(including VAT)
Brake pads/shoes (2 wheels) — front £10.60
Brake pads/shoes (2 wheels) — rear £8.80
Silencer(s) £223.56
Tyre — each (typical advertised) £66.12
Windscreen £198.72
Headlamp unit £18.00
Front wing (nose section complete) £358.56

Warranty Period
12 months unlimited mileage

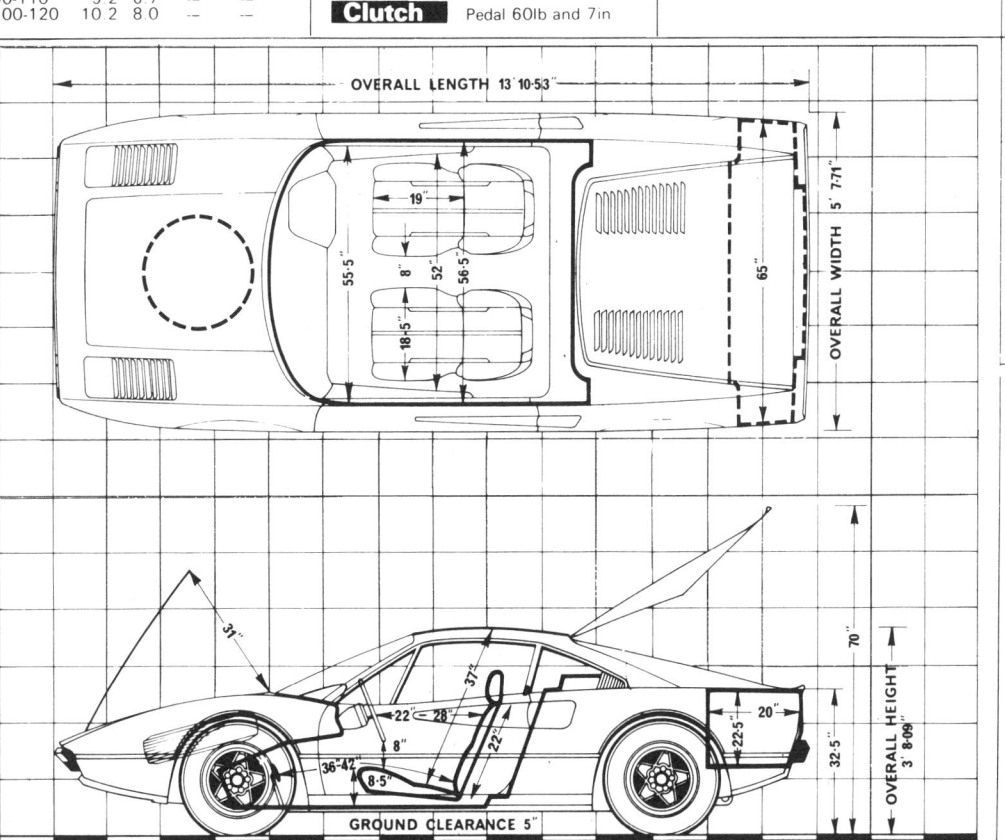

Weight

Kerb, 25.6 cwt/2,870 lb/1,300 kg.
(Distribution F/R, 41/59)
As tested, 28.7 cwt/3,220 lb/1,460 kg
Boot capacity: 8.6/1.4 cu. ft.
Turning circles:
Between kerbs L, 40ft 8in; R, 40ft 6in
Between walls L, 42ft 6in; R, 42ft 4in
Turns, lock to lock 3.3

Test Scorecard

(Average of scoring by *Autocar* Road Test team)

Ratings:
6 Excellent
5 Good
4 Better than average
3 Worse than average
2 Poor
1 Bad

PERFORMANCE	4.83
STEERING AND HANDLING	4.58
BRAKES	4.00
COMFORT IN FRONT	3.75
DRIVERS AIDS	3.75
(instruments, lights, wipers, visibility etc.)	
CONTROLS	3.75
NOISE	4.83
STOWAGE	3.00
ROUTINE SERVICE	3.10
(under-bonnet access, dipstick etc)	
EASE OF DRIVING	4.00
OVERALL RATING	**3.97**

Comparisons

Car	Price £	max mph	0.60 sec	overall mpg	capacity c.c.	power bhp	wheelbase in.	length in.	width in.	kerb weight	fuel gall	tyre size
Ferrari 308 GTB	11,997	154	6.5	19.2	2,927	255(SAE)	94	156½	67¾	25.6	16½	205/70x14
Lamborghini Urraco S	9,434	143	8.5	18.7	2,463	200	96½	167¼	69½	25.8	17½	205/70x14
Jaguar XJS	10,507	153	6.9	15.4	5,343	285	102	191¾	70½	34.8	19	205/70x15
Porsche Turbo 3.0	17,500	153	6.1	18.5	2,993	260	89½	169	69	22.4	17½	205-225/50x15
De Tomaso Pantera	11,500	159	6.2	13.0	5,763	330(gross)	99	168	72	27.8	19	185-215/70x15
Maserati Merak	9,430	135	8.2	17.4	2,965	190	102¼	171	70	28.5	18½	185-205/70x15

Ferrari 308 GTB

On the facia-top, there are three rotating air vents that have flaps which allow the air to be directed. These air vents are supplemented by two vents on the central console that can be used for conditioned air only. In normal Ferrari practice, the air distribution to each side of the car via the car's heating system is individually adjustable, though there is only one fan to boost either or both sides. There is a single temperature control for the car's water-valve heating system and two rotating controls for the air-conditioning, one to select temperature and the other to control air-flow speed. All the controls for heating, ventilation, and air-conditioning are positioned on a panel between the seats formed by the top of the central tunnel. At its front, the tunnel has the gated gear change, an illuminated ashtray, a cigarette lighter, and the choke control lever. The front three of the centrally-mounted switches on the tunnel control the heated rear window, the heater fan, and the speed of the windscreen wipers — intermittent and continuous wiping being available on high or low speeds.

On rhd Ferraris, the releases for the front and rear compartment lids are both brought to the driver's side of the car — both releases have an emergency pull should the main one break. Also unique to Ferraris sold in the UK is a Maranello-converted circuit for the electric windows, which allows them to be operated with the ignition switched off.

Other sensible details that show care for rhd owners includes windscreen wiper conversion, a footrest for the left foot alongside the clutch pedal, and transfer of the operating switches for both side windows to the padded roll that forms the top of the driver's door trim.

The big foot pedals are arranged to allow heel-and-toe action with an organ pedal-type accelerator. The driving position is typical Italian with a steering wheel set fairly flat and the need to adopt a knees-up/arms stretched driving position.

Living with the Ferrari 308 GTB

Cars with high cruising speed potential are wasted if they are not used for long journeys. Too often, such potential is spoilt by inadequate comfort, noise, and tiring controls. In the Ferrari's case, only two serious drawbacks take the pleasure out of long-distance touring. The least serious of these is the clutch pedal effort which, at 60 lb, is very much in the heavyweight class, and is made more wearing because full clutch travel of 7in. is needed to engage the gears smoothly and quickly, and to avoid grating them when selecting reverse. The worst drawback concerns the choice of covering for the facia top. In either strong

Above: With the get-you-home spare tyre removed, the battery can be reached. With the wheel in place, there is some space for small luggage items

Right: Opening the engine cover also reveals the main luggage boot beneath its zipped cover. The space is carpeted and holds a comprehensive toolkit as well as the jack and wheelbrace

Below: the fuel filler is neatly hidden behind a louvred flap on the left rear quarter panel. The plastic bib avoids paint damage through careless filling

overhead lighting at night, or when heading towards the sun during the day, the reflections in the windscreen are bad enough to warrant being called a hazard.

All other aspects of comfort, like seats that do not cause aches, good visibility (except at angled junctions), and a wealth of on-board stowage space, may be taken for granted. The optional air-conditioning really spoils the occupants and, by mixing warmed air from the heating system, it is possible to get that elusive combination of warm air to the feet and cool air to the face.

Only short sections of the front and rear of the car are out of sight when manoeuvring, and the GTB is only slightly worse than its more prosaic GT4 2+2 brother in this respect. The wipers are not fast enough on the higher of their two speeds, and the contents of the washer bottle are soon used up. The halogen headlamps give splendid light on main and dipped beam, and are well up to the performance of the car. However, since the headlamps are stowed in electrically-retractable pods, they cannot be used for daylight flashing, there being no auxiliary lamps.

Starting the Ferrari's stirring engine should present no difficulties, provided some simple rules are observed. The choke need only be used in the coldest of weather, three dabs on the accelerator usually sufficing when starting from cold. When the engine is hot, just a third of the accelerator travel should be used to get started, being careful to avoid further movement of the accelerator before the engine has fired up. Despite its lengthy run, the accelerator linkage is excellent, giving smooth, progressive response that helps to avoid jerkiness in driving. Until the engine is warmed thoroughly, and especially when it has just come off the cold-start accelerator mechanism, full throttle at low revs should be avoided as there can be a sudden gulp and some hesitancy. There is a lesser, similar effect noticeable even when the engine is hot, when some hesitancy occurs in the transition from idle to main jets.

One imagines that engine access will not be too much of a problem to most owners of this rather expensive car. This is as well since the access is not good. To change the plugs of the front bank of cylinders would take all day, and there seems to be twice as much of everything as there needs to be. Two coils, two timing belt tensioners, two fuel filters, and two contact breakers all compound to frighten the life out of the average garage mechanic. In fact, with the help of the excellent handbook, all relevant service items could be handled by the enthusiastic owner/driver, and there would be great satisfaction in avoiding the high service costs shown in Service Details.

Such obvious items like the dipstick, brake fluid reservoirs, and battery are all easily accessible in either the front or rear of the car, and routine weekly checks pose no problems.

Little has been said so far in this Test about the glass-fibre bodywork. The reason is simply that, if you did not know what the car was made of, you would never guess. There is no smell of glass-fibre, the doors shut with a delightful dull thud, there is no sign anywhere of star-crazing of the gel coat, and the finish of the paintwork is as good as that which can be expected on steel. There is the obvious advantage of saved weight, avoided minor parking blemishes, and absence of rusting. For these reasons as well as many others, the Ferrari Dino 308 GTB is guaranteed a place in treasured collections in the years ahead.

Conclusions

The Ferrari is so good in most respects that the areas of shortcomings stand out like sore thumbs. Maranello Concessionaires are working on an alternative facia covering to avoid the screen reflection problem, but there are still some areas that need improvement. The clutch pedal effort is unjustifiably high and out of character with the light weight of other controls. The large turning circle and length of the 308 do not allow for parking easily in confined spaces. When into a space, the absence of a check strap on the doors could allow them — and other cars — to be dented too easily while one struggles from the car. On the performance side, the loss of high speed stability in crosswinds is regrettable and hard to understand.

All round, one is lost in admiration for the superb quality of the mechanical engineering, the standard of finish of the body and interior details, and the all-round efficiency. It is the best Ferrari we have yet driven. □

MANUFACTURER:
Ferrari S.E.F.A.C.,
Maranello, Italy.

UK CONCESSIONAIRES:
Maranello Concessionaires Limited,
Egham By-Pass,
Surrey.

PRICES	
Basic	£10,254.00
Special Car Tax	£854.50
VAT	£888.68
Total (in GB)	**£11,997.18**
Seat belts	Standard
Licence	£40.00
Delivery charge (all UK)	£35.00
Number plates	£8.50
Total on the road (exc. insurance)	**£12,080.68**
Insurance	Group 7 (on application)

EXTRAS (inc. VAT)
Air conditioning*	£374.40
Metallic paint	£152.10
7½in. wide wheels	£222.30

Fitted to test car

TOTAL AS TESTED ON THE ROAD	**£12,455.08**

308 GTB
FAZZAZ ON THE FLY

Merry Christmas, and here comes Bob Bondurant with your present

BY JOHN LAMM

WHEN IT WAS all over and Bob Bondurant had put 40 hard laps on Rick Schrameck's Ferrari 308 GTB at Sears Point International Raceway, he pulled off the track, shut off the 3-liter V-8 and climbed out to the accompaniment of, "What's she like, Bob?"

He grinned and said, "It was a nice car when it was new."

If most of us had even tried to drive the Ferrari as quickly as Bondurant had, the car would most likely have been used up, to say nothing of ending up broken and possibly destroyed. But driving cars quickly and cleanly is what separates the better racing drivers from the rest of us and the very reason we asked Bondurant to drive the European version 308 and give us his impressions of this newest Ferrari.

He began: "The main thing I like about the 308 is that the car is so neutral you can do anything you want with it. In other words, it doesn't understeer badly or oversteer badly. It is an easy car to drive so you can go through a turn, lift off to make the car oversteer and then just really plant your foot in it and it takes off.

"This is so much better than a Daytona. That car steers heavy, sits high and has a bag of power, but you can't really use the power like you can in this car. This is more like driving a Ford GT40 when they first came out. The Daytona oversteers a bit and it understeers a bit and it's just not as easy to drive as the 308. You also have to be careful when you're driving fast in a Daytona, while with the 308 you can just drive and not worry about the car doing something unpredictable in the middle of a turn. I would say this would be a much safer car for the average guy to drive. It has very smooth weight transfer and transition in cornering, which is important for the driver and passenger as then everything seems sort of nice when you're going fast.

"You get into a Porsche and they work fine up to a certain speed and then they can go into an understeer or an oversteer; kind of a snap oversteer. With the 308, even the guy who isn't quite as adept could take it that much closer to the limit.

"The 308 has a nice smooth acceleration; it doesn't explode all of a sudden. It just comes on nice and strong, but smoothly. And it has a good power range, with not only a decent bottom and top end, but also a good mid-range. I only ran it to 7500 rpm, mostly around 7200 . . . you don't need any more than that, really.

"The brakes are phenomenal, because normally if we were out there in a street car as long as we were with the Ferrari, the brakes would have gone away and started to smell. They have a nice soft feel—no, not soft but predictable, with good pedal pressure. They don't grab or pull one way or the other, but just squat the car down very nicely and stop it.

"This is an easy car to heel-and-toe. Whoever designed it took his time and put together a car that really fits the driver. It is very comfortable to drive as far as seating position goes; the leather-covered steering wheel is also comfortable. And the car has enough headroom.

"The steering is very nice, with a light feel to it—like a Dino. It's not only very light, but also predictable, and it transmits back to the driver what the car is doing, which I feel is important. It's so light that for all I drove out there I never got tired. If I had driven a Daytona that far I would have really been pooped.

"About the only thing I can say about the car that is negative at all is that it has a typical kind of Ferrari gearbox—somewhat stiff. I think after it's driven for a while even that would loosen up.

"Of course, the car looks sleek as hell; like it's already moving quickly. The 308 has a nice finish and is pretty well detailed. I think the BBS wheels really add a lot to it (Our test car was shod with intermediate rain/dry 225/60VR15 Pirelli P7s with hand-cut treads mounted on 15 x 8.5-in. BBS wheels, supplied by Intermag—Ed.) although the stock wheels also look good on it."

How does the 308 GTB rank among all the cars Bondurant has driven?

"I like that Mercedes of mine very well (a 450 SEL), but then we're talking about a sedan that handles like a sports car. The 308 is probably the best sports car I've ever driven, particularly bearing in mind that this is a street machine and it would be unfair to compare it to pure race cars.

"I want one."

But then who doesn't, the only barrier between most of us and a fiberglass 308 being the $29,525 dealers are going to require before they deliver one. Schrameck, of Ray Ramsey's Ferrari of San Francisco, had hoped to have a U.S. version for us to test, complete with anti-smog mufflers and bumper mustache. The boat arrived on time, but was a bit tardy to the docks, so you'll have to hold your prancing horses for a couple of months until you can start checking acceleration times and skidpad results.

It will be interesting to compare the U.S. version to the 308 European model Paul Frère tested for our February, 1976 issue. Then he got a 6.4-sec 0–60 time, with a 14.6-sec quarter-mile and a top speed of 154 mph.

Oh, and if you can scrape together the down payment for a 308 (you should know most customers pay cash), you can be sure it won't depreciate as soon as you drive out the front door. With the GTB's already excellent reputation, the disappointment of the wedgy 308 GT4 model and the discontinuation of the Daytona, you won't lose a cent.

The way some Ferrari folks talk about the potential value of the 308 GTB, perhaps they should list it on the New York Stock Exchange . . . or offer Ferrari futures at the Chicago Board of Trade.

They're certainly more interesting than soybeans.

Ferrari 308 GTB

A successor to the Dino and little brother to the Boxer Berlinetta

by JOHN CHRISTY

Road Test

Of all the automotive confections that have rolled out of the doors of the Ferrari plants in northern Italy, perhaps the most pleasurably sensual was the Dino 246 GT. Other than the Barchetta 212 and the later Berlinetta 250 GT, there had not been a road-going Ferrari, a Ferrari that could be driven directly from the dealer's door, that had a closer kinship to the cars that the Ferrari factory raced. Small, 2-seated, mid-engined, with sensuously curved bodywork, the 246 GT bore a direct relationship to the 206 SP, 275 LM and 330 P, racing sports cars of the purest sort and all-conquering in their time.

Yet it did not bear the Ferrari name, and it was at first looked upon by the "real" *Ferraristi* with a certain amount of condescension as not being a true Ferrari. However, realization of the Dino's honest worth came quickly, and it has become one of the most sought-after, at least in the U.S., of all the works of the master of Maranello. In short, it became a classic in its own time and, save for the redoubtable and virtually unobtainable Boxer Berlinetta, is the only Maranello product that has never sold, used, for less than the price it brought when new.

So, it came as a distinct shock when, with almost no warning, the Dino 246 stopped production two years ago and was replaced by an angular 2+2 device called the Dino 308 GT/4. About the only relationship the new offering bore to the 246 was the fact that it carried the same name and was mid-engined. Even this last was different, in that the delightful, high-revving, Dino Ferrari-conceived V-6 was replaced with a 3-liter V-8, albeit still 4-cammed and multi-carbureted.

An unwelcome model change by a manufacturer who produces as few cars as Ferrari would be of little moment, were it not for the fact that Ferrari is unique. The uniqueness lies in the fact that a host of automotive enthusiasts around the world, most of whom will probably never be able to own a Maranello product, *care* about Ferrari and all his works. Perhaps it is simply that it's comforting to know that in our overprotected Western society there is still such a thing as a Ferrari automobile. In any event it exists, and when an unfortunate design rolls out of the Maranello

49

plant the public reaction is much as if a beloved master artist suddenly began turning out tasteless daubs.

Thus it was a matter of moment when rumors and then sneak photos began filtering out of Italy indicating there would be a new 2-seater version of the 308 and, most important, that it would be done by Pininfarina, who had designed the original Dino 246. The photos that followed the rumors revealed bodywork that was an attractive cross between that of the fierce Boxer Berlinetta and the original 246.

The new car, designated 308 GTB, is now here and there's good news and bad news. The good news is that, fiberglass bodywork notwithstanding, it is all Ferrari, even unto the nameplate and prancing horse insignia. It is also everything a racebred Ferrari should be: lean, taut and full of sound and fury, a little brother to the redoubtable Boxer and a definite descendant of the 246. The bad news is that, at $30,000, it costs almost exactly twice as much as a Dino 246, thus putting it well beyond the reach of the moderately prosperous enthusiast who might conceivably scratch up the price originally commanded by the latter car. At least one could dream of rescuing the earl's daughter from a sticky wicket and being suitably rewarded with the price of a Dino. However, what with inflation being what it is, maybe the rescue price of earls' daughters has suitably risen—or perhaps we must raise our sights a bit and confine rescue efforts to daughters of dukes, industrialists and heads of state.

Price aside, the new 308 GTB begs comparison with its Dino 246 forebearer. They are almost identical in all dimensions, the differences in most cases being read in fractions of an inch. The 308, despite its glass fiber bodywork, is 300 lb heavier than was our test 246 (December 1972) but the 45 additional horsepower of the V-8 can be expected to, and does, make up for weight penalty. The chassis design is virtually identical in the two cars, consisting of a stout tubular frame with independent double A-arm, coil-sprung suspension at each corner. The major difference between the two chassis designs is in the extra structure around the engine bay of the 308 which wasn't needed in the steel-bodied 246.

Though the chassis and suspension designs are nearly identical, the execution and suspension tuning are distinctly different. The springing on the 246 was surprisingly soft, and the shocks were calibrated to give moderate control on jounce and very strong damping on rebound. The ride was beautifully soft and supple with superb control over rough surfaces. In spite of the soft ride, the road-holding was impeccable, the car clinging fiercely in a turn with no perceptable lean.

The 308 is quite different in that the springing is harder and the damping would appear to be more equal on jounce and rebound. It lacks the silky suppleness of the 246, and while still quite comfortable, the feeling is somewhat harsh. The effect is more like a racing car than a road machine and, on smooth surfaces, feels as though it is sticking better and that, *in extremis,* it could be set into a drift. On the 200-ft circle the Dino 246 pulled a near-race-car .87 g lateral acceleration reading. We have rated the 308 marginally better at .879 g, which was the average of several laps around the circle. The strip chart shows that we actually reached into the .9 region, but we were afraid to keep it up because the lateral force pulled the engine oil away from the pickup, causing a momentary drop in oil pressure. There is no doubt that it clings like a limpet on smooth surfaces, but we were led to wonder if the 246 might not do better on rough secondary roads.

In terms of straight-line performance the 308 and 246 could hardly be a better match. With the exception of the Dino's slightly lower shift points, the two acceleration curves are almost identical. The Dino, hampered with considerably less emissions equipment, did the standing quarter mile in 15.25 sec at 92.11 mph. The 308 does the number in 16.9 sec at 90.8 mph. It could be that the major difference was in the shifting. The 308 gearbox was stickier or more balky than was that of the 246, and each shift in the 308 took possibly a half-second more time than did the shifts in the Dino. Given that extra second lost in the two shifts it took to get to the quarter mile mark, the times would be more nearly identical. We had heard that the 308 tended to be a bit delicate when being taken off the line and that it wouldn't take to hard launches, so we motored off the line and then nailed it as the revs got to around 3000. However, the speed traces showed that it broke loose at that point, so we tried winding it up to 3000 rpm and then let in the clutch— we didn't side-step the clutch but just let it in smartly. The result was a satisfying strip of rubber and a g-force trace that rocketed up to .65 as the fat Michelins bit the pavement. The 308 *will* take off smartly when so bidden. However, it still isn't a good idea to savage it on take-offs. No Ferrari is a drag machine, and the

Ferrari 308 GTB

308 is no exception.

The real thrill of the 308 does not lie in how quickly you can launch it but in how quickly it can be motored once launched. This car, like the Dino, will quickly let you know that, unless you are very good indeed, it exceeds your capabilities. You can easily do things with it that you wouldn't dream of doing in lesser machinery, and what might seem merely brisk motoring to you can look absolutely suicidal to an outside observer. Thanks to the fact that the engine is set laterally, acceleration from any moderate speed is deceptive. There is no lateral rocking couple in the engine to clue you to your rate of acceleration. The effect is visual and visceral, and a hard poke at the throttle in an intermediate gear can bring the future toward you very quickly indeed.

A Dino driver would not feel out of place in the 308. The cockpits and the general layout are much the same, but there are subtle differences, by no means all of which are in the newer car's favor. The seats of the 308 are an improvement over those of the 246, at least to most people, in that they are more adjustable, having a provision to change the rake where the Dino seat did not. They are also more resistant to the inevitable wear that is a built-in problem with the mid-engine exoticar. The steering wheel of the 308 is a touch larger than was that of the Dino, and though the actual turns lock-to-lock are nearly the same, the effect of the larger wheel is slower steering.

The major difference lies in the dash. The Dino's dash was pure Ferrari, while that of the the 308 shows a definite Fiat influence, that firm now controlling Ferrari. In the case of the Dino, all of the instruments were under a deeply recessed, hooded pod, with the small controls for heater, defroster and vent fans to the right. The whole dash was covered

continued on page 106

Specifications:

GENERAL
Importer	Modern Classic Motors Reno, NV 89505 and Chinetti-Garthwaite Imports Paoli, PA 19301
Number of U.S. dealers	40
Warranty	1 yr/10,000 miles
Base list price POE	West Coast: $29,525 East Coast: $28,500
Options on test car	None
Price as tested	$29,525

POWER UNIT
Location	Ahead of rear axle
Type	Liquid-cooled V-8
Valve gear	4 OHC
Bore & stroke	3.19 x 2.80 in.
Displacement	179 cu. in./2926cc
Maximum net power	240 bhp at 6600 rpm
Maximum net torque	195 lb/ft at 5000 rpm
Compression ratio	8.8:1
Recommended fuel	Premium
Carburetion	4 Weber 2-bbl
Emissions control	Air pump, thermal reactor
Ignition	Distributor

DIMENSIONS
Wheelbase	92.1 in.
Track, front	57.87 in.
rear	57.48 in.
Length	166.5 in.
Width	67.75 in.
Height	44.09 in.
Ground clearance	3.5 in.
Fuel capacity	21.1 gal.
Luggage capacity	5.3 cu. ft.

CHASSIS
Body/frame	Tube frame/fiberglass body
Suspension, front	Unequal length A-arms, coil springs, tube shocks, anti-roll bar
rear	Unequal length A-arms, coil springs, tube shocks, anti-roll bar
Steering system	Rack and pinion
Ratio	NA
Brake system	Hydraulic, vacuum-assisted
Front brakes	Vented disc, 10.8 in.
Rear Brakes	Vented disc, 11.0 in.
Wheel rim size	Alloy, 14 x 7.5
Tires	Michelin XWX, 205/70VR x 14

DRIVETRAIN
Transmission	5-spd transaxle
Gear ratios:	
Fifth	0.95:1
Fourth	1.24:1
Third	1.69:1
Second	2.37:1
First	3.58:1
Final drive ratio	3.71:1
Differential	Integral
Drive wheels	Rear

Data:

TEST CONDITIONS
Weather	Clear
Temperature	75° F
Altitude above sea level	1008 ft
Pavement	Tarmac
Test car odometer reading	5600

SPEEDOMETER ERROR
Mean velocity error	1.6%
Odometer error	.3%

WEIGHT
Curb weight, full tank	3090 lb
Distribution, front/rear (%)	42/58
Test weight, half tank	3280 lb

SPEEDS IN GEARS
First	39 at 7500 rpm
Second	63 at 7500 rpm
Third	90 at 7500 rpm
Fourth	120 at 7500 rpm
Fifth	150 at 7000 rpm (est.)

ACCELERATION
0-30 mph	3.0 sec
0-40 mph	5.1 sec
0-50 mph	6.6 sec
0-60 mph	8.2 sec
0-70 mph	11.1 sec
0-80 mph	13.4 sec
0-90 mph	16.9 sec
0-100 mph	20.3 sec
Standing quarter mile	17.0 sec
Speed at end of quarter mile	90.8 mph

BRAKING
30-0 mph	33.3 ft
60-0 mph	137.3 ft
Fade rating	nil

MANEUVERABILITY
Steering wheel diameter	14 in.
Turns lock-to-lock	3.3
Turning circle diameter	39.5 ft
Tire pressures, front/rear	28/34 (normal)
Max. speed on 200-ft-dia. circle	37 mph
Maximum lateral acceleration	.879 g
Speed at onset of tire noise	30 mph
Time through slalom	11.8 sec

FUEL ECONOMY
Mileage on 73-mile test loop	15.2 mpg
EPA weighted average mileage	NA
EPA city mileage	NA
EPA highway mileage	NA

NOISE LEVEL
Interior at idle	67 dB
Interior under full acceleration	86 dB
Interior at 30 mph	74 dB
Interior at 60 mph	77 dB

ROAD TEST:

Ferrari 308 GTB

A radical new message from *El Commendatore*.

Here is a car that's going to make the purists grind their teeth. The thing says Ferrari all over it, prancing horses everywhere you look. You can see six of those noble little Ferrari trademarks right from the driver's seat and that is without even craning your neck. Walk around the outside, and you'll see more—front, back and both sides—proclaiming for all the world that this is the latest message to the faithful from *El Commendatore*.

But the purists, bless their over-informed hearts, know perfectly well that this little thigh-high sweetie officially called Ferrari 308 GTB is nothing but a social-climbing Dino. Look underneath if you don't believe it. The engine is the same all-aluminum, sidewinder V-8 bolted to the same five-speed transmission that you'll find in the Dino 308 GT4. The suspension pieces are identical too. And so are the wheels. What's old Enzo trying to pull here anyway, sticking his rampant horses to a bogus Ferrari?

It must be hell being a purist—fogs the mind and blocks the vision. You can't see the truth for the facts. Now, to be completely correct, this new Ferrari shares virtually all of its major mechanicals with the Dino 308 GT4. And it has a V-8, not one of the glorious V-12s, with belts instead of chains driving the four camshafts. So there is none of the old clatter, nothing to cock an ear to two blocks away. As a final indictment, the body is made of—can it really be true?—fiberglass. Like a Corvette. Or a boat. If you're the sort who weighs the merits of an automobile by the specifications printed on the back of brochures, you're going to conclude that the EPA and the DOT and the old rocking chair have taken a firm grip of Mr. Ferrari, and he is easing himself into retirement on the strength of past accomplishments.

And that, in one short paragraph, is why it is so dangerous to judge a car by the ancestry of its parts. Because this latest Ferrari is a remarkable machine, a truly happy collection of ideas and components that really work well together.

Visually, it's a first-round knockout, just the right blend of old Pininfarina Curvaceous and Modern Wedge. The fenders swoop with undisguised joy while the nose droops in the best aerodynamic tradition. And there are enough slots, scoops and vents to satisfy those who want the Functional Look. In red, the effect is stunning. Ferrari's red seems to get more intense each year: The test car made fire trucks look pale. It also made fire trucks with all of their chrome fitments look frivolous. The Ferrari is trimmed entirely in black—bumpers, taillight surrounds, door handles, everything. Which makes the red even redder.

If you are going to spend big money for a car, $28,780 in this case, it's just assumed that the car will have visual flare in proportion to its price. But with rare exceptions—the 250 GT Berlinetta Lusso and the GTO 64, to name two, and these are more than 10 years old—Ferraris generally have had frumpy styling. So even at first glance, this new GTB is a radical departure from the norm. But then this car goes against the Ferrari tradition in a number of ways. It would appear that its designers approached the problem of a mid-engine GT coupe with an open mind. And as a result, the GTB is one of the most logi-

53

FERRARI 308 GTB

cal, comfortable and easy-to-drive mid-engine cars ever built.

If you think back a bit, you will observe that although mid-engine cars have been the dominant theme in racing for about 15 years, the concept has not fared well on the road. Usually, designers got the proportions wrong. DeTomaso's efforts, the Mangusta and the Pantera, are prime examples. He tried to have a V-8 engine between the wheels and at the same time maintain a wheelbase typical of front-engine cars (both the Pantera and the Corvette measured 98 inches). The result was a short cockpit with little legroom and vertical seats positioned very close to the floor. Calling this a torture chamber would be a rather severe description, but you get the idea. Another source of discomfort was the combined effect of the steeply raked windshield and the short roof. The top of the glass was right over your forehead, and all of the sun's energy beamed right in full force on your lap. It was like sitting in a Radarange.

With the GTB, Ferrari has handled all of these dimensions very nicely. The wheelbase is even shorter than that of the Pantera—only 92 inches—but the transverse engine mounted just forward of the rear wheels is a very compact arrangement, and it does not intrude on cockpit space. So the GTB's interior dimensions are much more generous than a Pantera's. Your legs must angle slightly toward the center of the car to clear the arches for the front wheel, but this is less bothersome than in, say, a Porsche 911. And although the windshield angles back severely, it starts from a cowl that is well forward and finishes accordingly, leaving plenty of roof for shade. Overall, the GTB is a very compact car. It's just over a foot shorter than a Corvette and strictly a two-seater. Baggage space

> **You have the feeling you'll be well taken care of, come what may, even if that means 500 miles in one afternoon.**

provides about the same volume as in a Corvette, but the shapes of the compartments in the two cars are different enough to make you use different bags. All of the Ferrari's luggage space is rearward of the engine, and it runs the full width of the car. You get to it by lifting the engine lid and then unzipping a vinyl tonneau cover. The space is modest, but at least one golf bag should fit, for those of you who accept or reject cars on that basis. But regardless of what you like to carry along on your travels, if it won't fit in this trunk, chances are it won't go anywhere. The front compartment is entirely filled with the spare tire and brake cylinders, and there is barely enough room behind the seats for this magazine. All that's left is a pocket in each door and a locking compartment between the seats.

But if the GTB is obviously not a pack horse, it still has practical aspects that may have escaped your first notice. The fiberglass body should be easier to repair after a crunch than limited-production steel body panels. Chances are that some enterprising shop will make a mold from which replacement body sections could be lifted for the cost of dune-buggy bodies. Considering the brand name we're dealing with here, the markup might be a bit higher, but the net should still be less than factory-fresh equivalents from across the sea. Fiberglass is really quite a logical body material for this sort of machine. Manufacturing costs for a short run of cars should be less than with steel, and the finished product won't rust. It's also fairly dent resistant as well. A light hit on fiberglass may scratch the paint, but the body is resilient enough to resume its former shape with no permanent damage.

To be sure, fiberglass brings its own problems. You don't have to ride one block in the GTB to know its body material. There is a sort of dull clunking sound, a sensation you feel as much as hear, whenever you hit bumps. This Ferrari, all Corvettes and some Lotuses have it. There is nothing annoying about it. It's just there. And in our minds, a fair trade for the rust resistance.

When you mention fiberglass, people usually conjure up visions of fenders rippling like mill ponds. This isn't the case with the GTB. We inspected the body surfaces quite carefully and found them to be generally within the range of what you'd expect of steel cars. Interestingly, the Ferrari's front deck lid is made of aluminum, and the surface finish of it is no smoother and no rougher than the glass nosepiece surrounding it.

Although the GTB's entire body, with the exception of the front lid, is made of fiberglass, the construction of the car itself is not much different than metal-bodied Ferraris. All of them use a frame welded from simple brackets and steel tubing that's usually rectangular. The tubes are always sprayed black, and you can always see rust around the welds, even on new cars. This is typical of all the Italian exotic cars, and if the sight of rust blemishing your handwrought frame makes you nervous, it might be best if you didn't look so close.

Certainly the GTB offers other visual attractions. The interior is done up in the Michelin Man style now coming into vogue with the Italian coachbuilders. Most surfaces in the cockpit are covered with huge pillowy rolls of vinyl. Pininfarina gave the padded-cell look to the Lan-

Oh, the Advantage of Owning a Ferrari Dealership

• You might think that the Ferrari dealers of this nation would be stumbling over themselves to lend this magazine a test car. They are not. The only one to step up to its solemn duty was International Motorcars Corporation, which also happens to be the newest Ferrari dealership in the country. But don't think that International Motorcars came through because it was too green to know any better. Instead, the reason had to do with geography—because the newest Ferrari store just happens to be located smack in the heart of Jackson, Mississippi, and business is done differently in that part of the country.

As you would expect, there is not a surfeit of Ferraris in Mississippi. Until just recently, the exact count was seven. And four of these were owned by two men from Jackson: Charlie Kemp, better known for his exploits in road racing, and Sam Scott, a lawyer of no modest skill. Together they started International Motorcars, not to populate the state with more Ferraris but as a way to get dealer tags for their own cars. In Mississippi, the license plates for a Ferrari run about $300 per car per year, but as a dealership, they can get all they want for the lump sum of $160. Scott drew up the papers, and they were in business.

The idea of actually becoming a Ferrari store came later. "To be honest, we just thought it would be a neat thing to do," Scott says. "But we didn't know if we could qualify. When we called the importer, they had heard of Kemp through racing and said they would consider us."

After a few months of negotiations, the deal went through. International Motorcars bought $2000 worth of genuine Ferrari parts and a pair of new cars and hung its sign out in front of a remodeled Chrysler-Plymouth dealership in downtown Jackson, a location that is within an exhaust-shriek of the state capitol.

Having a Ferrari store and actually selling Ferraris are two different matters. Kemp and Scott had no way to know if they could move cars. But they soon discovered that the Ferrari business, although it may move in strange ways, definitely moves. Scott remembers the first car they sold. "A kid arrived out front in a taxi with an airplane ticket in his hand. He was wearing a T-shirt and jeans with holes in the knees. Turned out he was from New Orleans and had seen our ad in the *Times-Picayune*. We had a used 275 GTS on the floor. He walked around the car twice—didn't even sit in it—and said he'd take it. The price was $14,500, and he gave us a check for $8000 to hold it."

That was last July. Since then business has been well above expectations. They've sold 11 cars, six of them new ones. And poised on the showroom floor right now is one brilliant-red Ferrari 308 GTB certain to become a much sought after collector's item. It is the *Car and Driver* test car. If you tell Charlie Kemp that you are a regular reader, he might even make you a deal.　　　　　　　—*Patrick Bedard*

Scott and Kemp: The only car dealers in Jackson who don't wear double knits.

ACCELERATION standing ¼ mile, seconds

FERRARI 308 GTB	16
FERRARI/NART 365GT4/BB	14
FARRARI DINO 308 GT4 (1975)	15
PORSCHE TURBO CARRERA (1976)	13.5

Scale: 13–21

BRAKING 70-0 mph, feet

FERRARI 308 GTB	220
FERRARI/NART 365GT4/BB	207
FARRARI DINO 308 GT4 (1975)	153
PORSCHE TURBO CARRERA (1976)	178

Scale: 150–230

FUEL ECONOMY C/D mileage cycle, mpg
Tested by Automotive Environmental Systems, Inc.

- City driving / Highway driving
- FERRARI 308 GTB (NOT AVAILABLE)
- FERRARI/NART 365GT4/BB — ~12 hwy, ~8 city
- FARRARI DINO 308 GT4 (1975) (NOT AVAILABLE)
- PORSCHE TURBO CARRERA (1976) — ~20 hwy, ~14 city

Scale: 6–38

PRICE AS TESTED dollars x 1000

FERRARI 308 GTB	28
FERRARI/NART 365GT4/BB	52
FARRARI DINO 308 GT4 (1975)	22
PORSCHE TURBO CARRERA (1976)	28

Scale: 0–80

INTERIOR SOUND LEVEL dBA

- 70-mph cruise / Full-throttle acceleration
- FERRARI 308 GTB — ~75 / 90
- FERRARI/NART 365GT4/BB — ~85 / 95
- FARRARI DINO 308 GT4 (1975) — ~75 / 92
- PORSCHE TURBO CARRERA (1976) — ~75 / 82

Scale: 60–100

Heinz Maurer

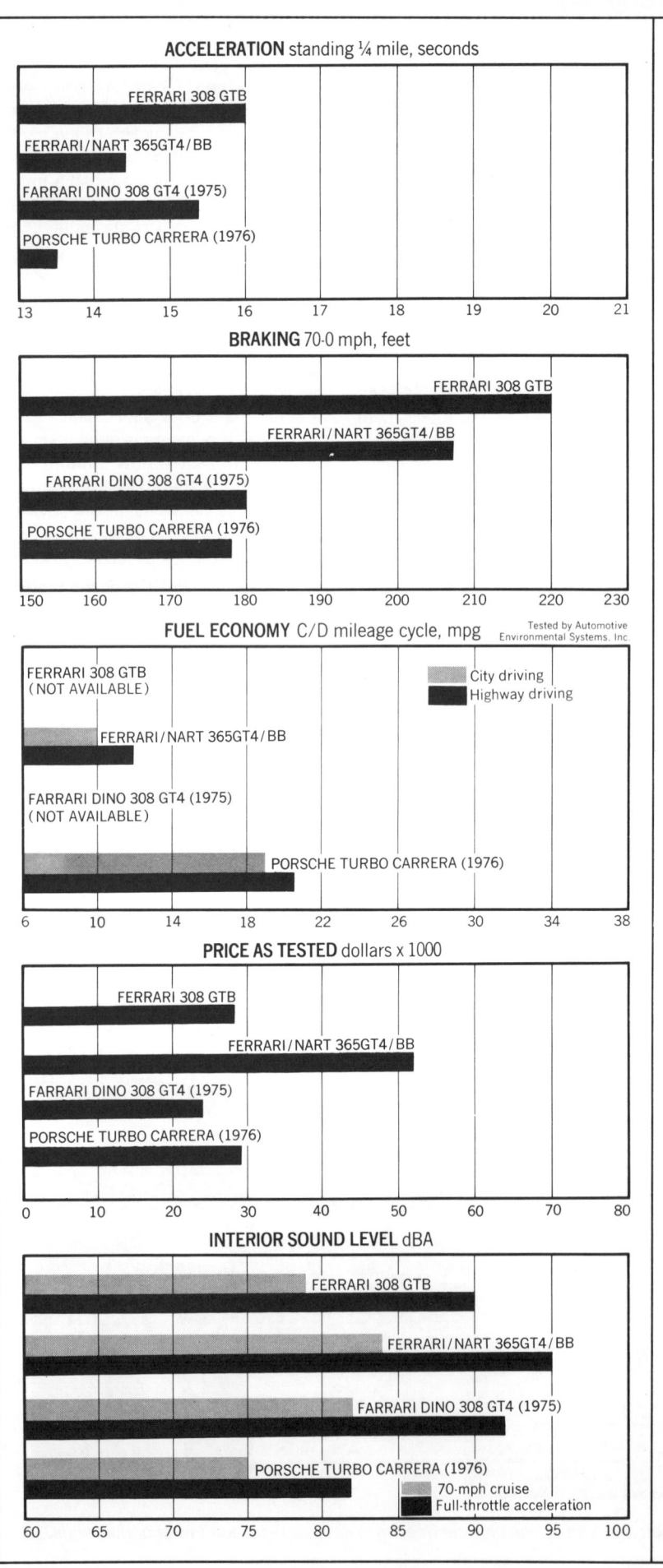

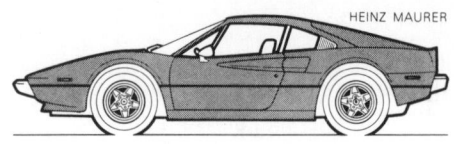

FERRARI 308 GTB

Importer: Chinetti-Garthwaite Imports Company
1100 West Swedesford Road
P.O. Box 455
Paoli, Pennsylvania 19301
Vehicle type: mid-engine, rear-wheel-drive, 2 passenger coupe

Price as tested: $28,780
(Manufacturer's suggested retail price, including all options listed below, dealer preparation and delivery charges, does not include state and local taxes, license or freight charges)

Options on test car: Base 308 GTB, $28,580; dealer preparation, $200.

ENGINE
Type: V-8, water-cooled, cast aluminum block and heads, 5 main bearings
Bore x stroke . . . . 3.19 x 2.79 in, 81 x 71 mm
Displacement . . . . 178.6 cu in, 2927cc
Compression ratio . . . . 8.8 to one
Carburetion . . . . 4 x 2-bbl Weber 40DCNF
Valve gear . . . . belt-driven double overhead cam
Power (SAE net) . . . . 240 bhp @ 6600 rpm
Torque (SAE net) . . . . 195 lbs-ft @ 5000 rpm
Specific power output . . . . 1.34 bhp/cu in, 82.0 bhp/liter
Max. recommended engine speed . . . . 7700 rpm

DRIVE TRAIN
Transmission . . . . 5-speed, all-synchro
Final drive ratio . . . . 3.71 to one

Gear	Ratio	Mph/1000 rpm	Max. test speed
I	3.59	5.6	43 mph (7700 rpm)
II	2.35	8.4	65 mph (7700 rpm)
III	1.69	11.9	91 mph (7700 rpm)
IV	1.24	16.0	110 mph (6900 rpm)
V	0.95	21.0	110 mph (5250 rpm)

DIMENSIONS AND CAPACITIES
Wheelbase . . . . 92.1 in
Track, F/R . . . . 57.9/57.9 in
Length . . . . 172.4 in
Width . . . . 67.7 in
Height . . . . 44.1 in
Ground clearance . . . . 4.1 in
Curb weight . . . . 3110 lbs
Weight distribution, F/R 42.0/58.0%
Battery capacity . . . . 12 volts, 66 amp-hr
Alternator capacity . . . . 770 watts
Fuel capacity . . . . 18.5 gal
Oil capacity . . . . 9.4 qts
Water capacity . . . . 19.0 qts

SUSPENSION
F: . . . . ind, unequal-length control arms, coil springs, anti-sway bar
R: . . . . ind, unequal-length control arms, coil springs, anti-sway bar

STEERING
Type . . . . rack and pinion
Turns lock-to-lock 3.2
Turning circle curb-to-curb . . . . 39.3 ft

BRAKES
F: . . . . 10.6-in. vented disc, power-assisted
R: . . . . 10.9-in. vented disc, power-assisted

WHEELS AND TIRES
Wheel size . . . . 7.5 x 14-in
Wheel type . . . . cast alloy, five bolt
Tire make and size . . . . Michelin 205/70VR-14 XWX
Tire type . . . . steel-belt radial, tubeless
Test inflation pressures, F/R . . . . 28/34 psi
Tire load rating . . . . 1490 lbs per tire @ 36 psi

PERFORMANCE
Zero to — Seconds
30 mph 3.1
40 mph 4.3
50 mph 6.9
60 mph 7.9
70 mph 10.3
80 mph 12.8
90 mph 16.3
100 mph 21.1
Standing ¼-mile . . . . 16.0 sec @ 89.0 mph
Top speed (estimated) . . . . 140 mph
70-0 mph . . . . 220 ft (0.74 G)

FERRARI 308 GTB

cia Beta Scorpion first, and now it has done a variation on the theme for Ferrari. The effect is altogether different than what the world has come to expect of cars, particularly sports cars. There is no wood, no engine-turned aluminum, no black crackle to act as background for dozens of gauges, levers and knobs. Instead, the dash, doors, roof and seats all seem to run together in a padded cocoon of vinyl. There is but one interruption. Directly before the driver, right where he would hope it would be, is the instrument cluster. All the dials are round. The tachometer shows a 7700-rpm redline. The speedometer reads to 180 mph. The stylists have fancied up the markings a bit more than they should have for maximum legibility, but the message is by no means lost. Curiously, the cluster ended up too small for all the dials, so two—the clock and the oil temperature—are tucked away below the dash on the left, half buried by the voluptuous padding.

The controls you must reach for either sprout from the steering column or are located on the console. In Italian cars, reaching for the steering wheel tends to be an Olympic-level task. At first glance, the GTB looks as difficult as they come. The wheel angles forward at the top in that awkward way. But then when you slip into the seat, you find the steering is more manageable than you anticipated. The reach is reasonable, and the low steering effort, one of the delightful aspects of a mid-engine car, makes minimal demands. Parking requires some exertion, but you'll never be discouraged from any maneuver on the road.

In fact, there is very little about the GTB that is discouraging. Sports cars and GT cars traditionally have been specialized machines, and frequently you were asked to make major sacrifices in comfort or convenience just to have the look of speed and, if you were lucky, the performance to match. But the GTB is wonderfully accommodating. It has plenty of room inside for two adults. Your head won't rub the molded, one-piece headliner, and you won't have to tuck your elbows in your belt loops. The seats are exceptionally good. They are finished in leather, very firm and deeply contoured. They give you the feeling that you'll be well taken care of, come what may, even if that means 500 miles in one afternoon. And the combination lap-and-shoulder belts fit comfortably, something that couldn't be said of their equivalent in the Dino 308 GT4.

This new Ferrari also sets high marks for tractability. It seems happy enough chugging around at 1000 rpm, and there is no need to slip the clutch to move away from a start. Just let the pedal out as you would in a Pinto. This car doesn't care. You can drive it as a little old lady would, and it will just motor along. If your style is more vigorous, you'll have to muscle the lever a bit—and the brake and accelerator pedals are far apart for heel-and-toeing.

No matter what your pace, you'll always be aware of the engine. The old chain clatter is gone, but you can still hear the rest of the machinery churning around back there just behind your shoulder. The largest single component of the noise, particularly under full throttle, comes from the air intake. You'll notice a scoop on each side of the car just forward and above the rear wheels. These are functional. The one on the driver's side leads to the oil cooler, the other to the air cleaner. A fair amount of induction roar seems to escape from the latter. Fortunately, it's not located on the left side just behind the driver's ear as it was on past models. The engine has four Weber two-barrels, and this arrangement is a challenging one to silence in the best of arrangements. Besides, who wants to muzzle a Ferrari?

The tractability of the engine may be unexpected to those who have not driven recent Ferrari models, but the GTB's soft ride is completely unprecedented. The springs, particularly in front, are very soft, resulting in low-frequency ride motions, lower than any car of this sort we can remember. This car floats down the road more like a Buick than a Ferrari. And this has its benefits. Expansion strips and broken pavement are easily absorbed, and you don't cringe at the sight of normal road bumps as you might in other cars of this stripe. But there is a penalty when you are in a hurry. The nose bobs up and down like that of a Showroom Stocker. This sensation, however, is enjoyable for the same reason that Showroom Stockers are fun—you have the feeling that it is only your skill that is controlling an errant machine. But there also is a loss of precision, a quality you expect from a Ferrari more than a soft ride. And it could lead to trouble. When the pace quickens on undulating roads, you occasionally hear one of the tires scraping the bodywork and feel the suspension come to the end of its travel. There is nothing like bottoming the suspension to upset your balancing act. The way the GTB is now, it's perfect as an executive commuter, but we think it'll need a little tighter control in the suspension department before it can be a hero on the rural blacktop.

Taken altogether, the GTB really is quite a radical Ferrari. The styling is nothing short of inspired, and the fiberglass body and soft suspension must have caused a great deal of soul-searching among the engineers before they finally granted their approval. Compared to this, the idea of packaging Dino mechanicals under the Ferrari label is a relatively minor shift in marketing strategy. The purists may think that Ferraris and Dinos spring from altogether different sources, but in fact both roll out the same door of Ferrari's Maranello factory.

Just in case you are interested, the next new model to roll out of that door will be a spider version of the GTB with a steel body. It will be called the 308 GTS. Look for it this summer. All of the GTB's grace and sun on the top of your head at the same time should make it a marvelous package.

•

Ferrari's little red rocket

MOTOR Test

The Dino 246 is alive and well and living in the 308 GTB...

THE COMMENDATORE'S latest road rocket is not so much a design, but an amalgamation of existing ideas, married together with a view to satisfying the increasing varied demands of the luxury sports-car market.

The 308 GTB Ferrari looks for all the world like the Berlinetta Boxer from the nose to the A-pillar, and reminds of the classic Dino 246 from the door to the tail.

The styling is by Pininfarina, but once again it's not so much an original design, but a consolidation of proven popular styling features. Whether it's as classic a design as the original 246 remains to be seen, but the GTB is certainly an innovative and unusual change of concept for the Modena magicians.

It's a payoff for all the Ferrari lovers who never really warmed to the 2+2 308 GT4 in the way they did to the Dino 246. It's also a good way for those would-be owners who can't quite run to $60,000-odd to get themselves into a more reasonably priced coupe. The Boxer wasn't sold in Au-

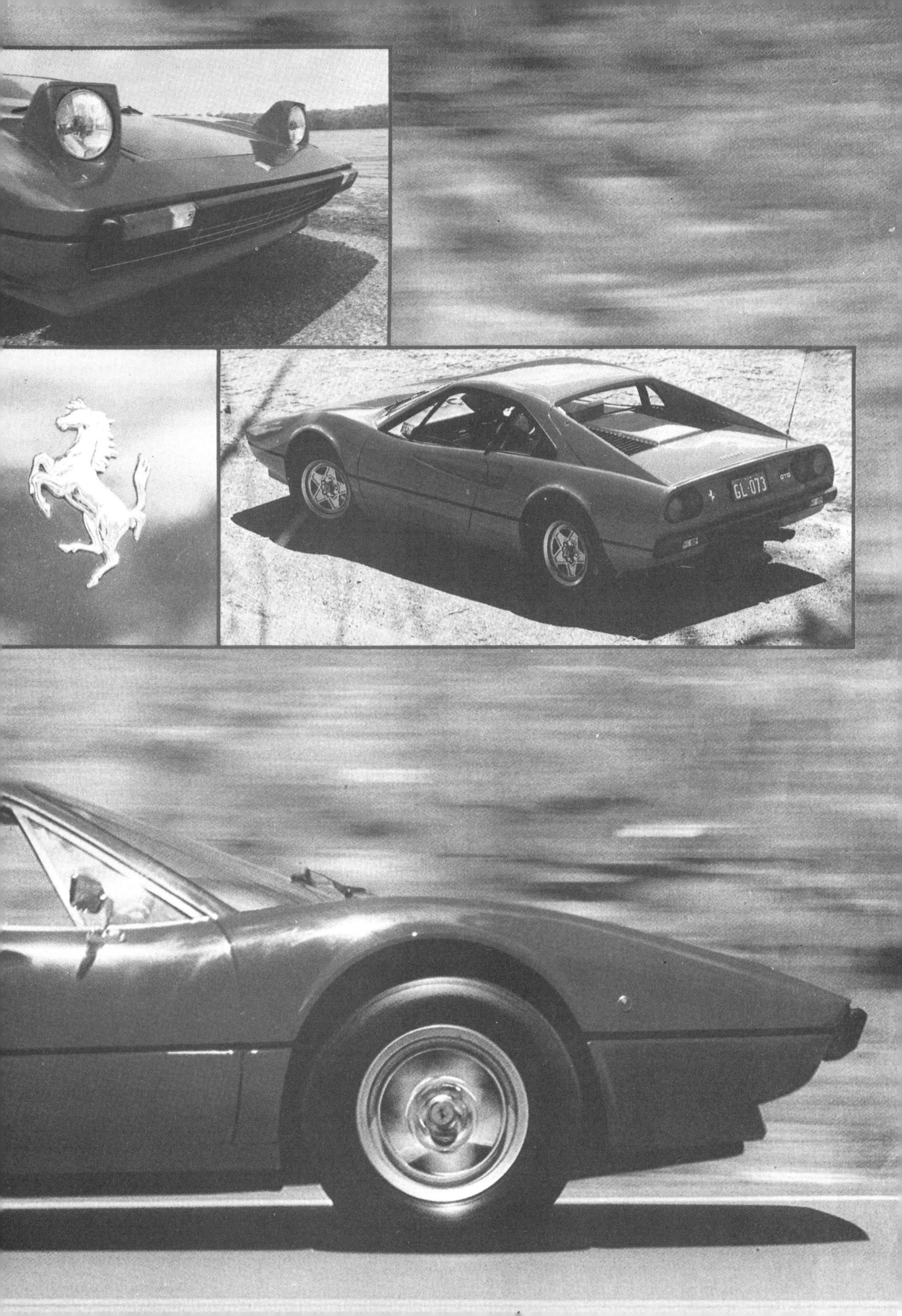
GL 073

stralia after January 1976 when it cost $54,700 — and the GTB's current price is $41,274, which is some saving.

There's no way it can be any more than a two-seater — it doesn't pretend to be able to shoe-horn any extra bodies into the cabin but, surprisingly, there is quite a deal of luggage space in both front and rear boots. That means it's quite a reasonable touring car on European roads. But as a Grand Tourer for Australian conditions, the fibreglass car's design makes it instant nervous breakdown territory. It has only 120 mm of overall ground clearance, and the overhang at the front is such that you'll scrape it on just about anything taller than a dead match.

If the insurance premium doesn't frighten you, the car's unsuitability for our roads will give you graphic evidence that transplanting luxury European exoticars to Australia doesn't work at all.

The 308 GTB's wheelbase is 210 mm shorter than the GT4, and the two-seater is 90 mm shorter and 80 mm narrower than the 2+2. Surprisingly, despite fibreglass panels and smaller dimensions, the GTB is only 60 kg lighter than the GT4. And, interestingly, the fibreglass panels are segmented and this allows speedy and economical repairs to be affected if the car is pranged.

Access to the car is easy, through wide opening, and quite light, doors. The rear hatch is unlocked by a lever mounted in the right-hand C-pillar and when the lid is raised there is reasonably good accessibility to the transverse-mounted V8 engine plus a roomy luggage compartment, covered by a zippered, fabric top.

The Ferrari's 3-litre V8 is a superb piece of Italian craftsmanship. The double overhead camshaft engine uses twin belt-driven camshafts, four 40 DCNF Weber carburettors, and is mounted transversely just ahead of the rear wheels, driving through Morris Mini-style transfer gears to the excellent five-speed manual gearbox which is mounted behind the engine.

The transmission, despite complications which arise when designing shift mechanisms of this nature, is quite smooth and easy to change. The shift from first to second takes quite a lot to get used to, but this aside the gate is easy to move through and the pressure required is quite a lot less than we've experienced on other Ferraris.

The clutch has extremely long travel, which necessitates compromises in both seat design and travel. The pedal

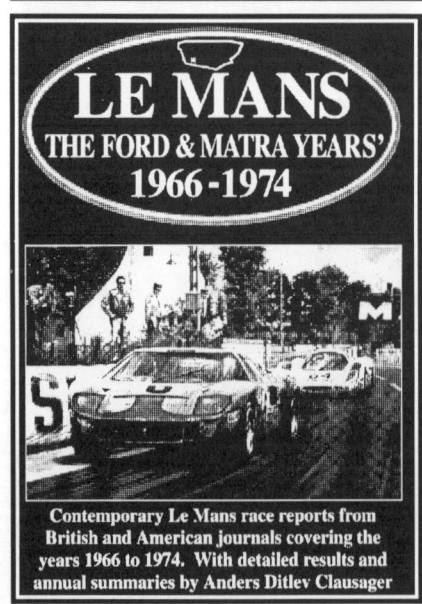

LE MANS 'THE JAGUAR YEARS' 1949-1957

Contemporary Le Mans race reports from British and American journals covering the years 1949 to 1957. With detailed results and annual summaries by Anders Ditlev Clausager

LE MANS 'THE FERRARI YEARS' 1958-1965

Contemporary Le Mans race reports from British and American journals covering the years 1958 to 1965. With detailed results and annual summaries by Anders Ditlev Clausager

LE MANS 'THE FORD & MATRA YEARS' 1966-1974

Contemporary Le Mans race reports from British and American journals covering the years 1966 to 1974. With detailed results and annual summaries by Anders Ditlev Clausager

AVAILABLE FROM BROOKLANDS BOOKS

must be pushed all the way to the floor for proper engagement, and the pedal pressure is extremely high. It's uncomfortably high in the city, but we can't help feeling it's a lot lighter than the 246 Dino, which was almost unbearably heavy.

We have always felt the Ferrari-type gearshift gate was a cumbersome piece of design, but it never fails to operate super-smoothly on the road. Once out on the highway, you forget about the wide separation of the gear positions — the GTB's shifts are made easily, and without excess concentration being required. The box is a delight at higher cruising speeds, especially between fourth and fifth.

The chassis for the GTB is completely different to the GT4 and the outer skins are all-new, though the two-seater uses the same all-independent, coil-wishbone suspension of the larger 2+2 — and, of course, the identical powertrain.

The GTB also uses the same upper and lower wishbones as the GT4, but the springs are slightly different and shock absorber settings have been modified to alter the ride characteristics.

A detailed look at the difference in shock absorber settings between the 246 Dino and the 308 GTB reveal an

The four-wheel disc brakes are ventilated, mounted outboard, and fitted with an automatic pressure compensating valve which manages to be extremely accurate in all circumstances. On test, the brakes operated on exactly the right amount of pressure and we only locked the front discs on one occasion. The GTB uses the same size tyres all round, and the test car wore 205/70 VR14 Michelin XWX radials mounted on the very attractive Ferrari light alloy wheels with 6.5in rims. The pattern was obviously the inspiration for South Australia's Globe Wheels when it designed the Globe 'Bathurst' mag for Touring Car racing.

Though the 308 GTB sits very close to the road, the cabin gives the feeling of great spaciousness. The little coupe is only 112 mm high, but access to the cabin is easy and the seats are set at just the right height to give a commanding view of the road.

The seats aren't the most comfortable we've tried, but they do offer reasonable support and are finished in expensive-smelling leather facings. They adjust for length and angle, and headroom is more than adequate for taller drivers. The doors feature a stylish arm-rest mounted at window-sill level which follows the line of the dashboard,

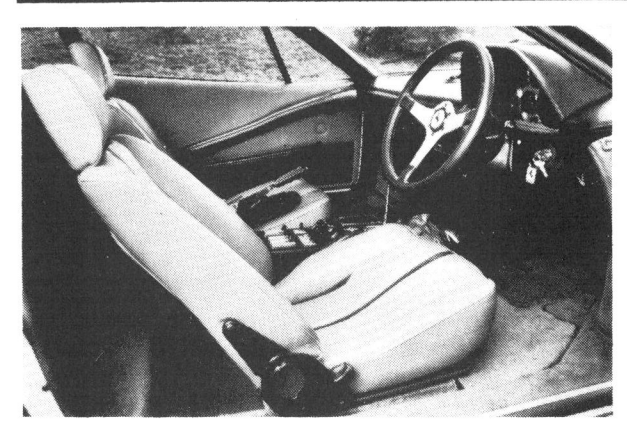

interesting approach to the challenges of providing a ride and handling compromise suitable for a today's buyer.

At the front, the expansion/compression rates between the 246 and 308 show the GTB to be significantly softer, while the rear settings are slightly stiffer.

Spring rates on the 308 are harder, to offset the increased weight and to counter the softer shock settings.

For those interested in slightly more precise handling, if you should ever feel the need, then the front-end geometry and the toe-in of the rear wheels can be modified to provide a stiffer ride, but even more pin sharp handling, akin to that of the 246.

The rack and pinion steering is very positive, but not at the expense of lightness. At parking speeds, the car is easy to manoeuvre, but on the highway the steering is carefully loaded to provide the degree of precision you expect in a Ferrari. It's quite low geared, giving just over three turns, lock to lock, but it provides excellent feedback. However, on our bumpy roads there were times we wished this was not so. The car doesn't seem to suffer unduly from steering deflections, but needs concentration when driving quickly over indifferent surfaces.

giving the cabin the appearance of a truly integrated design.

The GTB comes with air conditioning and electric windows and this helps make the car very comfortable in Australia's climate. The electric windows are painfully slow in operation and at either end of their travel give forth with a horrible graunching noise as the winding gear continues to engage against the ratchet.

The controls are located in the true GT style. A centre console houses shift lever, a large ashtray, air conditioning controls and knobs for choke, heater fan, heater temperature and additional wiper speed. The remainder of the controls are column mounted, with two levers on the left side of the column for lights and trafficators, and a third stalk on the right-hand side for the wipers/washers.

The gauges feature very attractive graphics and are mounted in a compact panel with white numbers on black faces. In the binnacle ahead of the driver, the 180 km/h speedo and 10,000 rpm tacho stare back at you, flanked by fuel, oil pressure and water temperature gauges. The oil temperature gauge and clock are almost hidden from view in a sub-panel mounted under the dash and to the right-hand side of the ignition lock.

There are vents on the top of the dash for demisting and the only other outlets are two adjustable grilles in the centre console under the radio and above the gear lever. These provide a hefty draft of air-conditioned atmosphere to the occupants' faces. The leather bound Momo steering wheel is set at a good height with the rim just rising above the line of the instrument binnacle.

The dash is finished in plain black vinyl, which didn't seem to fit very well on three different GTBs which we've inspected. However, the whole cockpit appeals to us overall because of the simplicity of design and the efficiency of the layout. There's a lockable compartment mounted between the seats and this is sufficient to hold directory, cassettes, smokes, and of course driving gloves.

The speakers for the radio/stereo are mounted behind tastefully-designed grilles in the doors, ahead of capacious map pockets.

The GTB does suffer from a couple of problems in the 'reflections' department — because of the 'fast' angle of the front windscreen, there are annoying reflection nozzles and the instrument binnacle and this can be very obstructing on a very sunny day. Also, the flat glass lenses on the instruments produce another series of reflections which shouldn't be so in such a thoroughbred.

Despite these two criticisms, we loved the overall cabin layout. The cockpit is generous in its amount of useable space (for two) and it's well laid out, offering the driver all the driving aids in easy reach and excellent vision for a low slung coupe.

Apart from other aspects — like noise — we don't think it at all unreasonable to use the GTB as a long-distance touring car — albeit on European roads, thank you.

We were surprised to find the GTB was not a great deal quicker than the GT4. In fact, in some cases it was slower. However, the surge with which the power comes, and the excellent torque characteristics, gives you the impression that you're travelling quite fast enough.

As you ease the GTB off the line and accelerate up through the gears, it's a marvellous feeling to see the tacho needle climbing smoothly toward the 7500 rpm redline. As it nears that mark, the V8 is howling just behind the rear window and — with the driver's window down — you can hear the sucking of the four Webers and the scream of the various gear drives in the all-alloy engine.

Unfortunately, due to its alloy construction and the number of gears whirring and spinning around inside the engine, it's quite a noisy powerplant. The fibreglass body panels don't help the situation at all, allowing a free path for any noise generated in the engine compartment.

With the windows up, you can still hear — though much subdued — a buzzing which signifies you're doing with a Ferrari what everyone would like to do to a Ferrari: driving it to its limits.

Even outside, as the car flashes past down the highway, there's a magnificent howl as the car approaches, peaking in the gears, a slight pause, then another roar as the next gear is engaged and the road-hugging Red Flash rockets away out of sight. The emissioned GTB we drove was not much quicker, overall, than our most recent drive in a GT4, but the torque comes in from around 1800 rpm and pulls smoothly up to its peak at 5000, making the GTB a very easy car to drive in town.

There is obviously nowhere in Australia where you can comfortably — or legally — put the GTB's top speed of 252 km/h to the test, but at high touring speeds the car accelerates easily in fourth and fifth, giving clear evidence that Ferrari completely understands the demands of today's traffic-clogged driving patterns and has reacted suitably.

Due to its relatively light weight, the GTB handles top gear cruising easily, running from 100 to 130 km/h in a mere 7.5 seconds. And it will accelerate in fifth from 145 km/h to 180 km/h in just under 7.0 seconds — that's quick!

However, engine and transmission noises at this speed tend to be quite high, despite windows up. There are, thankfully, no unwanted vibrations and rattles from the powertrain at all. The Ferrari engineers seem to have taken great pains to eliminate such obtrusive noises.

The Lamborghini Urraco 3000 is a quieter and more refined car in this area, but it fails to achieve the same sensations and in its defence we must say the GTB is one of the most refined cars Ferrari has ever produced.

The GTB feels an immensely safe car from the moment you drive it through your first corner. It's stable, it disregards crosswinds, the broadness of the track tells you that it's sitting squat on the road, and it will go willingly where you point it. There is a hint of initial understeer, but easing the throttle back solves the problem and the responsive steering allows you to guide the car carefully through any corner, whether you're doing 50 km/h or 150 km/h. In fact, the 150 km/h corner in the GTB is pure delight. You know this car has more adhesion and precision than anything else you've driven and you point it with confidence and calmness.

It is a beautifully balanced car with a progressive ride and no undue harshness from the suspension. We felt at first that the spring rates might have been a little too soft for absolute control, but believe us this car gives you Control (note the capital C). We remember our drive session in the 3-litre Urraco and how smooth and refined the ride was, still with a high degree of control. The Ferrari is not as comfortable, but at the same time it is not uncomfortable.

It always seems as if there is a gear for every situation — a feeling generated by the smoothness of the torque curve, and the tautness of the drivetrain and suspension. The steering has just the right amount of travel between locks and lets you almost 'conduct' the GTB through a corner like a concert-master leads his audience. And, it doesn't matter whether you're driving sedately or thrashing around at 7000-plus in the gears — the GTB always does what it's told.

In terms of control, placement of controls, flexibility of the engine and the ease of steering, the GTB is not a tiring car to drive at all. In fact, we don't know many people who would baulk at taking it on a 1000-kilometre trip.

But it does have some fatigue-inducing qualities — like the engine and transmission noise and the resonance which seems to come when you're cruising at high speeds for long periods.

There is an absence of wind noise, but the XWX tyres did generate some road roar and this leads us to think the GTB has a glorious future when Ferrari gives some development time to a little more refinement of the suspension and suppression of some of the more obtrusive noises.

Of course, there are noises which must, and should, accompany any drive-time in a Ferrari. The 308 GT4 features a four-branch exhaust system, which allows the almost syncopated pulse of the V8 to growl at passers-by, but the GTB's four extractor tubes are forced into a large exhaust silencer, emerging through a single exhaust pipe with a throttled, innocuous note reminiscent of more domesticated cars.

However, since the 'real' noises are generated forward — just behind the driver's ears — it is this which convinces you that you're in a real Ferrari. We appreciate the howl of the oil pump, distributor and other assorted gear drives, even the sucking and blowing of the Webers, but some of those sounds give you a real headache after 1000 kilometres and maybe the engineers should look at a little more sound insulation in the future.

The GTB came along when all of Europe was worrying about the long-term effects of the fuel crisis. Pundits said there'd be no more cars like it, that new V8-engined coupes were already a thing of the past. Enzo Ferrari has proved them wrong. Ferrari's recipe for a sports car is absolutely perfect — Ferrari lovers will desire it, and sports car freaks will blow their minds. This is a real sports car in the accepted sense: it's fast, safe, beautifully balanced, exotic, Italian and sheer exhileration to drive.

FERRARI 308 GTB ROAD TEST DATA

ENGINE
- Location: Transverse, mid-mounted
- Cylinders: V8
- Bore x Stroke: 81 mm x 71 mm
- Capacity: 2927cc
- Compression: 8.8 to 1
- Aspiration: Four twin-choke 40 DCNF Weber carburettors
- Fuel pump: Electric
- Valve gear: Single dry plate
- Maximum power: 190 kW at 7700 rpm
- Maximum torque: 284 Nm at 5000 rpm

TRANSMISSION
- Type/locations: Five-speed all-syncro manual
- Driving wheels: Rear
- Clutch type: SDP
- Gearbox Ratios
  - 1st: 3.418
  - 2nd: 2.543
  - 3rd: 1.693
  - 4th: 1.244
  - 5th: 0.918
- Final drive: 3.7 to 1

SUSPENSION
- Front Suspension: Independent by coils and wishbones with anti-roll bar
- Rear suspension: Independent by coils and wishbones with anti-roll bar
- Shock absorbers: Direct-acting telescopic
- Wheels: Light alloy 6.5 x 14
- Tyres: 205/70VR14 radial

STEERING
- Type: Rack and pinion
- Turns lock to lock: 3.1
- Turning circle: 12 m

BRAKES
- Front/rear: Disc
- Servo assistance: Vacuum standard

DIMENSIONS AND WEIGHT
- Wheelbase: 2340 mm
- Overall length: 4230 mm
- Overall width: 1720 mm
- Overall height: 1120 mm
- Track, front: 1460 mm
- Track, rear: 1460 mm
- Ground clearance: 120 mm
- Kerb weight: 1090 kg

CAPACITIES AND EQUIPMENT
- Fuel tank: 80 litres
- Cooling system: 13 litres
- Engine sump: 18 litres
- Battery: 12V 60Ah
- Alernator: 55A

CALCULATED DATA
- Weight to power: 5.7 kg/kW
- Specific power output: 64.9 kW/litre

FUEL CONSUMPTION
- Average for test: 18.9 litres/100 km
- Best recorded: 18.4 litres/100 km

PERFORMANCE
ACCELERATION

	Manual
0-60 km/h	3.1s
0-80 km/h	5.4s
0-100 km/h	6.7s
0-120 km/h	10.8s

OVERTAKING TIMES (holding gears)

km/h	4th	5th
50-80	5.1s	9.8s
60-100	4.8s	7.8s
80-100	4.7s	7.2s
100-130	4.9s	7.5s

STANDING 400 METRES
- Average: 14.7s
- Best Run: 14.5s

SPEEDS IN GEARS

	Max. km/h	rpm
1st	66	7700
2nd	95	7700
3rd	133	7700
4th	180	7700
5th	252	7700

THE FIVE-STAR TEST
- Comfort: ★★★
- Handling: ★★★★
- Brakes: ★★★★
- Luggage capacity: ★★
- Performance: ★★★★★
- Economy: ★
- Quietness: ★

COMPARISON CHART

	FERRARI 308 GTB	BMW 633 CSi	DE TOMASO PANTERA GTS	JAGUAR XJ-S	LAMBORGHINI URRACO
PRICE	$41,274	$38,675	$40,000	$29,950	$42,500
CAPACITY	2927cc	3120cc	5763cc	5354cc	2996cc
POWER	190 kW	147 kW	260 kW	213 kW	197 kW
TOP SPEED	252 km/h	220 km/h	260 km/h	240 km/h	265 km/h
LITRES/100 km	18.9	14.8	20.0	20.1	15.6
WEIGHT	1090 kg	1470 kg	1420 kg	1682 kg	1300 kg
WEIGHT/POWER	5.73 kg/kW	10.00 kg/kW	5.46 kg/kW	7.89 kg/kW	6.59 kg/kW

You Can Afford a Ferrari

Little-known facts about how the Internal Revenue Service can help you buy a 308 GTS

by William Jeanes

Ferrari ownership carries with it many of the obligations that attach to mistress-keeping. Both projects require huge cash outlays and are psychologically risky, but they are guaranteed to make you the envy of your fellows. Neither activity suits the average man, the faint of heart or the happily married, let alone the financially disadvantaged or the terminally dull. Both undertakings have an undeniable allure, but we are compelled to recommend the Ferrari. Not for any moral reason, but because in the long run it will be cheaper.

A Ferrari can provide almost as much pleasure as a mistress and offers the additional advantage of being unlikely to call your home. Consider then the new Ferrari 308 GTS, the spyder version of the 308 GTB. This latest ingot produced by the precious metal artisans at Modena radiates all the qualities expected of Ferraris, including sure-footedness, power, excitement and an aura of jet-set fashion. Thanks to the removable targa top, the 308 GTS also offers wind-in-the-wallet motoring. The car costs $35,000 or thereabouts, a sum capable of producing a tax-free lifetime income of $200 a month if invested in municipal bonds. But, keep heart; the $35,000 price tag represents the only bad news about the GTS.

The good news is that by using some common sense you can own the $35,000 1978 Ferrari 308 GTS for a piddling $225 a month and probably make money on the deal.

To do this, you will need the following: an understanding banker, a top-drawer credit rating and a trade-in worth $5000. But before going into the specifics of this revolutionary—and eminently possible—transaction, you should be told a bit about the Ferrari you will soon be driving.

Secreted in the Pennsylvania hamlet of Paoli, a brief $30 cab ride from the Philadelphia train station, lies Algar Motor. (Any sharp-eyed descendant of Natty Bumpo will be able to ferret it out in no time.) Algar Motors, from this remote location, serves as the nation's importer of Ferraris. If a dealer has sent you, or if you bring along a cashier's check for $35,000, Algar's bright young manager, Randy Turney, will give you a GTS—if he

Ferrari 308 GTS

has one, which isn't likely these days. The other way to get your hands on one of these gems is to let a Ferrari dealer, who is also an old friend, talk you into driving one down to Atlanta for him.

(There is a third way but it is not recommended. Even the Ferrari people provide a test car for the automotive magazines. But the chances of this car surviving abuse from the road testers are miniscule at best. It is far better to find a car that has been neither trashed nor melted down to make a graven image.)

The old friend was Charlie Kemp, an owner of International Motorcars in Jackson, Mississippi. Kemp is one of the few moderately successful road racers to learn that selling cars can be more profitable than racing them. His latest profit producer is the GTS, and the one I drove to Atlanta for him was the first for his firm, and one of only a handful in the country. Already sold, it impatiently waited out its imprisonment in Algar's frigid warehouse.

Allowing for my wrong turn out of Algar's driveway, the distance between there and Atlanta stands at 872 miles. The trip divided nicely into three legs: Paoli to South Hill, Virginia; South Hill to Charlotte, North Carolina; Charlotte to Atlanta. Not a bad ride, considering the horse.

The horse in question came in gleaming black with a black vinyl interior. For those of you who feel Porsche owns the black-on-black franchise, be warned that serious disillusionment awaits. Maybe it's the little yellow Ferrari emblem on the nose that makes it work. Whatever, it does work.

Before covering two dozen miles, it became apparent that I held the controls of a genuine emotional experience. Sitting there on a black vinyl seat, designed by an Italian with soul rather than by an anonymous Teutonic chiropractor, my hands on the black leather Momo steering wheel, and peering into the night over green-numeraled instruments set into the black padded dashboard, I reached a potentially costly decision: Not having lost contact with reality altogether, I decided *not* to buy the car right on the spot.

What I *did* decide to do was drive the car to Atlanta. Really *drive* it. Lacking Fuzzbuster, Smokey Spray, Cop Off, CB radio, foil-filled hubcaps or any of the sissy stuff that supposedly fends off the forces of law and order, I figured the worst I can do is blow a couple of hundred dollars in fines. Balancing this possibility against the psychic satisfaction of just laying the Quicksilvers to the cocoa-colored mats, there was no contest. So, exhibiting the kind of will power that made Wilbur Mills a household word, I led myself into Temptation's bar and ordered a round for the house.

Located immediately behind the two seats and before you get to the rear axle, a transverse-mounted 3-liter Ferrari V-8 lies in conspiratorial wait. It contains four camshafts, four double-barrelled Webers, and enough valves to stock a parts store. It is shamelessly pleased that you have hurled caution and the fuel crisis out the electric windows. (In fact, the fuel crisis remained within the purviews of my conscience. Before the Cleveland Amorys of the energy world suffer gas pains, may I offer by way of expiation the news that I averaged 18.2 miles per gallon over 872 miles of high-speed travel. Furthermore, the catalytic converters attached to each bank of the V-8 kept my exhaust fumes well within legal bounds. When the average car on American roads can do that, the doomsayers can come after me with the dogs and I'll go quietly.)

So there I went, tooling southward at speeds that the Ferrari and I—if no one else—found mutually agreeable. Once you've make up your mind to risk arrest, tickets and money, a feeling of underwater-like calm descends. You stop worrying. I've always imagined the same feeling applied to parachute jumping... the worst part is waiting for the green light to go on.

The temptation to reflect upon Great Motoring Truths as you zoom elegantly in your Ferrari through a North Carolina springtime is irresistible. Targa top stowed neatly behind the seats, the sun shining and the compelling sing of the engine trailing behind, one realizes the first of several truths: Your fellow motorists—the ones who know what you're driving—are devastated by envy. People driving Mercedes 450SLs *hate* you, their narrowed eyes blistering the paint of your GTS as you pass. Another truth lies in the difficulty of gracefully explaining to service station attendants that the car costs more than their building and grease rack. But the salient nugget of motoring lore that emerges concerns rearview mirrors: Most people are unaware that their car is equipped with such a device. Time after time I hummed up behind a citizen who was flying along the passing lane at speeds up to 40 mph. Perversely, I would sit there, two or three car lengths back, and wait for the inevitable reaction. First comes the head jerk. Then the wheel jerk. Then the realization that he doesn't know if the right lane he is pulling into is free of traffic. Utter, pathetic, confused ineptitude. He would have been no more shocked to find a cobra crawling out of his console than he is to find a car behind him. It's the kind of rolling stupidity that sends me shrieking into the woods. If the National Safety Council, Ralph Nader and Joan Claybrook would combine forces to teach Americans how to use rearview mirrors, the world would be considerably safer.

Despite the hindrance of the occasional blindered motorist, the intensely seductive spirit of the Ferrari had no difficulty asserting itself. It's entire interior evokes a racing car atmosphere, which is what Grand Touring cars are supposed to be about. The steering is quick and precise, yet its no-play feel is reassuring rather than intimidating. The shifter on the 5-speed transmission does its work as easily as a slot machine handle, but the gearbox itself functions with the precision associated with metal lathes and bank vault doors. The 4-wheel Ferrari disc brakes are, of course, legendary, and deservedly so. In fact, every last piece of the car worked with the authority that naturally comes with excellence. No tinny slam to the doors. No gaping cracks in the trim—inside or outside. No tacky wood-grain vinyl. In short, a thoroughgoing expression of automotive quality that soon aligns you with the zealots whose fondest dream is to see Enzo Ferrari issuing the noon benediction at the Vatican.

But enough of expensive sensual thrills. Just how are we going to make this $35,000 vision of excitement available to those of you who have not written multi-million-dollar screenplays, struck oil under the patio or successfully landed a planeload of dangerous drugs on Sunset Strip? With less difficulty than you might think.

First, you will need the friendly banker and the excellent credit rating already mentioned, because what you are about to do is borrow $30,000. Do not panic; people borrow such sums every day. You may require some collateral, perhaps the equity in your home or the 500 shares of AT&T you've got stashed away, but always remember this critical point: Banks are in business to lend money. They *want* you to have their $30,000 and they do not really care what you do with it. They *do* want reasonable assurance that you intend to repay it some day and that, until that time, you will pay interest regularly and on time.

You will also remember that we credited you with an automobile worth $5000 as a trade-in. You must use this car to strike a deal with your local Ferrari deal-

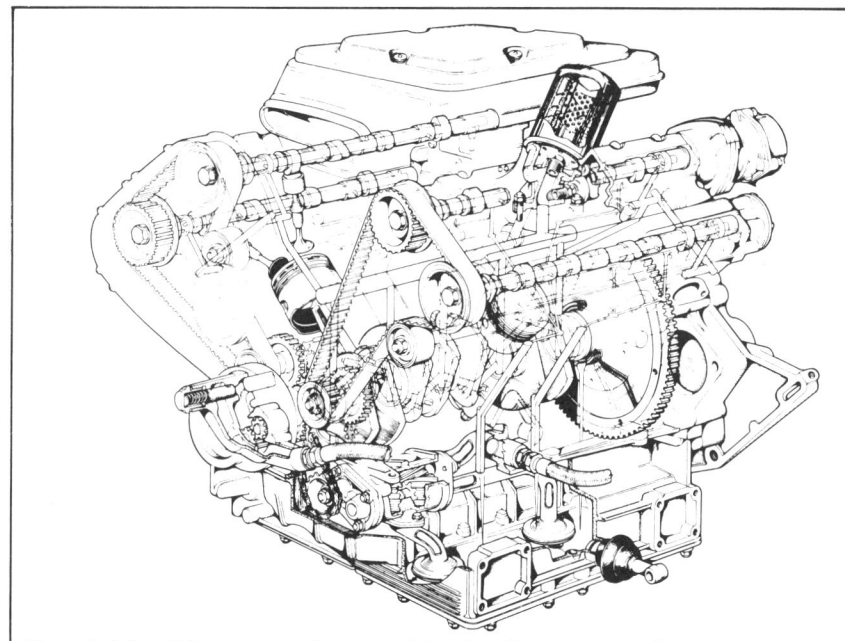

Ferrari's 3-liter V-8 powerplant features a light alloy block with shrunk-in cast iron liners. Four overhead cams are driven by toothed belts with tensioners. U.S. versions have wet sump lubrication system.

er whereby you buy your GTS for the cash difference of $30,000.

Now go to your banker and borrow the $30,000 on an open note. An open note has no monthly payments; you owe the entire amount at the end of a specified period, usually 90 or 180 days. At the end of the term, you pay the accrued interest and renew the note for another 90 or 180 days. It is by no means necessary to reduce the principal, though an occasional reduction will have a positive effect on your banker's mental attitude.

Under this system, at prevailing interest rates (about 9%), you will pay $2700 annually for the privilege of owning your Ferrari. That works out to $225 a month, or $8100 over a 36-month term. Not bad when you consider what $225 on a conventional car payment buys you these days. But the low monthly outlay isn't even the good part. How's this: It is tax deductible in its entirety, because it is interest expense! This means that your indulgence in a grand touring car has not only pumped your ego but has also resulted in a fat tax write-off.

And there's more. Ferraris have, over the past decade, appreciated with stunning regularity. A Ferrari Daytona that sold five years ago for $19,000 will sell in minutes today for $30,000. A Daytona spyder that sold new for $21,000 in 1973 brings as much as $90,000 today. Those figures are easily supported and will exercise a calming effect on your banker.

There's another point to mention. While the subject under discussion here is the 308 GTS, there are plenty of exotic cars selling for much less that you may have thought were beyond your reach. The same financing gambit works equally well on a $20,000 Ferrari, a $14,000 Pantera or a $6500 Shelby GT350.

All of this information ping-pongs around in my head all the way to Atlanta...which I reached, miraculously, without the first traffic ticket. It's actually *possible,* I keep telling myself. I am transported by visions of Jimmy Carter's reaction to tax-deductible Ferraris. If he likes the three-martini lunch, he'll love this. In fact, this procedure might even be the final prod that sends Ralph Nader 'round the bend.

You will by now have realized that owning a Ferrari is a far better investment than the mistress who figured in our initial ponderings. You could, depending on her demands, fork out $4800 a year in apartment rent alone. That's a $2100 loss right there. And then there are the requisite gifts, dinners and out-of-town trips to consider. God knows what those will cost. And the possibility of losing half your income through the divorce courts, should you be so tied, poses an ever-looming threat. Furthermore, mistresses have a depressing tendency to depreciate. And if you try to write any of it off as a legitimate business expense, you'll wind up like Wayne Hays.

Clean living, 18.2 mpg, big tax deductions and a Ferrari 308 GTS may not represent the one true way to good citizenship, happiness and a rising financial statement, but they do point to a life style that has distinct merit. Who knows, with a stunning Ferrari, a tax refund check and an air of genteel self-rightousness, you might attract some moneyed Eurasian temptress whose only goal in life is to bring a smile to your lips. It's certainly worth the try...and the worst that can happen is that you get 24-hour-a-day use of what has to be one of the world's great automobiles. [MT]

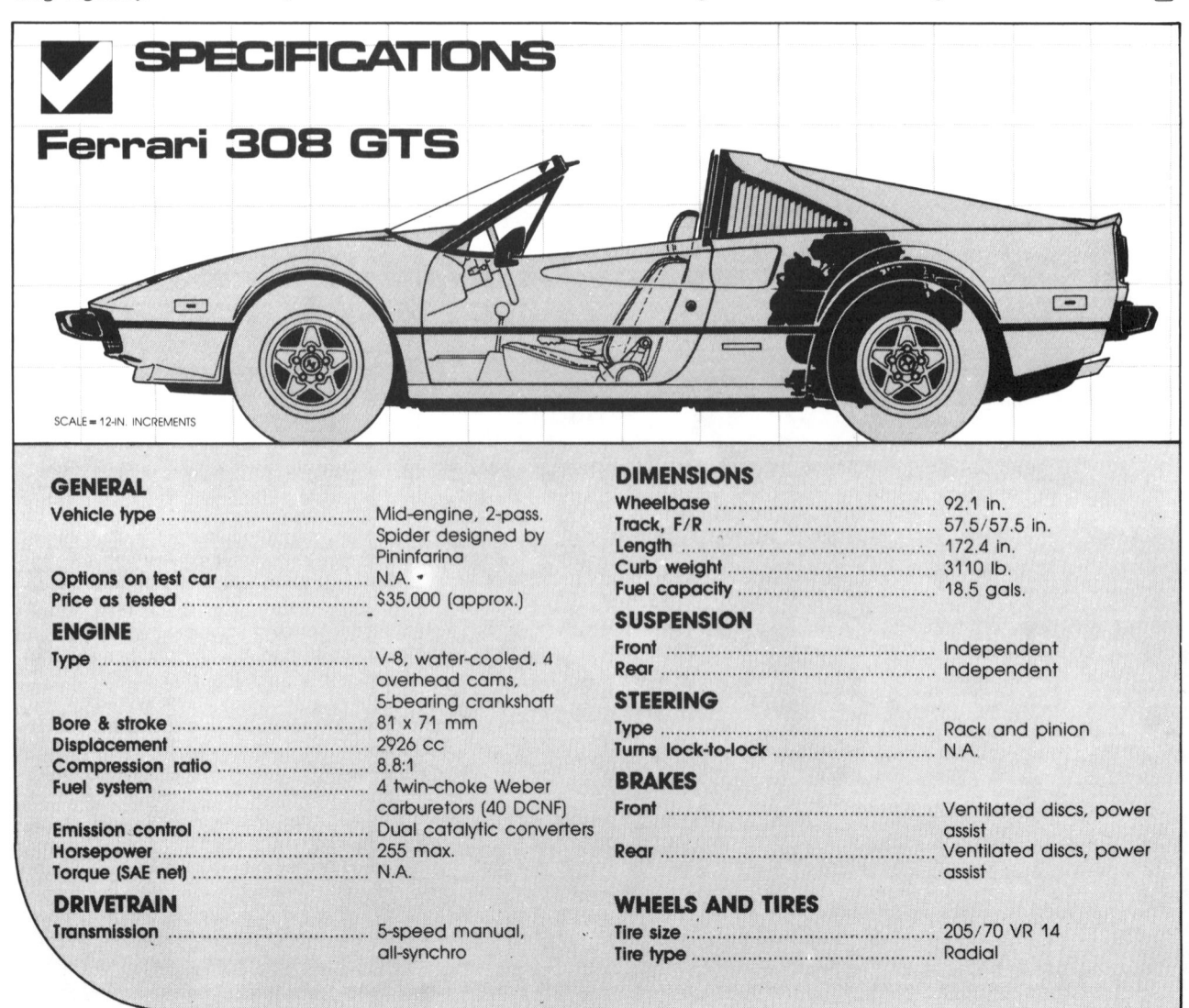

SPECIFICATIONS
Ferrari 308 GTS

SCALE = 12-IN. INCREMENTS

GENERAL
Vehicle type	Mid-engine, 2-pass. Spider designed by Pininfarina
Options on test car	N.A.
Price as tested	$35,000 (approx.)

ENGINE
Type	V-8, water-cooled, 4 overhead cams, 5-bearing crankshaft
Bore & stroke	81 x 71 mm
Displacement	2926 cc
Compression ratio	8.8:1
Fuel system	4 twin-choke Weber carburetors (40 DCNF)
Emission control	Dual catalytic converters
Horsepower	255 max.
Torque (SAE net)	N.A.

DRIVETRAIN
Transmission	5-speed manual, all-synchro

DIMENSIONS
Wheelbase	92.1 in.
Track, F/R	57.5/57.5 in.
Length	172.4 in.
Curb weight	3110 lb.
Fuel capacity	18.5 gals.

SUSPENSION
Front	Independent
Rear	Independent

STEERING
Type	Rack and pinion
Turns lock-to-lock	N.A.

BRAKES
Front	Ventilated discs, power assist
Rear	Ventilated discs, power assist

WHEELS AND TIRES
Tire size	205/70 VR 14
Tire type	Radial

CAR and DRIVER

JUNE 1978

ROAD TEST:

FERRARI 308 GTS

Putting the Grand back into Grand Touring.

BY STEVE THOMPSON

FERRARI 308 GTS

Almost the first thing you learn about fast cars after you've stopped listening to the old men dispensing automotive wisdom down at the corner gas station is that not many cars deserve the label "GT." As soon as the moto-moguls of Detroit and Japan figured out that some kind of mystical energy grew around a car if they called it a "Grand Touring" car—they were certainly never sure what the name meant or why the aura existed—any highline-trim option got to be known as a GT. It usually denoted "gage paks," "road wheels" and the ever-racy "stick shift." That all started in earnest about a decade and a half ago, and it's still going on today.

Considering that the popular progenitor of the GT fad was the Pontiac GTO, that's not surprising. The GTO wound up selling ten times as many cars in its first year as had been predicted, so marketing guys all over the world scrutinized the elements of its success. One of them was its name: GTO. Gran Turismo Omologato—homologated grand touring car. (Homologation is the process by which the FIA, the international motor-sports governing body, determines whether a car is acceptable for classification in one of the racing groups. Thus, to be homologated means to be accepted for class racing.) That the Pontiac did not also take the awesome character from the most famous carrier of the title—the 1963–64 Ferrari GTO—was one of the bitterest blows the "purists" of the Sixties had to endure. And, incidentally, this was one of the opening shots fired in the ensuing decade-long war between them and the hated "musclecar" drivers.

All this is by way of explanation for the following: the Ferrari 308 GTS *is* one of those rare cars that deserve all that goes with the title. (Unlike the original Ferrari GTO, which was a long-distance race car that, when used on the street, was not exactly the smooth, comfortable automobile that "GT" would indicate.) You won't find the reason for our praise of the new GTS on the spec sheet, where it demonstrates competent exoticar performance but no more. Nor will you find it simply in the photographs, where anyone can see its breathtaking styling. Nor, of course, does its equally breathtaking price—over $35,000—explain it. The reason is in the way it works.

Mind you, the spec sheet is pretty impressive. The sidewinder three-liter twin-cam V-8 rides amidships in the welded steel-tube frame, and a five-speed gearbox dispenses the motor's 205 hp to the fat Michelin 205/70VR-14 XWX radials, which themselves ride on beautiful cast magnesium-aluminum alloy Campagnolo wheels. The body, which was all fiberglass on the previous 308 car, the GTB (tested in March '77), is a composite of steel, aluminum and fiberglass panels. The floorpan, removable top panel, and lower body extremities are 'glass, the front hood is aluminum, and everything else is steel. This all results in a weight gain of 180 pounds over the GTB, but as you can see from the performance data, speed has not been affected. The all-plastic GTB managed a quarter-mile in sixteen seconds at a speed of 89 mph, while the heavier GTS did the same in 15.8 and 90 mph. Observed top speed in the GTS was 139 mph, while the GTB maximum was calculated at 140.

Still, your average well-tuned Corvette will turn in similar numbers, and though it lacks the eyeball-popping styling of the Pininfarina body, you could easily be forgiven for thinking that the Vette, at a fraction

70

of the cost, will do all that the Ferrari can. Which is where the ideal of Grand Touring comes in, where the immense differences between simply driving somewhere and touring show up.

The Ferrari is a complete car. No aspect of its performance or its personality overpowers any other aspect. So while it goes very fast, it doesn't make you *feel* as though you're going fast. And while it has a low, almost painfully pure body style, it demands no sacrifices from its driver or passenger in room or comfort. The same is true of the steering, brakes and suspension. The car steers lightly and positively at high speed, but is free of that brutally direct feel that can sap the joy of driving the minute heavy traffic is encountered. Indeed, the feel of the car is more like that of a Porsche 911 than of, say, a Pantera. While not whippet-quick, the steering, at 3.3 turns lock-to-lock, seems perfectly natural, and it is immediately easy to learn its characteristics. It holds no surprises.

Neither does the suspension. With a large amount of travel built into each wheel's movement, the car seems at first to have an almost Buick-like ride. But the impression soon evaporates as you discover

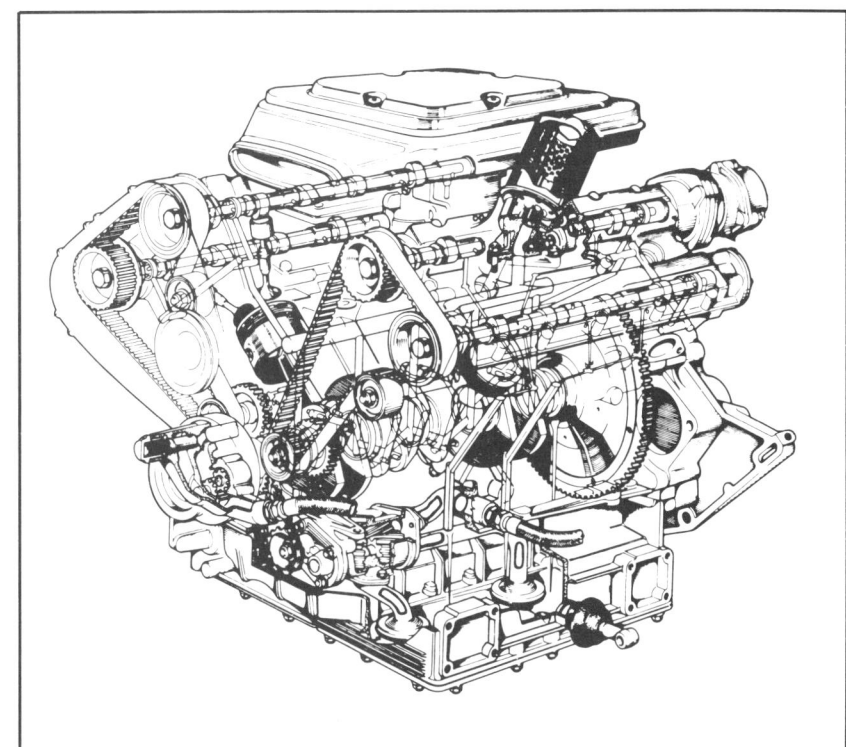

Better eight than never: Ferrari fanatics turn their noses up at it, but there's nothing wrong with a twin-cam three-liter V-8 with over 200 horsepower and all the right noises.

FERRARI 308 GTS

the capabilities of the suspension. Although it will allow lateral accelerations of 0.80 g before one end or the other breaks out of line, it is far more competent than that number (admittedly a middling figure for an exoticar) would have you believe. Again, it is the *way* you can drive it that counts, not the final limit. And unlike something like a tail-happy Porsche, the 308 GTS can be driven with an ineptness of style that would have other machines going backward into the wall. It is, above all, a forgiving steerer: understeer at first, with tail-out attitudes available anytime under power, tending toward slight oversteer at the chassis limit. All that translates to a safe, predictable handler with superb roadholding and ride. It goes where you want it to with just the right amount of effort in inputs, and, maybe more importantly, *won't* go where you don't want it to. *You* drive *it*, not the other way around.

None of that should make you imagine for a moment that it isn't a complete ball to throw around on back roads. If you think of it as a big, well-tuned Fiat X1/9, you'll get the picture. You can go howling into tight little hairpins with the V-8 bellowing out behind you and jump on the brakes, bang the close-gated shifter down a couple of gears, listen to the Cosworth-Ford noises as the engine burbles and growls on the overrun, and then get back on the power anytime you want, the chassis and tires having corrected any errors you may have committed going in. It lacks the razor-sharp definition of something like a Lamborghini Countach or a Lotus Esprit, but its responsiveness is still head and shoulders above the bulk of the so-called GTs you can buy today.

Of course, the fact that the engine makes those V-8 noises instead of the "ripping canvas" howl of the twelve-cylinder cars is more fuel for the purists' fire. To a rational enthusiast who's never been involved in the fierce intertribal wars of the marque freaks, it may seem strange that a car like the GTS would ignite such passions. It certainly seems strange to us. Especially since the car is so good that not a whit of its value would be lost if the prancing horses and lettering were removed from its bodywork.

There are, naturally, some flaws in the car. The two gauges that got left out of the central dash nacelle—the oil-temperature and clock—are stuck over to the left, under the thick roll of padding on top of the dash. They are hard to see, and since proper procedure for cold-morning warm-up in our brand-new test car required an oil temperature of 160 degrees before fancy stuff with the gearbox was permitted, seeing them was important. Moreover, the stylists who had a hand in the layout and design of the gauges themselves only hit about .500; the needles and lighting are somewhat less clear than could be the case. And the auxiliary control switches—for the heater, wiper speed, backlight defogger and electric antenna—are placed somewhat awkwardly on the center console between the seats. They are therefore switches you must look at to manipulate—hardly the thing called for when you're blazing down a rain-swept mountain road at eight-tenths at midnight.

In all other respects, the interior is a masterpiece of clean and simple ergonomics. The ventilation system—which blows not only through the traditional Ferrari triple vents on the center dash but also through side vents and air-conditioning ducts smack in the center of the dashboard at chest height—is one of the best, if not *the* best, yet encountered in a Ferrari, and possibly in any exoticar. The mixture and direction of the airflow are simple and direct—and you can get the air to circulate in any way you like, which in turn makes long-distance driving far less wearying. (And for those believers in the Italian-Cars-Always-Overheat myth, the GTS never once notched its temperature needles past normal, even in full-on air-conditioning mode while stuck in a simmering Southern California traffic jam. Sorry about that, boys.)

Popping the top lid makes the car even more fun. And best of all, getting it off and on again is a matter of only about fifteen seconds and less effort than it takes to lift a couple of grocery bags into the back of a station wagon. Two simple swivel clamps secure the padded fiberglass top. You release them (they're in back), slide the front backward (the locating pins are in the top, with holes in the windshield) and slip the light panel behind the seats, where it can be completely encased in a vinyl waterproofed cover. It takes up what little incidental package space there is behind the seats, but you can always put your magazines and newspapers into the beautifully crafted lockable pouches built into the GTS's door panels.

With the top off, the GTS seems far more roadster-like than its narrow roof opening would suggest. The air flows so well across the gap that at speeds over 60 mph there is virtually no annoying burble to tickle scalps and drag paper scraps out of the ashtray. Indeed, so efficient is the

FERRARI 308 GTS

ACCELERATION standing ¼ mile, seconds

Car	Seconds
FERRARI 308 GTS	~15.8
CHEVROLET CORVETTE L82	~15
LOTUS ESPRIT (1977)	~17.2
PORSCHE TURBO	~13.5

BRAKING 70-0 mph, feet

Car	Feet
FERRARI 308 GTS	~189
CHEVROLET CORVETTE L82	~175
LOTUS ESPRIT (1977)	~210
PORSCHE TURBO	~167

FUEL ECONOMY C/D mileage cycle, mpg

(City driving / Highway driving)

- FERRARI 308 GTS (NOT AVAILABLE)
- CHEVROLET CORVETTE L82 (NOT AVAILABLE)
- LOTUS ESPRIT (1977): ~22 / ~27
- PORSCHE TURBO: ~10 / ~17

PRICE AS TESTED dollars x 1000

Car	Price
FERRARI 308 GTS	~37.6
CHEVROLET CORVETTE L82	~11
LOTUS ESPRIT (1977)	~16
PORSCHE TURBO	~34

INTERIOR SOUND LEVEL dBA

(70-mph cruise / Full-throttle acceleration)

Car	Cruise	Full-throttle
FERRARI 308 GTS	~80	~91
CHEVROLET CORVETTE L82	~75	~86
LOTUS ESPRIT (1977)	~78	~88
PORSCHE TURBO	~73	~85

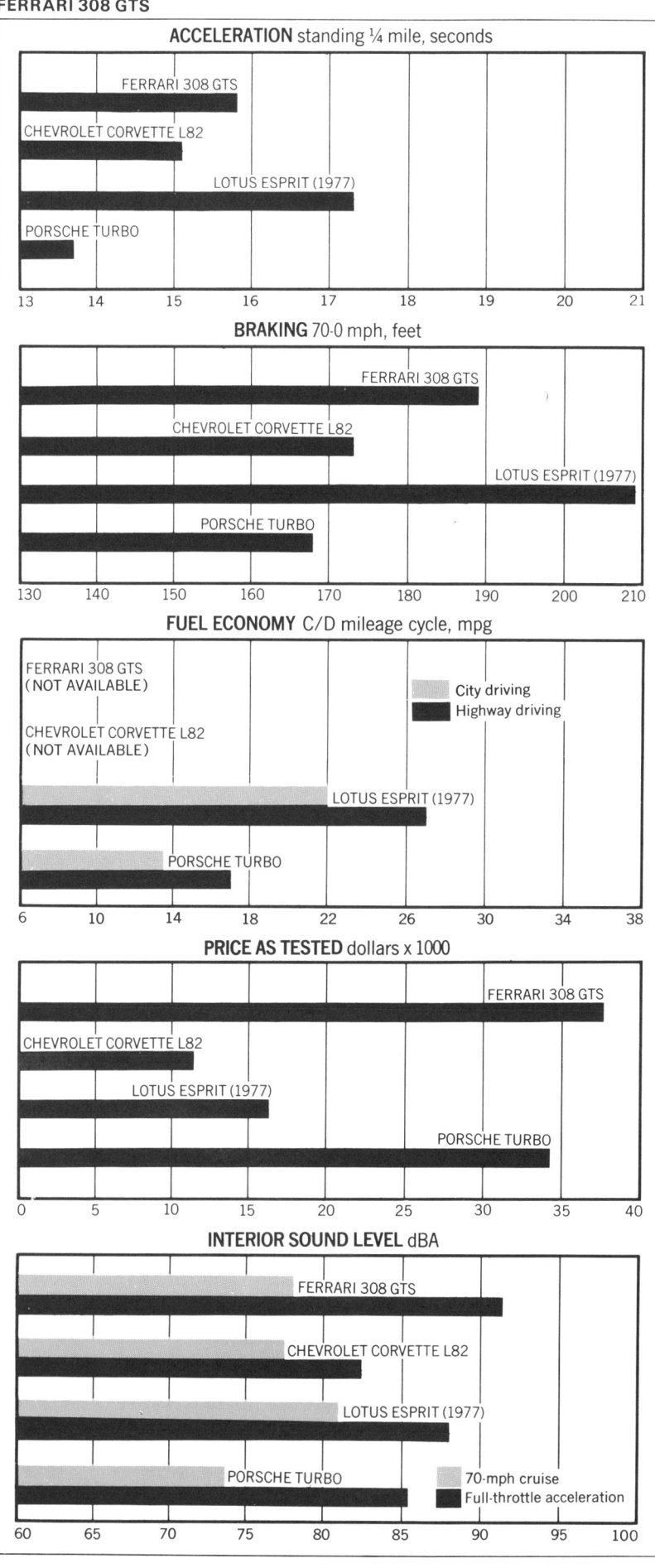

FERRARI 308 GTS

Importer: Modern Classic Motor Cars
3225 Mill Street
Reno, Nevada 89510

Vehicle type: mid-engine, rear-wheel-drive, 2-passenger, coupe/convertible

Price as tested: $37,649
(Manufacturer's suggested retail price, including all options listed below, dealer preparation and delivery charges, does not include state and local taxes, license or freight charges)

Options on test car: base 308 GTS, $36,411; Blaupunkt radio, $800; metallic paint, $438.

ENGINE
Type: V-8, water-cooled, aluminum block and heads, 5 main bearings
Bore x stroke .... 3.19 x 2.80 in, 81.0 x 71.0mm
Displacement .... 179 cu in, 2926cc
Compression ratio .... 8.8 to one
Carburetion .... 4x2-bbl Weber 40 DCNF carburetors
Valve gear .... belt-driven double overhead cams
Power (SAE net) .... 205 bhp @ 6600 rpm
Torque (SAE net) .... 181 lbs-ft @ 5000 rpm
Specific power output .. 1.15 bhp/cu in, 70.1 bhp/liter
Max. recommended engine speed .... 7800 rpm

DRIVETRAIN
Transmission .... 5-speed, all-synchro
Final drive ratio .... 3.71 to one

Gear	Ratio	Mph/1000 rpm	Max. test speed
I	3.56	5.6	43 mph (7700 rpm)
II	2.37	8.4	65 mph (7700 rpm)
III	1.69	11.8	91 mph (7700 rpm)
IV	1.24	16.1	124 mph (7700 rpm)
V	0.95	21.0	139 mph (6650 rpm)

DIMENSIONS AND CAPACITIES
Wheelbase .... 92.1 in
Track, F/R .... 57.5/57.5 in
Length .... 172.4 in
Width .... 67.7 in
Height .... 44.1 in
Ground clearance .... 4.6 in
Curb weight .... 3290 lbs
Weight distribution, F/R ...... 42.2/57.8 %
Battery capacity .... 12 volts, 66 amp-hr
Alternator capacity .... 770 watts
Fuel capacity .... 18.5 gal
Oil capacity .... 9.4 qts
Water capacity .... 19.0 qts

SUSPENSION
F: ind, unequal-length control arms, coil springs, anti-sway bar
R: ind, unequal-length control arms, coil springs, anti-sway bar

STEERING
Type .... rack-and-pinion
Turns lock-to-lock 3.3
Turning circle curb-to-curb ... 39.3 ft

BRAKES
F: .... 10.7-in dia vented disc, power-assisted
R: .... 10.9-in dia vented disc, power-assisted

WHEELS AND TIRES
Wheel size .... 7.5 x 14-in
Wheel type .... cast magnesium-aluminum alloy, 5-bolt
Tire make and size .... Michelin XWX, 205/70VR-14
Tire type .... steel-belted radial ply, tubeless
Test inflation pressures, F/R .... 30/34 psi
Tire load rating .... 1490 lbs per tire @ 36 psi

PERFORMANCE
Zero to	Seconds
30 mph	2.5
40 mph	3.7
50 mph	5.5
60 mph	7.2
70 mph	9.6
80 mph	12.2
90 mph	15.8
100 mph	20.6

Standing ¼-mile .... 15.8 sec @ 90.0 mph
Top speed (observed) .... 139 mph
70-0 mph .... 189 ft (0.86 g)

FERRARI 308 GTS

design that we thundered through a fierce rain squall without putting the top on and did not get wet. (Later on we discovered that the car *with* the top on is watertight, except when the electric windows are rolled up and forced into the rubber seal of the top and don't quite slide up all the way into their intended grooves; the resulting tiny gap collects water and occasionally allows a drop or two to enter the cockpit during hard cornering.)

With the top off, too, the beautiful music pouring out the four exhaust pipes provides a heady counterpoint to driving. Still, even topless, the GTS is not so noisy that it is ear-breaking. Unlike the 308 GT4, it can be driven quite hard for extended periods without any distress from vibration or noise, although its 78 dBA at 70-mph cruise is loud, and fullthrottle is enough to let the surrounding countryside know that you're driving something special. This is undoubtedly due to the fact that even with the eight-inch difference in their wheelbases (92.1 for the GTS, 100 for the GT4) the GTS allows more room for sound deadening and vibration damping than the fourplace car, which attempts to cram a lot more passenger space into a decidedly smallish package. In addition, the intake howl that used to echo through the long scoop molded into the driver's side of the car has been quieted by using the passenger's duct for feeding the air cleaners.

On the other hand, feeding and clothing the passenger and driver on a long trip with the GTS may be something of a problem, in that the luggage space is essentially all in the zippered and carpeted rear compartment. And *it* only displaces 5.3 cubic feet in a sort of oblong, rectangular way. Fitted or soft luggage would be appropriate, as would a luggage rack for serious journeys. We discovered that the hard way, by trying to get an American Tourister pullman suitcase aboard. Eventually, the passenger's seat was sent forward the requisite amount of space to allow the bag to ride behind the seat, but it was an untidy solution and points out the uncompromising nature of even the best exoticars.

And exactly how does the GTS fare when stacked up against more formidable opposition than the Corvette? Well indeed. Besides its simple beauty, it provides its driver with a kind of dynamic, rhythmic balance that allows perfectly confident high-speed driving for several hours at a time. Indeed, the only thing that ever made us stop was the need to refill the GTS's fuel tank, which was on the small side and would allow only about 175 to 200 miles of genuinely quick travel before its sensors would activate the low-fuel light. The seats and cockpit provide such a pleasant driving environment that a tank twice as big would have resulted in a stay on the road twice as long.

But of course, there *are* GT cars with plenty of luggage space, with larger fuel tanks, as distinguished in pedigree as the GTS, and even, sometimes, a hefty pile of *lire* cheaper. All that is undeniable. But somehow, as delightful as driving those cars is, as rewarding as it must be for a Porsche 928 owner to step into his garage at midnight to gaze proudly at his assemblage of Teutonic cold steel, as much as a Countach can spin heads as it passes, neither one—nor their counterparts—have that special flavor unique to the GTS. Just as Grand Touring is a matter of subtleties that set the act itself apart from merely driving somewhere, so the Ferrari 308 GTS, as a classic example of the best of Grand Touring cars, is set apart from the rest by its own subtleties.

It is a distinction the gas-station sages will never understand, and maybe that's for the best. After all, they still haven't figured out what GTO stood for. ●

Tale of two Targas

If you've got £15,000 or so to spend on a high performance sports car with an open top, your choice is limited to just two: Ferrari and Porsche. We took them both — and pitted them against each other on road and test track in a classic confrontation of Germanic efficiency versus Italian flamboyance to find out which, if either, is the superior Supercar

IT WAS the stuff that dreams are made of. A Porsche and a Ferrari, perfect weather, the open road, and the whole day before us.

Our task? To drive both cars in convoy, swapping every 30 miles or so, to the MIRA test track; to test them to their limits on MIRA's acceleration straights and steering pads; and to repeat the convoy run back to our starting point.

The purpose? To take two of the world's most desirable driving machines, to drive them as they should be driven, and so to compare them in the only truly valid way that two cars can be compared: on the road, on the track, and back-to-back.

They made a natural pair. Porsche and Ferrari both belong to that club of just a half-dozen builders of sports cars that are to be truly reckoned with — cars with pedigrees as long as your arm. And among that select band, no two are closer in concept than the Porsche 911 SC Targa and the Ferrari 308 GTS.

Both have two-seater cockpits; both have three-litre engines; both cost within spitting distance of £15,000; and both, alone among cars of similar conception, have detachable roof panels.

The Ferrari is most expensive in a range of three models based on the same running gear. At the bottom is the 308 GT4, with four seats, ungainly (by Ferrari standards) appearance, and a price tag of £14,754. Then there is the two-seater GTB, arguably the most beautiful of modern supercars, originally of glass-fibre but now of steel construction, and with a price of £15,500. At £16,499 the GTS is essentially just a Spider version of the GTB — £999 is a lot to pay for the privilege of a detachable roof panel.

Porsche's 911 range is simpler than it once was with, apart from the Turbo, just the one basic 911SC model. This costs £13,499 whether in coupé form or, as tested here, as a Targa. Additionally there is the Sport option — basically a handling package plus wings, seats and stereo/radio cassette thrown in and adding another £1500 to the price: again, the Targa option is available at no extra cost.

Superficially, then, the two protagonists couldn't be more similar. Delve beneath their skins, however, and they are poles apart. Both cars have rack and pinion steering, both have braking by ventilated discs on all four wheels, but that is all that they do have in common.

They reflect different national characteristics, and divergent business philosophies. The Porsche is a product of basic parameters and components that were finalised a long time ago, enabling basic tooling costs to be amortised over a long production run, while the details have been refined and refined again. Thus it has a steel monocoque shell first introduced almost 14 years ago, with the same air-cooled flat six engine mounted anachronistically behind the rear wheels. It is supported by torsion bar springs all round, with MacPherson damper struts and (effectively) wide-based lower wishbones at the front, semi-trailing arms at the rear, and an anti-roll bar at each end.

The Ferrari reflects the Italian firm's empirical approach; its ability to chop, change and redesign at short notice; its handbuilt construction. At the heart of it is a separate, tubular steel chassis, clothed in a shell that was originally all glass-fibre when the 308 GTB was first introduced, but is now a mixture of plastic and metal. In the modern idiom, its engine is transversely mounted ahead of the rear wheels: a light alloy, water cooled V8, it has a classic specification with a carburetter choke for every cylinder and two overhead camshafts per bank. Suspension is by wishbones, coil springs, and an anti-roll bar at each end.

Compared with the Ferrari engine's all-singing, all-dancing spec, the Porsche's is subtle understatement, with just a single overhead cam per bank, and Bosch fuel injection as opposed to rorty Weber twin-chokes.

Knowing the Germans, it seems likely that the claimed power output is also an understatement: 180 bhp (DIN) at a mere 5,500 rpm and 189 lb ft of torque at 4,200 rpm. Compare that to the Ferrari's claim of 255 bhp (DIN), developed right on the rev limit at 7,700 rpm, and 210 lb ft at 5,000 rpm, and it is easy to see which ought to be the quicker car.

But wait a minute. If the Porsche's kerb weight of 22.9 cwt seems a lot for a two-seater, the Ferrari's 26.0 cwt is absurd. That's heavier than a Rover 3500.

Also allow for some optimism and some pessimism in the respective manufacturers' power claims, and you end up with two cars having remarkably similar performance overall. The figures tell it all. One may be a little quicker at some speeds, the other gets ahead in other circumstances. Both are fully fledged Supercars.

Porsches are not always the easiest cars to get off the line. Many past test cars have suffered a surfeit of traction over engine torque and clutch grip. No such problems with the latest 911, however. Dropping the clutch at 5,000 rpm sends it hurtling off the line with just the right amount of wheelspin to reach 30 mph in just 2.2 sec, a stunning performance even by Supercar standards that makes the Ferrari's 2.5 sec look positively pedestrian. This head start gives the 911 a lead that it maintains up to over 110 mph. With a 0-60 mph time of 6.6 sec the Ferrari is fabulously quick, but the Porsche beats it by ½ a second. By the quarter mile post the Porsche is some 25 yards ahead of the Ferrari, but the Italian is just about to start hauling in the German.

The Porsche attains 100 mph just 16.3 sec after departure from rest, while the Ferrari gets there only 0.3 sec later. By the kilometre post the gap is narrowing fast: the Porsche gets there first, but is only travelling at 118 mph, whilst the Ferrari is just nudging 120 mph. Soon it will overtake the German, and given enough space we would expect the Ferrari eventually to leave the Porsche dwindling in the distance. There isn't the space to prove it on this island, but rough checks suggest that 150 mph plus seems quite credible from the GTS (Ferrari claim 157 mph) while the Porsche seems to be good for 145 mph (the factory claims 140 mph plus).

On balance, the test track figures leave little to choose between them: we'd have to call it a dead heat. On the road, subjectively, the differences are surprisingly pronounced.

It's not what they do, but the way that they do it: the Ferrari is deceptive, it simply does not feel particularly fast, but the Porsche leaves you in no doubt. Perhaps it is the manner of the power delivery.

At the very lowest revs the Ferrari's carburation is a trifle patchy, while the injected Porsche is as clean as a whistle; hence it's 30-50 mph time in fifth of 7.9 sec as against the lower-geared Ferrari's 9.0 sec.

But once over that low speed rough patch the GTS gets into its stride sooner and the remaining 20 mph increments in top gear are quicker in the Ferrari all the way up to 90 mph. There is no stomach flattening wrench, no sudden climbing up onto the cam: instead engine power builds up steadily right through the rev range up to about 7,000 rpm. The red line is at 7,700 rpm, but, unlike the last 308 we drove, the GTS felt reluctant to go all the way.

By contrast, the Porsche's red line is set at a modest 6,300 rpm, yet the engine feels like it wants to go a lot further. There is nothing steady about the power flow from the Porsche's engine: while it pulls as cleanly in the lower rev ranges as the Ferrari, it does so less strongly. Then suddenly, at 3,500 rpm, it all

75

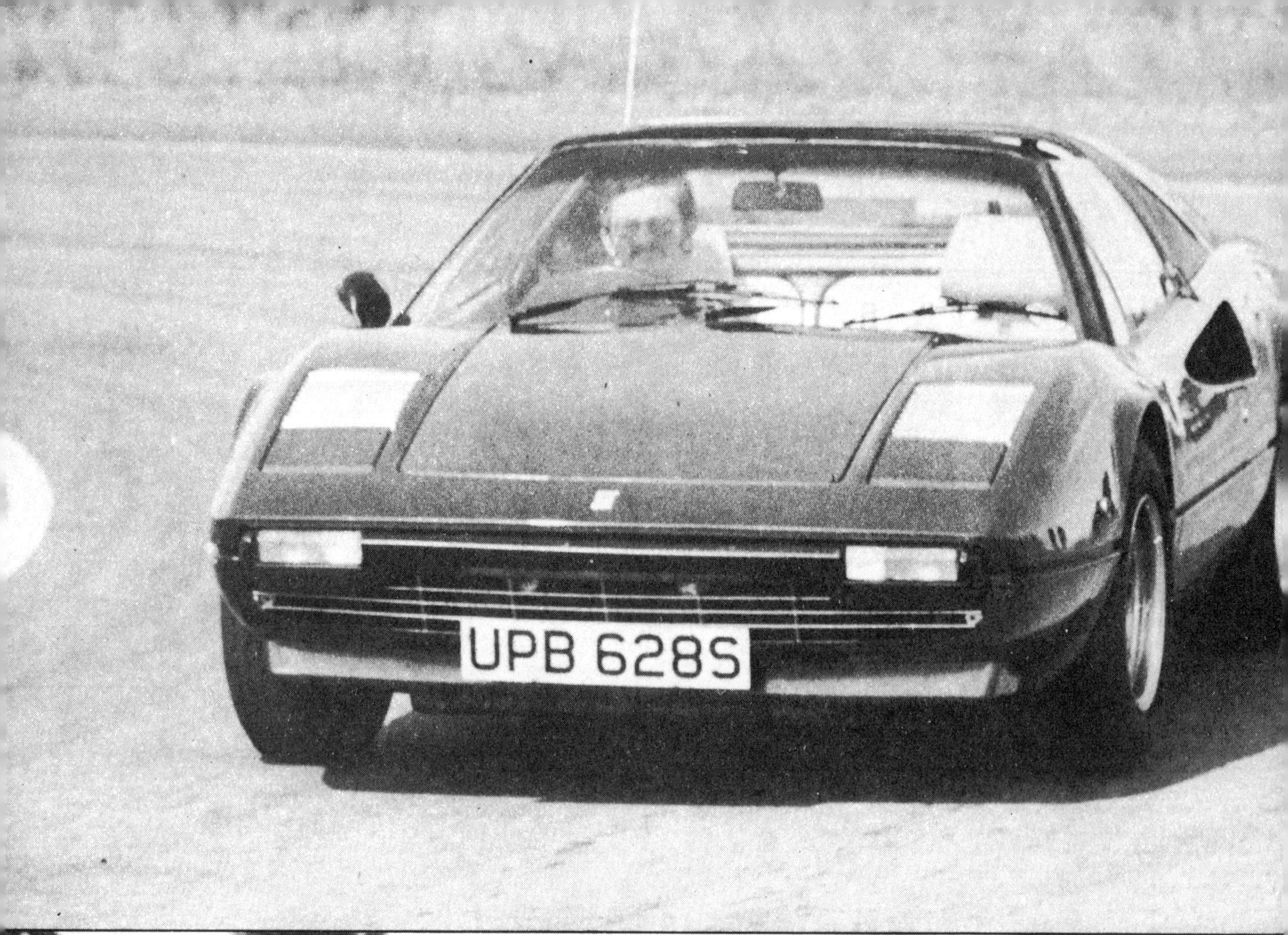
UPB 628S

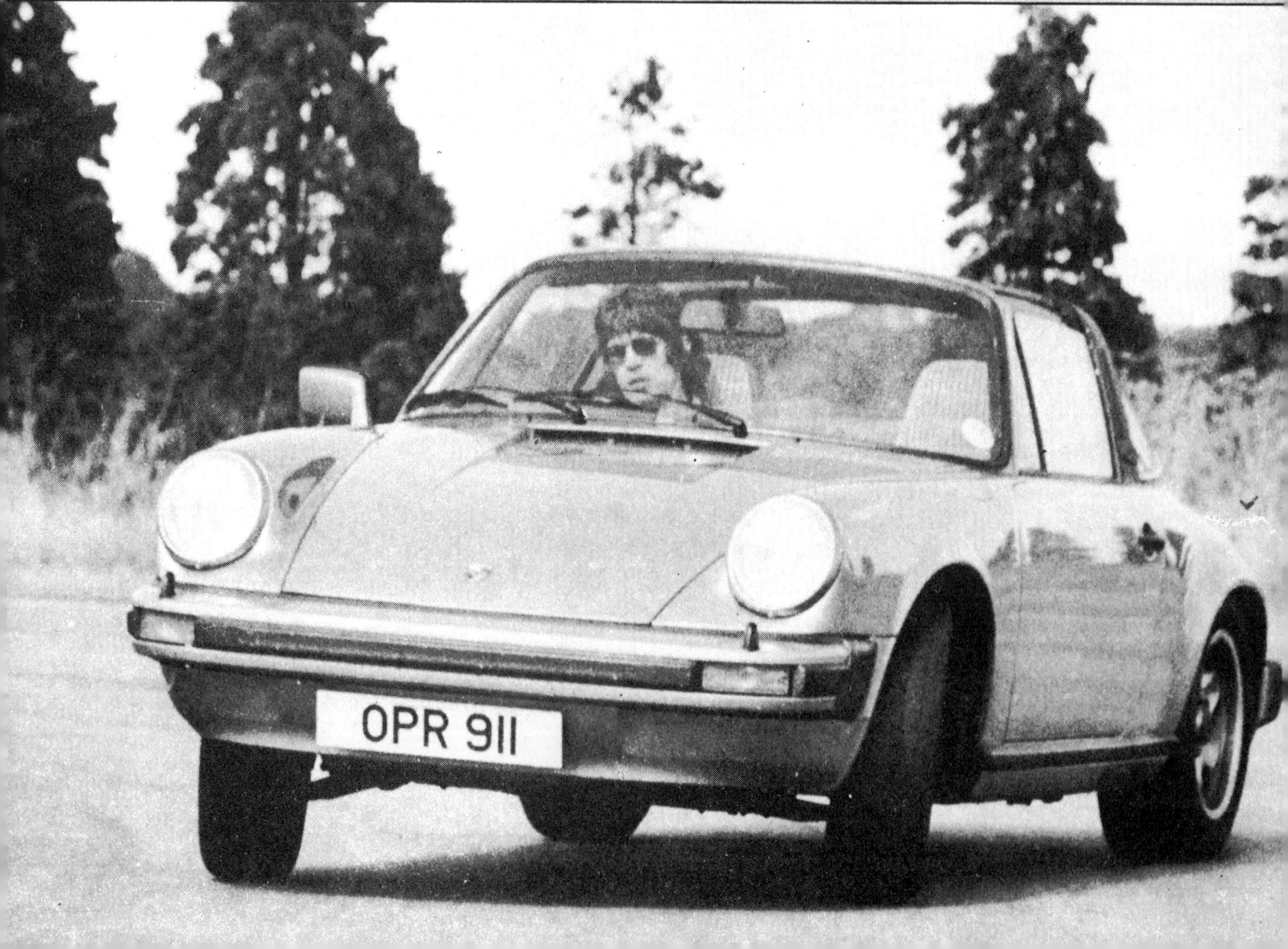
OPR 911

comes in with a rush and you have to snatch the next gear before you overshoot the red line. It is this 'coming on cam' effect, this vivid surge of power, that makes the Porsche *feel* so very much quicker than the Ferrari, in spite of what the figures may say.

Both engines are utterly smooth but neither is quiet. However, they differ in both the quality and volume of noise. The Porsche is easily the quieter of the pair, and what sound it does make is an enthusiast's delight — a combination of mechanical whine and whirr and burble from the exhaust. In volume it is never offensive, even under acceleration through the gears, and it gets "left behind" when cruising at speed in fifth.

In a different sort of way, the Ferrari has an equally satisfactory engine note. It is a very hard sound; it is loud throughout the rev range and becomes a positive howl at peak revs. Although engine noise is almost a prerequisite in a car of this sort, the Ferrari's is perhaps too obtrusive for all but the most avid enthusiast, to the extent that even cruising at modest speed on a motorway is a noisy business.

Each car has a gearchange that thrives on hard and positive driving. At first glance the metal "gate" that determines lever movements in the Ferrari looks intimidating, but any fears prove groundless. True, the change has a heavy, metallic quality, but the lever movements are short, the gate positive, the spring loading perfect. At town speeds there is some notchiness, but on the open road the lever is like a switch and can be slammed through as fast as the hand can move. What does limit the speed of gearchanges, however, is the clutch, which has a disconcerting tendency to "stick" on the floor on really fast changes: this may well have cost a couple of tenths of a second on acceleration times at the test track.

In its action the GTS's clutch is heavy, and it's engagement somewhat abrupt — a characteristic that is shared by the 911 but aggravated by the awkward operating angle of the latter's pedal.

Hence smooth changes on the Porsche need some concentration, especially in normal use when the gearchange can feel heavyish and notchy. Under very hard use, however, outstandingly quick changes are possible. The gate is clearly defined and the lever well sprung, but it is possible sometimes when changing from 2nd to 3rd to get lost in the no man's land between the 1st/2nd plane and 3rd/4th planes. These first four gears are on the usual 'H' pattern on the Porsche, with fifth to the right and forward: the Ferrari has the alternative pattern with a dog-leg 1st to 2nd change.

Ratios on both cars are suitably close and sporting, but whereas it is 1st, 2nd and 3rd that are stacked the more closely on the Ferrari, on the Porsche the ratios close up at the top end: both are more than satisfactory, though the Porsche has the advantage of usefully longer legs in fifth gear. Honours must also go to the Porsche for its lower level of transmission noise: an occasional moan from the gears on the overrun pales into insignificance compared with the high pitched whine from the Ferrari's gears, pronounced enough to be clearly audible over even the obtrusive engine.

Pitch a modern mid-engined Ferrari against the Porsche, with its long out of favour rear-engined layout, on a twisting road and it *ought* to be no contest. In reality, Porsche's engineers have so finely honed their car's handling as to effectively mask its fundamentally unsound layout. Thus the 911 can easily hold its own — 99 per cent of the time.

The sting in the tail is that *in extremis* the 911 will not forgive driver error: the Ferrari will.

At the sort of speeds forced upon us most of the time, the Porsche is easily the more satisfying car. Its steering is perfectly weighted and geared and positively alive with feel, whereas at town speeds the Ferrari's is too light and curiously dead in feel. That the steering is also somewhat low geared aggravates the quite pronounced understeer in the lower speed range. In such conditions understeer is also evident in the Porsche, but to a lesser degree: in either car you will only get oversteer under power if you deliberately break traction at the rear in first gear.

Push them harder under power, in 2nd or 3rd gear, say, and both cars will ultimately understeer prodigiously. Where they reveal their true colours is if you then lift off the throttle in mid bend. Do so in the Ferrari and it merely reverts to a neutral cornering attitude: it's safe: it's forgiving. Do so in the Porsche, and you've got a full-blooded tail slide. It isn't vicious, it doesn't happen too quickly, but your response has to be both prompt and carefully judged, because you only dare let that tail swing out so far. Catch it in time, and you'll live to spin another day. Let it go too far, and you'll probably never get it back again. It's a sobering thought.

But few owners are likely to find themselves experiencing it, simply because the limits are so very high, beyond the point at which most drivers' nerve runs out. To experience oversteer in the Porsche you've got to either deliberately provoke it, or do something very silly — or take it to the limit of adhesion at highly illegal speeds on the open road.

It is at such endorsement-baiting

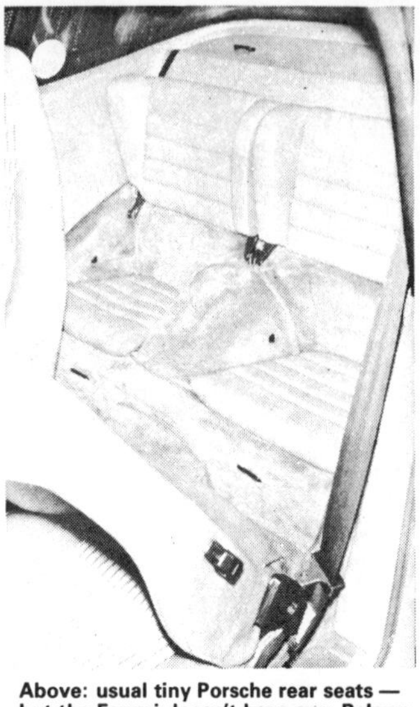

Above: pilot's seat and controls Porsche-style. If you're long-legged and/or tall then the position is ideal — if you're small it's still very comfortable

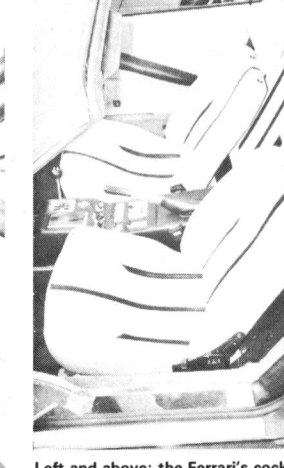

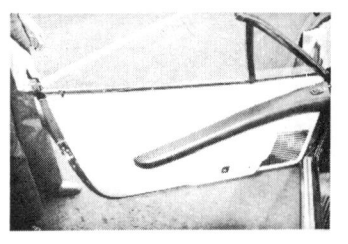

Above: usual tiny Porsche rear seats — but the Ferrari doesn't have any. Below: most Ferrari dials are easily visible. Bottom: futuristic armrest for Ferrari

Left and above: the Ferrari's cockpit. The driving position has a trace of the Italian long-arm/short leg syndrome but is still excellent

speeds that the Ferrari is transformed. With less torque available to the rear wheels it corners only just on the understeer side of neutral. The steering seems to gain in weight and feel, and it sits rock steady on the road. It is unequivocally a car for the open road. Quite apart from its more forgiving nature, the Ferrari is also more stable. At anything much over 110 mph the Porsche has a disconcerting tendency to yaw and goes light at the front end.

Thus the 308 GTS is best suited for that quick dash down to Nice for the weekend — it is just as satisfying as the Porsche but more reassuring. On the other hand in this England of crowded roads and silly speed limits the Porsche is the more rewarding — provided you treat it with respect.

On a wet road the Porsche demands more than respect, for the tail will all too easily step out of line, even under gentle acceleration. We experienced no rain in the Ferrari, but would expect it to be better if only by virtue of its greater weight and narrower rear tyres and past experience of those XWX Michelins.

When you spot that radar trap ahead, you want braking that is as undramatic as it is powerful. Both cars are quite fabulous in this respect. The Ferrari's pedal is appreciably harder in feel than the Porsche but both offer perfect progression and feel when braking from the highest of speeds. It would be nit-picking indeed to complain that the 911 is just a trifle unprogressive for check braking at town speeds, and that the handbrake on the Ferrari is not very effective.

Buyers of such cars are likely to drive them far, as well as fast. Nobody expects a limousine ride, nor space for kids and the kitchen sink too, but there must be plenty of space and comfort for two adults plus some luggage. Ride quality is relevant to comfort, and here it is the Ferrari that really surprised us. Within the first few yards of leaving the gates of Maranello Concessionaires we drove over a couple of vicious potholes at very low speed and there was such a severe jolt that we expected the worst from the GTS's ride.

Yet one hundred miles later we stepped out of the two cars to compare notes and suddenly realised that we had never given a moment's thought to the Ferrari's ride. And the next 100 miles confirmed what this fact suggested: that the Ferrari's is good by any standards, and quite remarkably so for a car of this character.

The Porsche's ride is exactly what you *do* expect from such a car. Very firm, especially at low speed, and accompanied by considerable bump thump, it is the price that you expect, and are prepared, to pay in return for handling precision. But the fact remains that the Ferrari handles to an equally high standard yet doesn't sacrifice ride comfort.

Inside, both cars have enough room for the longest-legged of drivers, but the Ferrari is a strict two-seater, and has less headroom, while the Porsche has two miniature rear seats. Two circus dwarfs or very small children can be carried in them, or tip them down and use the space for luggage overspill from the bonnet/boot. In the Ferrari there is a smallish but sensibly shaped boot behind the engine, but no room inside for anything bar what you can fit into a pair of door-mounted handbags.

In both cars the driving positions are angled to the left, but our two test drivers of very different dimensions were able to get comfortable in both. The very long one found the Porsche ideal, while complaining that the 308's gear lever was too close to the wheel, leaving insufficient room for his left knee when travelling in fourth gear. The shorter of the two preferred the Ferrari, which suffers to only a small degree from the long-arm/short leg Italian driving position syndrome; he complained that the Porsche's brake pedal was too high and too far to the left in relation to the throttle pedal.

In the Ferrari, column stalks cater for both lights and wash/wipe functions, while the Porsche's lights switch is on the facia. Both are comprehensively instrumented, the important dials on the Ferrari being well located in a nacelle in front of the driver, but the oil temperature gauge and clock are badly placed at the extreme bottom right of the facia. Both cars' instruments suffer from reflections, and while the Porsche's are better calibrated, half of them are hidden from the driver's view by the wheel.

Porsche's heating system now works pretty well, but proved redundant on the day we ran our convoy confrontation. It was gratifyingly warm and sunny, but it showed up the Porsche's ventilation as quite inadequate, all the more so in contrast to the outstanding optional (£600.21) air conditioning fitted to the Ferrari.

On such a day it was tempting to remove the Targa tops (a marginally easier operation on the Ferrari) but, although there was no buffeting, wind noise became intolerable on both cars if anything more than half their speed potential were used.

GENERAL SPECIFICATION

PORSCHE

ENGINE
Cylinders	6, Horizontally opposed
Capacity	2994 cc (182.7 cu in)
Bore/stroke	95.0/70.4 mm (3.74/2.77 in)
Cooling	Air cooled
Block	Light alloy
Head	Light alloy
Valves	Sohc per bank
Cam drive	Chain
Valve timing	
inlet opens	7° btdc
inlet closes	47° abdc
ex opens	49° bbdc
ex closes	3° atdc
Compression	8.5:1
Carburetter	Bosch Fuel injection
Bearings	8 main
Max power	180 bhp (DIN) at 5500 rpm
Max torque	189 lb ft (DIN) at 4200 rpm

TRANSMISSION
Type	5 speed, rear wheel drive
Clutch	8.9in
Actuation	Cable
Internal ratios and mph/1000 rpm	
Top	0.821:1/23.4
4th	1.000:1/19.2
3rd	1.261:1/15.2
2nd	1.833:1/10.5
1st	3.181:1/6.0
Rev	3.325:1
Final drive	3.875:1

BODY/CHASSIS
Construction	All steel monocoque
Protection	Floorpan, sills and wheel arches galvanised, PVC underseal, cavities Tectyl treated, 6 year warranty

SUSPENSION
Front	Ind. by lower wishbones, McPherson struts, longitudinal torsion bar springs, anti-roll bar
Rear	Ind. by semi-trailing arms, transverse torsion bars, telescopic dampers, anti-roll bar

STEERING
Type	Rack and pinion
Assistance	No

BRAKES
Front	11.1 in ventilated discs
Rear	11.4 in ventilated discs
Park	On rear, drum in disc
Servo	Yes
Circuit	Split front/rear
Rear valve	No
Adjustment	Automatic

WHEELS
Type	Light alloy 6 × 15 front, 7 × 15 rear
Tyres	185/70 VR 15 front, 215/60 VR 15 rear
Pressures	29/34 psi F/R

ELECTRICAL
Battery	66 Ah
Earth	Negative
Generator	980W alternator
Fuses	24
Headlights	
type	2 × Halogen H4
dip	110 W total
main	230 W total

FERRARI

ENGINE
Cylinders	V8, transversely mounted
Capacity	2927 cc (178.5 cu in)
Bore/stroke	81/71 mm (3.20/2.80 in)
Cooling	Water
Block	Light alloy
Head	Light alloy
Valves	Dohc per bank
Cam drive	Toothed belt
Valve timing	
inlet opens	30° btdc
inlet closes	50° abdc
ex opens	36° bbdc
ex closes	28° atdc
Compression	8.8:1
Carburetter	4 x Weber 40 DCNF
Bearings	5 main
Max power	255 bhp (DIN) at 7700 rpm
Max torque	210 lb ft (DIN) at 5000 rpm

TRANSMISSION
Type	5 speed manual, rear wheel drive
Clutch	9.5 in dia
Actuation	Cable
Internal ratios and mph/1000 rpm	
Top	0.918:1/21.7
4th	1.244:1/16.0
3rd	1.693:1/11.8
2nd	2.353:1/8.5
1st	3.418:1/5.8
Rev	3.247:1
Final drive	3.706:1, limited slip differential

BODY/CHASSIS
Construction	Tubular steel chassis with part steel, part glass-fibre bodywork
Protection	Underseal

SUSPENSION
Front	Ind. by wishbones with coil springs and anti-roll bar
Rear	Ind. by wishbones with coil springs and anti-roll bar

STEERING
Type	Rack and pinion
Assistance	No

BRAKES
Front	10.6 in ventilated discs
Rear	10.9 in ventilated discs
Park	on rear
Servo	Yes
Circuit	Split front/rear
Rear valve	No
Adjustment	Automatic

WHEELS
Type	Light alloy 6½ × 14 in
Tyres	205/70 VR 14
Pressures	30/34 psi F/R

ELECTRICAL
Battery	60 Ah
Earth	Negative
Generator	770W alternator
Fuses	18
Headlights	
type	2 × Halogen
dip	110 W total
main	220 W total

With the top on, engine noise in the Porsche is low enough to show up considerable road roar and wind roar from around the edges of the roof. In the Ferrari, noise from both the latter sources is lower, but this is academic as all is dominated by noise from engine and gearbox.

As befits their prices, both cars are assembled and painted to the highest standard, but inside it is the Porsche which looks far the plusher, the Ferrari being rather plain and plasticky.

So, when all is said and done, which would we choose?

At the risk of being accused of cowardice, or evading the issue, we're not going to say. We cannot say. One of our testers would have opted for the Ferrari, the other preferred the Porsche, but either could quite happily have lived with the other. Which just about sums it all up, for the choice between them really does boil down to personal taste.

They are, both, quite simply, outstanding cars and we count ourselves privileged to have had the opportunity to drive them. And we can only hope and pray that such anti-social, fuel-guzzling, noisy playthings of the privileged few will continue to be available for many years to come.

For the world really would be a poorer place without them.

Above: 2.9 litres of V8 Ferrari muscle, mounted transversely and producing a massive 255 bhp. Below: classic Porsche flat six, now of 3.0 litres, mounted behind the rear wheels, and churning out 180 bhp. The two cars' performance is similar because the Ferrari is much heavier

	Porsche	Ferrari
MAXIMUM SPEED		
Mph	145*	150+*
SPEEDS IN GEARS		
	mph at 6300 rpm	mph at 7700 rpm
1st	38	45
2nd	66	65
3rd	96	91
4th	121	123

ACCELERATION
mph	sec	sec
0-30	2.2	2.5
0-40	3.4	3.4
0-50	4.6	5.0
0-60	6.1	6.6
0-70	8.2	8.6
0-80	10.4	10.6
0-90	12.9	13.5
0-100	16.3	16.6
0-110	21.2	20.9
0-120	28.9	27.3
Stand'g ¼	14.5	15.0
Stand'g km	26.8	27.2

IN TOP
mph	sec	sec
20-40	8.2	—
30-50	7.9	9.0
40-60	8.4	7.9
50-70	8.8	7.5
60-80	9.1	7.8
70-90	9.0	8.3
80-100	8.9	9.3
90-110	10.5	11.3
100-120	13.6	—

IN FOURTH
mph	sec	sec
20-40	6.3	6.4
30-50	6.3	5.5
40-60	6.8	5.3
50-70	6.4	5.4
60-80	5.8	5.6
70-90	5.8	5.7
80-100	6.5	6.2
90-110	8.3	7.8
100-120	—	10.9
Twin test mpg	14.1	13.6

*Estimate

FERRARI 308 GT4
A driver's car in every sense of the words

ENZO FERRARI'S IMPACT on the high-performance automobile scene has been extraordinary. In the racing world, no other manufacturer has enjoyed the number of victories and the degree of success of Ferrari—especially over such a long span of time. On the street, the name Ferrari has become synonymous with sleek, beautiful cars with excellent performance and exceptional handling—cars that evoke images of Hollywood stars, jet-setters along the French Riviera and wealthy driving enthusiasts all over the world. But behind the fantasies are automobiles, pieces of machinery that have often been innovative, sometimes breathtaking, but always worthy of respect. The first car to carry the now common 2+2 designation was a 1961 Ferrari 250 GT, and this road test concerns the current U.S.-legal Ferrari 2+2, the Dino 308 GT4.

The 308 GT4 has been controversial among Ferrari devotees since its introduction because of its Dino appellation. Most *aficionados* of the marque already know that Dino was Enzo Ferrari's son, that he studied engineering in Modena and Switzerland and took an active role in his father's business until his untimely death in 1956. Some months later a tradition of smaller cars with V-6 and later V-8 engines was born, carrying Dino's name. Ferrariphiles have been reluctant, to say the least, in accepting these cars as Ferraris, pointing to Enzo Ferrari's oft-cited remark to R&T Contributing Editor Paul Frère some years ago that "A Ferrari is a 12-cylinder car." Be that as it may, the only three Ferraris legally available in the U.S. today are 8-cylinder cars. We don't intend to enter into the 8 versus 12 debate, because our experience has been that even among motoring journalists who have the opportunity to drive nearly every car in the world at one time or another, the chance to savor a Ferrari is cause for celebration. In any case, the Dino name has been dropped; now we have the Ferrari 308 GT4.

The 308 GT4 was unveiled to the world at the Paris Auto Show in the fall of 1973 and the Bertone styling was described as angular and somewhat unbalanced looking. Pininfarina had been responsible for the styling of all production Ferraris for the previous 20 years, so the choice of Bertone surprised many, but it appeared to be Fiat's (49-percent share in Ferrari at that time, 51 percent now) way of maintaining the balance of power between the two styling houses. Contributing Editor Jonathan Thompson presented a styling analysis of the 308 GT4 to accompany our comparison test of the Dino, the Maserati Merak and Lamborghini Urraco in September 1975, and noted that while most Ferrari fans were not impressed with the styling, "a close-up inspection reveals many virtues. It's a design that grows in appeal with familiarity and which will have a greater influence on future mid-engine cars than the Merak or Urraco." Prophetic words indeed, as we found ourselves considerably more impressed with the car's lines now than we were previously.

The drivetrain components of the GT4 are the same ones

found in the 2-seater 308 GTB and GTS models, starting with the 3.0-liter (2926-cc) 90-degree V-8 engine mounted crosswise and just ahead of the rear wheels. The engine is rated at 205 bhp (SAE net) at 6600 rpm and delivers 181 lb-ft torque at 5000, which is actually rather remarkable for less than 180 cu in., and is capable of delivering scintillating performance: 0–60 mph in 7.8 seconds, the quarter mile in 16.0 sec at 89.0 mph. For a car with a test weight of 3585 lb, the acceleration figures are commendable.

Acceleration is not the entire story, however, and the 308 rates top marks in driveability as well. In warm climates, such as ours in southern California, the manual choke is unnecessary and start-ups are a matter of switching on the ignition, giving the fuel pumps 10 seconds to prime, engaging the starter and listening to the delightful sounds as the engine purrs into life. The Editor noted that it's "lovely the way this V-8 sounds more like a 12-cylinder." The warm-up period is short and marred by only an occasional minor stumble or mild backfire through the exhaust during cold upshifts. Another staff member who spent a day with the Dino in very hot weather was very impressed with its ability to start hot with only a rare stumble, confirming our previous praise of the V-8's near-perfect driveability characteristics.

don't have the progressive buildup of g forces or the forgiving breakaway characteristics found in newer designs such as the Pirelli P7 or even Michelin's own TRXs. At the track during our slalom testing, the car reached the point where the rear end wanted to break loose, not smoothly but rather abruptly, and this doesn't inspire confidence near the limits. There is a slight warning to the driver as cornering adhesion reaches its ultimate point, and when that line is crossed the car can loop rather quickly.

The rack-and-pinion steering is characteristically notchy and rather stiff at low speeds, but it becomes wonderfully precise and sensitive at higher speeds. There is some kickback on severe jolts but most surface irregularities are nicely damped and the GT4 driver will feel very much at-one with the car during spirited driving. We were not especially impressed with the Dino's brakes in our 1975 test and we're gratified to find that stopping distances from 60 and 80 mph in our simulated panic stops have been reduced significantly: 167 ft (previously 208) from 60 mph and 270 (versus 316 ft) from 80. During these tests, there was moderate to severe locking of the front brakes and they were relatively insensitive to pedal modulation, but there was no

On the other hand, there was some fuel starvation evident in right-hand skidpad maneuvers, but none of the oil pressure drop we had experienced in high-g cornering of the previous 308 GT4 we tested. While on the subject of oil, the 308 GT4's lubrication system has a quirk in terms of checking the oil: The most accurate reading is obtained immediately after shutdown, rather than by letting all the oil settle into the sump.

The 5-speed gearbox is a model of compatability with the engine's output and the gated shifter is wonderfully precise in typical Ferrari fashion. The shifter and the clutch require a firm hand and foot, respectively, but the expert driver will find great pleasure in using them to their full potential, once the gearbox is warmed up and the slight balkiness of cold shifts disappears.

The 308 GT4's suspension, using unequal-length A-arms front and rear, provides a lovely balance of ride and handling characteristics. On the road, there is mild and progressive understeer that can be changed to oversteer if the driver works at it. The Dino has high cornering limits but the Michelin XWX radial tires

AT A GLANCE

	Ferrari Dino 308 GT4	Porsche 928	Lotus Esprit
List price	$38,460	$31,835	$33,220
Curb weight, lb	3405	3410	2350
Engine	V-8	V-8	inline 4
Transmission	5-sp M	5-sp M	5-sp M
0–60 mph, sec	7.8	7.0	9.2
Standing ¼ mi, sec	16.0	15.6	17.0
Speed at end of ¼ mi, mph	89.0	93.0	81.5
Stopping distance from 60 mph, ft	167	138	176
Interior noise at 50 mph, dBA	77	71	74
Lateral acceleration, g	0.779	0.811	na
Slalom speed, mph	57.9	59.7	na
Fuel economy, mpg	14.0	16.0	est 24.0

slewing and the control was very good. Pedal effort for a 0.5g-stop is a light 15 lb but fade during our six stops from 60 mph was nil.

Inside, the GT4 impresses in several areas including the seats, which offer very good support and also emit the lovely aroma of Connolly leather (standard), and the level of detail and finish is quite high. First-rate materials are used throughout the interior and the overall effect is one of tasteful simplicity. The instruments are logically positioned within one housing directly in front of the driver, most of the controls are there too, and the gearshift and handbrake levers are within easy reach. There are, however, several niggling problems inside the Dino: The seatbelts twist in their guides and hang up, and there is some glare off the flat instrument faces as well as reflections off the flat rear glass into the windshield at night. Our biggest complaint, however, is with the ventilation system, which is barely adequate. It seems that the Italian exotic carmakers have never understood that cooling the windshield with dash-top vents is not the way to win friends and influence comfort. In the 308, there are three such vents and they can be angled down so that you get some face-level flow, aided by a single-speed fan. The air conditioning, which is also standard equipment, deserves the same rating as the fresh air vent system: only adequate. There are two small central vents on the dash, mounted somewhat too low and not having sufficient adjustment, that simply aren't up to the job. The 3-speed fan that's part of the system is welcome for the options it provides in the volume of cooled air, but its top speed has a god awful sound like a jet engine approaching maximum rpm. One has to be near heat prostration to endure it.

Ferrari's attitude about stereo sound systems is that each owner will want to select his own, so while two speakers and the electrically operated antenna are included in the $38,460 suggested retail price, a radio and/or tape player is not. That price also includes electric window lifts, tinted glass, a heated rear window, a quartz clock and a limited-slip differential. Our test car, built late in 1978 but identical to the 1979 models, also boasted such optional equipment as a sunroof ($957), metallic paint ($670) and some $1000 worth of Alpine AM/FM stereo cassette player, Fossgate amplifier and six speakers.

Summing up the Ferrari Dino 308 GT4 is easy: It's an exciting driver's car. The handling is superb, the high-speed stability is exemplary, the ride is delightful and the performance is stimulating if not breathtaking. Outward vision is excellent, not a simple accomplishment in a mid-engine car, and scooting along a sinuous road is sheer delight. One can't ignore fuel consumption but even in this area the Ferrari 2+2 acquits itself reasonably well with a real-world average of 14.0 mpg (we measured 16.5 mpg on the highway and 10.5 during hard driving). At better than $40,000 as equipped, the 308 GT4 is well out of the reach of most of us, but it is a worthy member of Ferrari's long line of dream machines.

PRICE

List price, all POE	$38,460
Price as tested	$41,407

Price as tested includes standard equipment (air cond, elect. window lifts, Connolly leather interior, tinted glass, heated rear window, quartz clock, stereo speakers & wiring for radio, power antenna), sunroof ($957), metallic paint ($670), sound system ($970), pre-delivery inspection ($350)

GENERAL

Curb weight, lb/kg	3405	1546
Test weight	3585	1628
Weight distribution (with driver), f/r, %		42/58
Wheelbase, in./mm	100.4	2550
Track, front/rear	57.9/57.9	1471/1471
Length	176.7	4488
Width	67.3	1709
Height	47.6	1209
Trunk space, cu ft/liters	5.6	159
Fuel capacity, U.S. gal./liters	19.8	75

ENGINE

Type	dohc V-8
Bore x stroke, in./mm	3.19 x 2.80 ... 81.0 x 71.0
Displacement, cu in./cc	2926/179
Compression ratio	8.8:1
Bhp @ rpm, SAE net/kW	205/153 @ 6600
Torque @ rpm, lb-ft/Nm	181/245 @ 5000
Carburetion	four Weber (2V)
Fuel requirement	unleaded, 91-oct

DRIVETRAIN

Transmission	5-sp manual
Gear ratios: 5th (0.86)	3.53:1
4th (1.12)	4.60:1
3rd (1.52)	6.25:1
2nd (2.18)	8.96:1
1st (3.23)	13.28:1
Final drive ratio	4.11:1

CHASSIS & BODY

Layout	mid engine/rear drive
Body/frame	separate steel body/tubular frame
Brake system	10.7-in. (272-mm) vented discs front, 10.9-in. (277-mm) vented discs rear; vacuum assisted
Wheels	cast alloy, 14 x 7½
Tires	Michelin XWX, 205/70VR-14
Steering type	rack & pinion
Turns, lock-to-lock	3.3
Suspension, front/rear: unequal-length A-arms, coil springs, tube shocks, anti-roll bar/unequal-length A-arms, coil springs, tube shocks, anti-roll bar	

CALCULATED DATA

Lb/bhp (test weight)	17.5
Mph/1000 rpm (5th gear)	20.7
Engine revs/mi (60 mph)	2900
R&T steering index	1.30
Brake swept area, sq in./ton	251

ROAD TEST RESULTS

ACCELERATION

Time to distance, sec:
0-100 ft	3.0
0-500 ft	8.6
0-1320 ft (¼ mi)	16.0
Speed at end of ¼ mi, mph	89.0

Time to speed, sec:
0-30 mph	2.4
0-50 mph	5.7
0-60 mph	7.8
0-70 mph	10.1
0-80 mph	12.8
0-100 mph	20.5

SPEEDS IN GEARS

5th gear (6700 rpm)	138
4th (7000)	109
3rd (7000)	82
2nd (7000)	59
1st (7000)	38

FUEL ECONOMY

Normal driving, mpg	14.0

BRAKES

Minimum stopping distances, ft:
From 60 mph	167
From 80 mph	270
Control in panic stop	very good
Pedal effort for 0.5g stop, lb	15

Fade: percent increase in pedal effort to maintain 0.5g deceleration in 6 stops from 60 mph ... nil
Overall brake rating ... very good

HANDLING

Lateral accel, 100-ft radius, g	0.779
Speed thru 700-ft slalom, mph	57.9

INTERIOR NOISE

Constant 30 mph, dBA	75
50 mph	77
70 mph	77

SPEEDOMETER ERROR

30 mph indicated is actually	30.5
60 mph	60.5

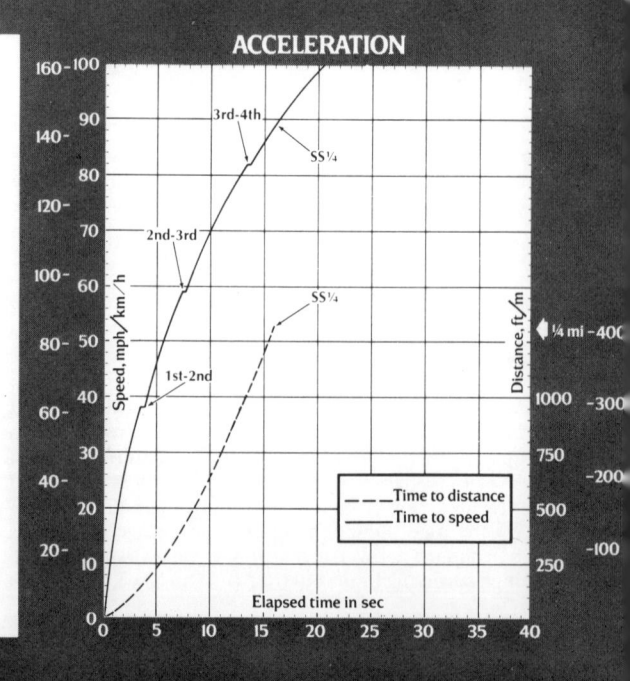

ROAD TEST

Ferrari 308GTBi

A specialist in punching definitive holes in the air.

• "Damn all, *that little sucker's a Ferrari!*" This piece of information crackles across southern Ontario's citizens'-band airwaves like the discovery of gold in the back yard. To truckers dulled by thousands of miles of chewed-up Chevrolets and proletarian Plymouths, a 308GTBi makes for a happy exclamation point in the early morning.

The Ferrari's windshield is wrapped around with a peach-colored sunup and splotched with the connect-the-dots remains of hundreds of early-rising Canadian bugs. The slope of the windshield has pasted them into an intricate streaking that shows how the air flows over the car. The pattern is interesting and the bugs did not suffer. We at *C/D* feel that it is our duty to make sure they never knew what hit them. To this end, the Ferrari is spewing out trans-provincial four-lane at a steady 115 mph. We know this only because 5500 rpm in fifth gear translates that way on the calculator.

While the speedometer refuses to talk to us uprange of 85 mph, the car is filling us in on its place in the universe with Italian sign language. Mainly, the car wants us to know it *is* Italian. It developed a whale of an oil leak, then sealed it up before we could figure out why. The first time we drove away, the driver's outside mirror unglued itself. Now the passenger's sun-visor snap has pulled out of the megabuck pseudo-burlap headliner, and the visor does a little curtsy over every bump. But these things are the way the car establishes its personality, and that is something you get plenty of for $48,000. With its shape and its height, which is about knee-high to the average hubcap, the 308GTBi organizes a sort of instant square dance of appraisal every time it rolls to a stop. People just naturally congregate around it. Not many ask if it's a kit car; the Ferrari's lines are too finely drawn for it to be misjudged, and it sits too nicely on its alloy wheels. If its Michelin XWXs are beginning to look a little skinny, Mr. Ferrari has a fix on the way. Only the earliest of 1980 cars will show up without sets of special wheels and Michelin TRX tires. The TRXs should give improved adhesion and they should also reduce steering effort. Change at Ferrari is almost always worth the wait.

FERRARI

The biggest change this year doesn't show up until you open the throttle or the rear deck lid. The muscular, dual-overhead-cam aluminum V-8 has been stripped of its fabled Weber carburetion. Bosch K-Jetronic Continous Injection System fuel injection feeds the hungry little cylinders these days. It meets current emissions standards with the help of a dual Marelli Digiplex electronic ignition system, and also with air injection, exhaust-gas recirculation, and controls for crankcase ventilation and evaporation of fuel and oil. A list like this no longer guarantees an asthmatic engine, and this is the big improvement indicated by the "i"-for-injection that's been added to the 308's moniker.

The car will stroke around town without loading up, overheating, or pausing even once when you nail the throttle. It simply runs, and runs freely. The Digiplex ignition, in spite of its computer memory, its choice of eight timing curves, and its claimed 15 percent improvement in highway fuel economy, is relatively crude electronically compared with some systems in use today, but we will not question its results because it works very well here.

Horsepower and torque remain the same, although the 308GTBi's 0-to-60-mph time of 7.6 seconds has added four-tenths since we tested the lift-top 308GTS two years ago. Top speed is up by one mile per hour, but braking from 70 takes 204 feet, about another car length, and that's too long for a car that will reel out to 140 mph. Braking feel, however, is exemplary.

Ferrari is worried about sustaining the integrity of both your body and Pininfarina's, so it sets the tire pressures at 27 front/34 rear in order to compen-

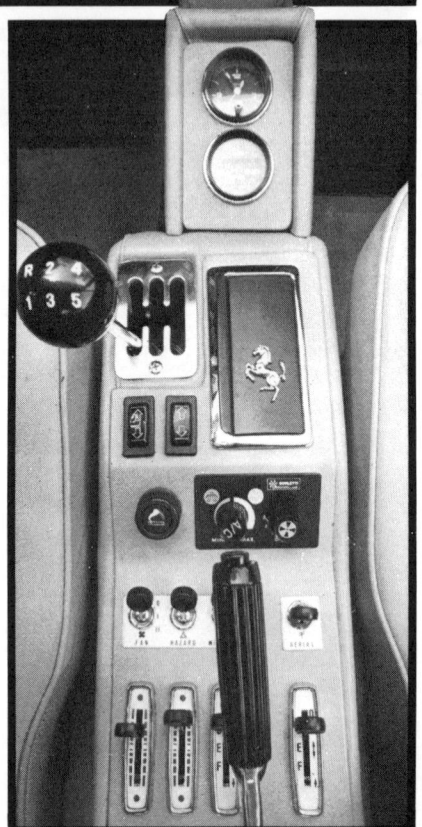

84

FERRARI

ACCELERATION standing ¼ mile, seconds

- PORSCHE 928
- FERRARI 308GTBi
- JAGUAR XJ-S
- LOTUS ESPRIT S2

BRAKING 70-0 mph, feet

- PORSCHE 928
- LOTUS ESPRIT S2
- JAGUAR XJ-S
- FERRARI 308GTBi

FUEL ECONOMY EPA estimated mpg

- LOTUS ESPRIT S2
- PORSCHE 928
- FERRARI 308GTBi
- JAGUAR XJ-S

CURRENT BASE PRICE dollars x 1000

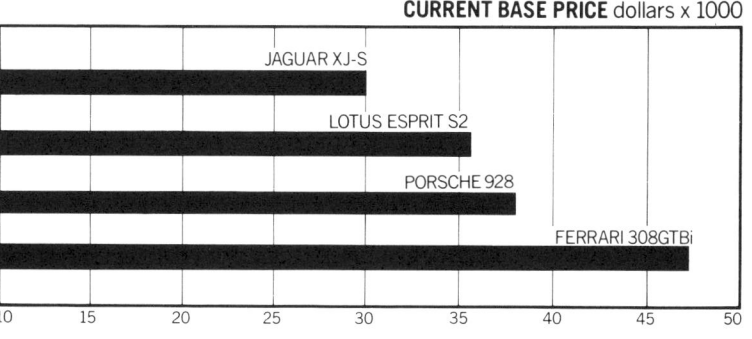

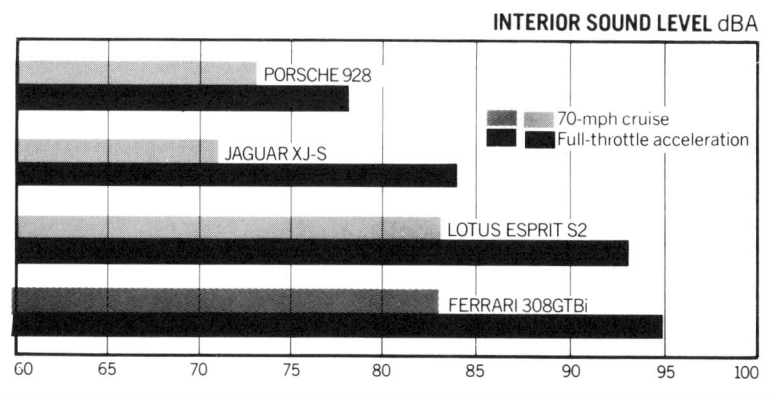

- JAGUAR XJ-S
- LOTUS ESPRIT S2
- PORSCHE 928
- FERRARI 308GTBi

INTERIOR SOUND LEVEL dBA

- PORSCHE 928
- JAGUAR XJ-S
- LOTUS ESPRIT S2
- FERRARI 308GTBi

(70-mph cruise / Full-throttle acceleration)

FERRARI 308GTBi

Vehicle type: mid-engine, rear-wheel-drive, 2-passenger, 2-door coupe

Price as tested: $48,064

Options on test car: base Ferrari 308GTBi, $47,440; metallic paint, $624.

ENGINE
Type: V-8, water-cooled, aluminum block and heads, 5 main bearings
Bore x stroke .... 3.19 x 2.80 in, 81 x 71mm
Displacement .... 179 cu in, 2930cc
Compression ratio 8.8:1
Carburetion .... Bosch K-Jetronic fuel injection
Valve gear .... belt-driven double overhead cams
Power (SAE net) .... 205 bhp @ 6600 rpm
Torque (SAE net) .... 181 lbs-ft @ 5000 rpm
Redline .... 7700 rpm

DRIVETRAIN
Transmission 5-speed
Final-drive ratio 3.71:1

Gear	Ratio	Mph/1000 rpm	Max. test speed
I	3.59	5.8	45 mph (7700 rpm)
II	2.35	8.9	69 mph (7700 rpm)
III	1.69	12.4	95 mph (7700 rpm)
IV	1.24	16.9	130 mph (7700 rpm)
V	0.95	21.0	140 mph (6700 rpm)

DIMENSIONS AND CAPACITIES
Wheelbase .... 92.1 in
Track, F/R .... 57.9/57.9 in
Length .... 174.2 in
Width .... 67.7 in
Height .... 44.1 in
Ground clearance .... 4.6 in
Curb weight .... 3280 lbs
Weight distribution, F/R .... 39.6/60.4 %
Fuel capacity .... 18.5 gal

SUSPENSION
F: ...ind, unequal-length control arms, coil springs, anti-sway bar
R: ...ind, unequal-length control arms, coil springs, anti-sway bar

STEERING
Type .... rack-and-pinion
Turns lock-to-lock 3.3
Turning circle curb-to-curb .... 39.3 ft

BRAKES
F: .... 10.7 x 0.9-in vented disc
R: .... 10.9 x 0.8-in vented disc
Power assist .... vacuum

WHEELS AND TIRES
Wheel size .... 7.5 x 14 in
Wheel type .... cast aluminum
Tire make and size .... Michelin XWX, 205/70VR-14

SAE INTERIOR VOLUME
Front seat .... 48 cu ft
Trunk space .... 5 cu ft

INTERIOR SOUND LEVEL
Idle .... 65 dBA
Full-throttle acceleration .... 95 dBA
70-mph cruising .... 83 dBA
70-mph coasting .... 77 dBA

PERFORMANCE
Zero to   Seconds
  30 mph 2.5
  40 mph 3.8
  50 mph 5.8
  60 mph 7.6
  70 mph 10.5
  80 mph 13.2
  90 mph 17.5
  100 mph 22.6
Standing ¼-mile .... 15.8 sec @ 88 mph
Top speed .... 140 mph
Braking, 70–0 mph .... 204 ft
Road horsepower @ 50 mph .... 15.0 hp
Roadholding, 282-ft dia skidpad .... 0.79 g
EPA estimated fuel economy .... 11 mpg
C/D observed fuel economy .... 15 mpg

FERRARI

sate for the mid-engined, rear-weighted tendency to spin at the cornering limit (a lowish 0.79 g on the skidpad). Ferrari likes you to have plenty of understeer to bull through before the car pivots around you, even though this puts a lid on steering precision. Frankly, though it works better overall with a little more air in front, the steering will always be truck-heavy. Every control feels massive. The 308GTBi is very heavy for its 92.1-inch wheelbase, and it lacks finesse until you're going fast enough to wind up in maximum security. But there are still places to drive that way and get away with it, and the GTBi will hurry the search.

This search is likely to be only semi-comfortable. The seats are very good, but mounted high enough off the floor that normally proportioned people butt up against the low roof. The wheel is raked away à la Continental Trailways, your feet are funneled down into an inboard cubbyhole for the pedals, and the gas pedal is so slickly covered that you'll execute a banana-peel ramp shot into the firewall until you get your shoes retreaded with Artgum. The interior has been reupholstered, again with Connolly hides, and this time with a greater appreciation for long-haul comfort. Only the hottest days match the air conditioner's capacity, but normal ventilation settings add mysteriously preheated air. The Ferrari gated shift is easy to manage, if heavy, but we are specifically annoyed that major controls (lights, wipers, etc.) work in directions opposite to world practice and common sense.

But why confuse common sense with the 308GTBi? Its purpose is to do smiles, and that it does. It is well fitted and nicely finished, and is fully capable of standing on its own merits as the best of factory-sold exotic cars in this country. But, although it will gall the purists of yesteryear to hear it, the 308GTBi can't hold a candle to the BMW M1. However, anybody who hasn't driven an M1 will never know the difference.

—*Larry Griffin*

COUNTERPOINT

• The 308GTBi is the most impressive Ferrari of the 308 series ever to pass through our hands. I only regret that it could not be stretched or bent to accommodate my size-46-long body more comfortably. Even so, once I'd managed the necessary slumping and compression of the physical me, the driving experience was an exhilarating one. If I tended to bump my head on the roof every time the car passed over an expansion strip or frost heave, what the hell, it kept me alert to all the fun I was having. What did I like best? Easy. The shift linkage and gate, Number One; and the clutch, Number Two. There's something solid and uncompromising about the way a Ferrari shift lever clicks through that gate. And as for the clutch, it's so heavy that when you first press your $36 Nike against its surface, you wonder if maybe you haven't managed to put your foot on the brake pedal or a frame member instead. This is not a car for the frail. Once you learn to cope with the very slippery accelerator pedal, you find wonderful throttle response, and the resultant noise is not unlike the aural effluvia of an old four-banger Offy if it'd been run through a Cherry Bomb, or maybe an old Smitty muffler. Nice car for short but conspicuous consumers. —*David E. Davis, Jr.*

Fifty Gs seems like a lot to spend for a car with no radio, at least to me. And at that price I think I'd like the air conditioning to work too. Maybe even a muffler that muffles—which this Ferrari doesn't have—would be nice (the sound meter had to be sent in for recalibration after the 308's cacophonous, 95-dBA acceleration runs). While we're at it, why not a lubrication system that maintains oil pressure at the limit of adhesion in both right- and left-hand turns? I could carp about the clutch pedal, which takes two feet to depress, and the Teflon-slippery seats, but what's the point? You'll go ahead and buy a 308 the instant you can afford one just because it's a Ferrari. What other brand can make a claim like that?

This is one of those cars hung up in the value warp, wherein dollars and cents beyond $20,000 are flipped upside down. The more you pay, the less you get. In this case, only the prancing horse could be worth that kind of money. The rest of the car is just a very convenient way to move the escutcheon around here and there for all the right people to see.

—*Don Sherman*

With all due respect to those of you who think Ferraris are assembled on hallowed territory, far above the clouds, and that each of the components has some sort of religious significance, I'd like to tell you that the 308GTBi isn't a very exotic car. Ferrari simply doesn't have the research facilities or the development capital to turn out a state-of-the-art automobile. An X-car is much more advanced structurally; the 308 is heavy for its size. And your average 1958 De Soto surpasses it in ergonomic technology. (In the De Soto you can at least reach the top of the steering wheel, and you never hit your head on the roof.) Except for a couple of cams, the pancake shape, and the breathtaking sticker price, there's really nothing here you can't find on an X1/9.

But that doesn't mean I don't like it. *Au contraire*, I love it. The pieces mesh with a rare sort of celestial harmony. A symphony of rich sounds caresses your ears, even when you're just trundling around town. The 308 will suck up Interstate at warp speed with astounding calm and assurance. And its sure-footedness is beyond reproach. No, the Ferrari's mechanicals aren't very exotic. But the way they work sure is. —*Rich Ceppos*

Dolce as well as presto

Ferrari's new fuel-injected Mondial 8 2+2

By Michael Scarlett

Fitting four seats into a relatively short wheelbase (for the type of car) and still contriving to keep a graceful aspect is not easy, but in most opinions Pininfarina has done outstandingly well on the new Ferrari. Bumpers are black resin type capable of absorbing 3 mph impacts without damage

BUILDING four-seaters, however genuinely so, when your name is one mainly associated with very high performance two-seaters, is a difficult trick. Ferrari have been doing various things of the sort for many years now, with varying success – more successfully than most of their few competitors or would-be emulators at any rate. Even so, cars like the Bertone-shaped Dino 308 GT4 and its ultimate replacement, the Pininfarina-styled Mondial 8 revealed at Geneva are inevitably a compromise. It is interesting to see how what Ferrari consider to be market forces and their (and Pininfarina's) own instincts have assisted or interfered in the making of the new car.

The company seek to widen the scope of their market by producing a more civilized and versatile machine, which will meet the needs of the modern sporting motorist, who in this price class is more often than not the type who, even when buying a Ferrari, increasingly expects more civilization and refinement than his sort 10 years earlier. Ferrari claim that many of their customers today are quite high mileage men, averaging up to 20,000 miles per annum, and not the owner who takes the car out only for the occasional pleasurable run. The modern Ferrari owner expects the same faultless driveaway from cold, the same reliability, even something of the same wider service intervals that is normal in the mass production car world. This, and the spread of government type approval interference across the motoring world, means the extinction of the fussy, coquette-ish Italian super-car, and maybe the start of a new generation of kinder, less demanding – if perhaps less exciting – products. This obviously applies very much in America, which is Ferrari's most important market, since it takes 35 per cent of their output.

Compared with the Dino 308 GT4, which appeared first at the 1973 Paris Salon, the Mondiale 8 is a softer, bigger car. It has 6½ per cent less power and nearly 15 per cent less torque, to drive a car which although it has virtually the same wheelbase and only 1.4 and 2.2in. wider track, is 11in. longer, has around 2½ per cent more frontal area (due to 1.6in. more height), and weighs nearly 10 per cent more. Its power-to-two-up-laden weight ratio, still a respectable 136 bhp/ton, is accordingly 14 per cent down on the Bertone model, assuming one goes by the maker's power figures – 215 bhp at 6,600 rpm and 179lb.ft. at 4,600 rpm against the older car's 230 bhp at 6,600 and 210lb.ft. at 5,000.

Fuel injected engine

In basic mechanicals, the engine is the same 81 × 71mm 2,027 c.c. 90-deg V8 as before, with no change to its compression ratio of 8.8-to-1. The big change is to the induction, which for the second time in a production road-going Ferrari, displaces multiple carburettors for fuel injection – Bosch K-Jetronic – one is tempted to add, "of course." The first case was on the big 400i model last June, where again there was a nine per cent reduction in power. Ferrari cite the advantages of injection here as a complete absence of flat spots, great operating smoothness, reduced consumption, simpler maintenancer over longer intervals and of course – the most important reason obviously – lower exhaust emissions. The Mondial 8 is in fact well ahead of European emission requirements, and complies with American ones up to 1982.

The other major change is the adoption of a new electronic ignition system developed in conjunction with Ferrari by Marelli and called Digiplex. It is not as sophisticated as Bosch's Motronic system used on some 7-series BMWs, but approaches it closely, with a hybrid solid state electronic memory which will automatically provide 512 ignition advance settings derived from eight levels of vacuum – in effect load – and 64 different engine speeds. The system includes an ignition cut-out which prevents over-revving beyond the 7,500 rpm limit. Again, the advantages are obvious – better ignition control, far better reliability assuming no electronic failure, improved consistency and no mechanical wear, which in turn means another item normally demanding regular service eliminated.

The four-camshaft, all-light

Ferrari's second production Bosch fuel-injection engine nestles behind cabin. For major engine work, the entire rear frame cum suspension can be wheeled away for ideal access

87

Car has pronounced overhangs front and rear; rear one hides an unusually generous boot for this layout

Ferrari Mondial 8 *continued*

alloy engine is of course arranged transversely, behind the driver and ahead of the back wheels, in unit with the five-speed transmission. This has nearly the same internal ratios except for a slightly higher indirect fifth – 0.92 instead of 0.952 – but a noticeably lower final drive – 4.063 (16/65) in place of the Dino's 3.706 (17/63). In conjunction with different tyres – very low profile Michelin TRX 240/55VR390 instead of 205/70VR 14in. – the result is a small reduction in overall gearing, from the 21.0 mph per 1,000 rpm of the Bertone car to 20.1 for the Mondial 8. Going by Ferrari's performance claims, as before the car is deliberately under-geared by 7½ per cent in terms of the amount it will rev past its engine peak at 143 mph (compared with 148 for the 308 GT4).

There are other differences within and without the gearbox. A circulating oil pump has been adopted for better cooling and more even heater distribution throughout the transmission, and it is claimed to lower transmission noise. Selectors and gears have been improved to reduce friction and gearlever vibration and to give better gearchanges. The countershaft provides the drive for the unusual electronic mileage recorder. More unusual still is the incorporation within the car's very comprehensive warning lamp system of an indicator which tells the driver if the gearbox oil level has dropped seriously.

Limited slip is standard for the differential, whilst the clutch has hydraulic control, automatic wear adjustment, and a constantly running release bearing.

Body and suspension

The frame is tubular as before, making much use of fabricated sheet box sections. It will meet current Federal crash safety dictates, and in European form has black resin bumpers good for 3 mph

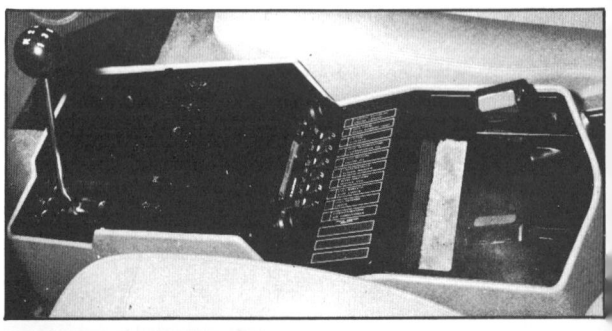

Right: Neat and tidy instruments layout gives driver all the information he might expect in a Ferrari. Digital distance recorder readout is on right of rev counter

Above: Panel of warning lamps behind radio informs on state of (from driver's right to left) engine temperature, brake hydraulics, bonnets open, transmission oil level, engine oil, screen washer fluid, service due, stop lamps, parking lamps, and air conditioning

Below: Buttons to left of speedometer deal with (right hand row from top) heated rear window, front and rear fog lamps, and (left hand row) releases for the lids of the engine compartment, front boot and car boot

88

parking lamps, engine and gearbox oil levels, coolant and washer fluid levels, brakes, and engine and boot lid closing. The standard specification includes leather upholstery by Connolly, air conditioning, electric mirrors and central locking (which includes locking details like the petrol filler cover), and loudspeakers wired for radio of the customer's choice. Options include a power sunshine roof and metallic paint.

The car's body panels are largely aluminium alloy. Where steel is used, it is either pre-zinc coated sheet (zincrometal) or zinc protected after assembly, with a final anti-abrasion pvc-based coat underneath. Special attention is paid to the vulnerable areas around spot welds.

parking collisions without damage. An interesting detail, a first for this maker, is the adoption of a rear end which, carrying engine, gearbox and rear suspension, can be removed complete for faster power plant assembly and maintenance. The suspension is little altered in principle, remaining all-independent with double wishbone geometry. The differences are much reduced kingpin offset, which comes down from 71 to 11 mm, low maintenance pivot bearings and anti-dive inclination of the upper front suspension arm. The improvements claimed include reduced steering kickback and effort. The steering column is adjustable for height and reach. Wheels are wider rimmed at 7in. instead of 6½in.

The body shape is as clearly Pininfarina as its Scaglietti-made predecessor was Bertone. Where the Dino 2+2 was almost arrogantly angular, the Mondial is curved. Its proportions aroused a certain amount of controversy on first sight at Geneva, perhaps in part due to the thrust-forward driving position and the quite noticeable amounts of overhang. At the back, this latter point is made surprisingly good use of in the quite sensibly shaped boot, which is 30 per cent more voluminous than the Dino's and which can be opened remotely from the passenger cabin. Headlamps are as usual the pop-out sort with an electrically powered movement.

The usual high Ferrari standard of instrumentation is preserved, with extras like an oil temperature gauge. Less usual is the adoption of an electronic digital display for both total and trip mileage indicators, plus the very elaborate control panel of warnings, carried in the centre of the car and telling the owner the condition of stop lamps,

Left: Seats are upholstered in leather. Rear ones are shaped to suit the position necessarily adopted by any extra passengers

Performance

The Mondial 8 – the name is an exact echo of an earlier Pininfarina Ferrari, the front-engined 1,984 c.c. open two-seater of 1953 – may not be quite as fast as the Dino 308 GT4, but it is still quick, with a standing quarter mile time of around 15 sec and a standing kilometre covered in 28 sec, passing the post at 130 mph, if one may take the maker's figures. It enters a highly competitive corner of the specialist market, with its most notable competitor coming from Germany in the shape of Porsche.

Above: A different Ferrari Mondial 8, also by Pininfarina – the classically beautiful front-engined 1,984 c.c. four-cylinder 170 bhp two-seater whose engine was based on the Formula 2 500 Ferrari

SPIN IN A SPYDER

It's a while since Ferraris went topless, so the opportunity to sample the latest 308 GTS was something Peter Anderson was unable to resist.

IF YOU'VE ANY doubt that there's still fun left in motoring in the 80's — try a Ferrari!

There may be a 100 km/h limit — but the red and black 308 GTS we tried recently can make a farce of it in second gear!

And we felt not a pang of guilt. Sure it used fuel, but no more than the Commodore 3300 Auto. And for sheer fun, the obedient Ferrari is hard to beat.

The 308 GTS is the latest in Ferrari's Dino series. Unravelling the coded model designation reveals it to be a 3.0 litre 8 cylinder Gran Turismo Spyder — the spyder meaning you can take off the roof.

Literally.

It is quite simply the easiest roof to remove and replace of any we've tried for at least a decade, the antithesis of the old "three boys and a trained monkey" days. That was the crew needed to erect the roof on the average British sports car during the 1960's.

Safety regulations did away with that, thank goodness. You can't roll one of those wire frame and canvas roofed machines over without all the bits and pieces falling down round your ears.

And with you held safely in your seat belt where you can't throw yourself into the passenger's lap as the roof caves in, you are much better off with the modern "targa" approach to open sports motoring.

The system of removing and stowing the roof panel, leaving the roll-over hoop in place behind the occupants, was initiated by Porsche. They called it the Targa for two good reasons: They were in the process of mopping up the Targa Florio race, then the world's trickiest long distance sports car event around Sicily. And the world Targa means shield, or protection, which is exactly what the design does.

In less time than it took you to read this far, you've unclipped the two catches of the GTS roof, lifted it off and stowed it behind the seats under its protective cover. It's really that quick and easy.

Getting in and out of the GTS is still a bit of a sports-car, acquired-art. Swing one leg in, then your body and then feed in the remaining leg.

Anyone trying the old "leap-over-the-doors" trick in typical Le Mans style and dropping into the leather bucket seat, may be surprised at the small length of the roof opening.

At least the big gear lever sprouting from its open, five-slot gate, is far enough out of the way not to snake up the leg of your trousers as you drop into the seat.

Marvellous how modern designers have avoided the old traps!

Two pumps on the accelerator, turn the key and vvrrumph, vvrrumph, the motor is running. There's a noticeable whine which can't be camshaft chain drive, for the four overhead camshafts, two per bank of the V8, are belt driven. There's also a bit of spitting and popping back through the four twin choke carburettors — one choke per cylinder — as the motor warms up.

Registration laws vary depending on the market, so Ferrari even has a 2.0 litre version of this three litre V8. There's less zip, less appeal for the foot to floor types — but a lot more economy. Still Australia has one of the world's lowest petrol price structures, so running costs in that respect aren't usually a factor in buying a Ferrari.

Performance and style are however. Extreme styles can render a car awkward to use, to see out of, or get into. The Ferrari avoids this pitfall despite appearance.

Side windows behind the seats have louvres to improve the car's lines and hide a fuel filler — but they don't obscure three-quarter rear vision.

Into first — which is, of course, competition style, cranked to the left — and you're off. About then you discover that second gear's synochromesh is most reluctant to let you near the ratio at all when the gear oil is cold

A quick dive into third and the remarkable torque of the motor pulls you firmly ahead even though the motor is still cold.

A brief look at fourth if you like and then into fifth for a 60-65 km/h doddle along with the traffic. The carburettor intake noise can be muffled with the roof on by shutting the electric side window.

Clearly, Enzo Ferrari's design team were thinking of left hand drive when they led the cold air from the left scoop into the oil cooler and the right scoop into the carburettor's air box. But then arranging anything at all for a mid-engined car is far from easy.

With the V8 tucked behind the occupants, and turned sideways to minimise length, the designers had to pick one side or the other for the twin side-scoop air-feeds to cool the oil and feed the motor. With left hand drive, the carburettor's intake noise would go unnoticed, mostly.

First traffic light. The motor idles without complaint at 1100 rpm even though neither of the temperature needles — oil and water as you'd expect in a proper sports car — had budged off their stops. Oil temp is hidden but there.

You sit so low, part of the car's speeding-at-rest style, that eyeball to eyeball confrontation with large pets is odds on at lights!

There is a very real danger, too, that the driver of the car ahead may not notice you tucked under his boot.

Spectacular though the Ferrari's styling might be, its diminutive size sometimes makes it almost inconspicuous.

Green light.

Off you go, second gear can now be found and the water temperature needle is edging off the stop. The oil pressure needle sits on 60 psi and the motor can be taken to 4000 rpm easily. The red line is 7700 rpm, the sort of limit you'd expect of a race-bred car.

The oil temperature needle hasn't budged. That race breeding has brought an oil cooler and dry sump system.

If you want the sort of oil temperatures to let the motor live all day at 7700 rpm, you need both an oil cooler and a dry sump system which will increase oil capacity, cut the height of the motor so it can be more easily wedged into the space behind the occupants, increase the oil capacity and help control aeration of the oil — worthwhile sophistication at a cost of more than $60,000 for the model.

With that healthy three litre V8 gurgling and whining out 179 kW, acceleration is the sort of shove-in-the-spine exhilaration it once was before Ralph Nader's air scrubbers got into the car act.

The Spyder, like all Ferraris, doesn't produce the tyre shredding torque of a big American V8. So when you put your foot down, there's a high thin whistle of air through the air box, plenty of Weber carburettor gobbling noises and lots of go. When you select the next gear, the same process continues right through to the end of fourth gear.

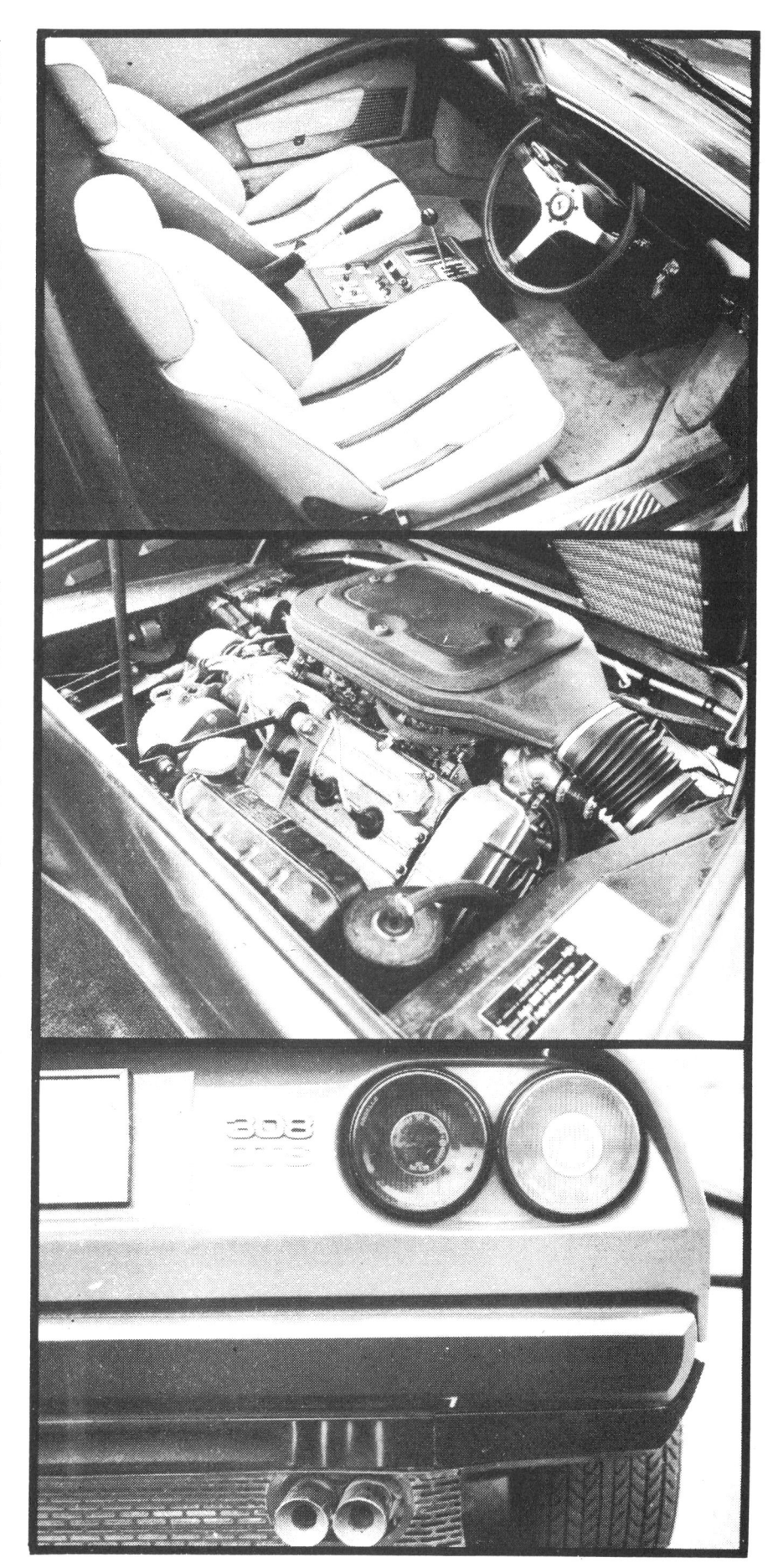

Maximum power in fourth comes at 6600 rpm which corresponds with a licence swallowing 171 km/h. But it will run to an even 200 km/h at the 7700 rpm red line.

After the shift into fifth; there's still an urgent push in the back but it isn't so dramatic. The car will look at 7200 rpm — 253 km/h — in favorable conditions which is more than you can say for society!

With its performance muzzled like a greyhound in training, the Ferrari might seem to have some of its use tarnished, its value diminished. Not so. It is still enormous fun to waft along at 100 km/h under the cascading autumn leaves, the sun's weak warmth filtering through the reds, browns and russets overhead, and the smell of burnt leaves mixing with the thin cool edge of the slipstream.

Straight line "zot" was always something American V8's could be relied upon to provide in gluttonous profusion. There again, a house brick on the American V8's accelerator could provide all the skill that was needed for it to do its thing.

This motor is a little finicky by comparison, but far more day-to-day useful than it once was. It is now endowed with a delightful flexibility. Fifth gear at 1000 rpm in traffic produces no snatches or coughs from the transmission or motor.

And there's not much lag in acceleration, either. For best effect, the car must be shuffled up and down the delightful open-gated shifter. There's a little arcane art in getting the shifts smooth and fast, but the reward is well worth it.

Cornering, roadholding and handling is what the car is all about, however.

It just turns. There's nothing dramatic about it apart from the fleeting thought that, if your outer ear stays stuck to your head, then your brains may not be splattered by the G-forces against the windows.

It is silly to go into under and oversteer with the GTS Ferrari. They are both there depending on how you drive it or what you want it to do — as things ought to be with a sports car. But for most situations neither is noticeable, just lots of neutral steer.

You turn the wheel and the car turns, no delays, no nonsense. And if you hit a bump in the middle of the corner, the behavior is the best of any we've driven. There's a movement of the body but no change in attitude or "sticktion."

If you rush up to a corner and crank on steering too quickly, jumping on the accelerator too early, you've got understeer. Turn it in gently, and just as the steering bites, jump on the accelerator and you've got oversteer.

Turn it in gently, aim for a late apex and feed the power in gradually, and the car comes out of the corner on rails going like a slingshot.

When it is wet, it pays to watch front end lightness turning into a corner. You can get more understeer than you bargained for.

There is only 41 per cent of the weight on the front wheels so they tend to get light and skatey on slippery surfaces. At all normal speeds, they are fine on the P7's and BBS mags fitted on the test car. Those are extra cost items but extra cost is relative when you are already over $60,000 lighter in your bank account.

If the front does get at all skatey — and the thought flashes that if things don't bite soon up front, then this expensive red gadget will join that cheap masonry in an expensive plasma — all the car requires is a gentle lifting of the accelerator or easing of the brake.

It is all so predictable.

What it boils down to is a car with staggering cornering limits, which pays big dividends to a kind and knowledgeable driver. Brutalise it and you'll still leave the lurchers swaying in your wake, but there is so much more potential there for those who seek it with experienced confidence.

It is interesting that so much understeer has been built into a sports car. It is the most easily accommodated handling trait when the car has an inexperienced driver. But it also can be used to advantage.

Competition drivers often seek a little understeer in their cars so they can get on the power earlier in a corner and come out of it faster. Oversteer is usually the slowest way through a turn — although, as rally drivers can show, it is probably the safest if you know how to manage it and you are driving on an unknown road.

That the Ferrari has been so finely tuned can be attributed to the firm's Formula One involvement. As part of that exercise, Enzo Ferrari had one of the world's finest handling circuits built beside his factory. Fiorano is on the outskirts of Modena — at the upper right hand end of Italy, as a geography master once described it.

There are varying turns uphill and down, together with high speed straights. While a car is being driven there, television cameras record its reactions, and timing lights record times through the bends. Small setting changes used for Formula One car test sessions can as easily be applied to road cars. And since the Formula One drivers themselves are usually hanging about waiting for the Formula One mechanics to fiddle with their latest Formula One toy, Ferrari uses the world's most expensive test drivers for his road cars!

Niki Lauda, who attended fee settling sessions with Ferrari armed with his own lawyer and calculator, is credited with sorting out the 308 Dino's handling.

That's all there really is to the 308 GTS. There's precious little room in the tail for bags under the zip cover, and none at all in the nose. The space in front takes a spare wheel and radiator — nothing else.

If you are wondering about the grilles behind the headlights on top of the mudguards, they are there for exhausting air which goes under the nose and through the radiator. It is typical of Pininfarina that all available air flow paths are used rather than random gouging of holes and louvres which some stylists seem to love. There's even an air spoiler lip over the windshield for breaking the air flowing over the open passenger space.

Louvres over the motor flatten a little at the rear of the left hand bank so they will not let rain fall on the rear cylinder bank's ignition system. Of course it has separate ignition. In many ways the V8 is two four cylinder, twin cam motors joined at 90 deg.

Workmanship on the body is particularly good — although the practicality of the low front spoiler came into question on test when it grounded on a couple of driveways.

In the out of the way spots such as underneath the panels, the finish isn't good as you'd expect of a robot-built, mass-produced nightmare on wheels costing a fraction of the price. It all comes back to the hand-finishing of cars which leads to comparative rarity, rising used values and often, rust! Special metals — single sided zinc coating for example — are available to the specialist makers now and, with underproofing, the exotic car, as Ferrari builds it at the Scaglietti works, is reasonably resistant to the rust-bug.

Rising used car value is an interesting point when discussing very expensive machines such as the Ferrari.

They hold their value. In 1973, a Berlinetta Boxer 365 buyer paid $36,000 for his bolide, then Ferrari's fastest. He's replaced it this year with the latest BB, the 512. The trade-in price? $65,000. That car has appreciated faster than real estate.

The original Dino came on the market for about $12,000 in the late 1960's and went out at $16,000 in the early 1970's. Today, it brings $25,000 to $33,000 depending on condition. And we prefer it to the BB, too, if that's of any consequence. You get the sort of power from the V6 which you can use most of the time.

The V8 Dino is much the same. The grip is a little scary. The handling superb. Probably the only real curb on using it to the full is the thought that if it ever lets go, your clothes will be out of fashion before the accident ends!

Who said motoring is dull, that cars are becoming all the same to look at and drive, and world car concepts promise wheeled boredom all the way into the future?

The Dino refutes all of that. Now as we enter the nervous 80's, Ferrari still stands for glamor, freedom behind the wheel and most of all, sheer fun.

IT WAS YEARS before "The Italian Job."

I was parked on the side of one of those Corniche offshoots looking down on Monte Carlo. It was mid summer and the typical south of France weather was there for the soaking as far as a pasty faced Brit was concerned.

It was quiet, that period soon after mid-day when continentals dally over lunch, following on with a brief nap in the sun — a distinctly civilised way of spending the middle hours of a summer day.

Flies buzzed around the wayside plant life, taking time out now and then to attack my long french loaf and chunk of fresh country cheese.

A new sound then entered my consciousness. It had been there for a short while, but my relaxed state hardly registered it until it could be ignored no longer.

Resonating off the rocky crags, it came closer, and louder, the sound of an engine exhaust being worked hard. At first the noise grew fairly slowly, but then, the closer it came its urgency surged and then died, accompanied from time to time by the sharp shriek of tyres.

Away to my left I saw the car for the first time. It was going like a bullet on the other side of a small valley. Best of all, it was bright red, standing out from the ochre and grey colored rock faces.

Undamped by intervening scenery, the exhaust sound was now blaring out across the countryside, rising to a crescendo and then popping and banging on the over-run.

The car disappeared behind yet another rocky outcrop and quite suddenly the sound disappeared too. It seemed like a long time before, from the interior of a tunnel over to my right those glorious decibels began to reverbrate through the confined space.

Finally, a blurr of red came out of the mouth, it's muzzle velocity like that of a big gun!

Then it was gone, the sound fading quickly at full speed on the straighter road leading to the Grand Corniche.

The feeling of silence was combined with an odd sense of loss. A Ferrari at Le Mans, or the Nurgurgring is one thing. Out there on the proper road, the environment for which it was designed, it was something else again.

And ever since that time the mere sight of a Ferrari has brought the memory flooding back, as clear and lucid as it was, there in the hills overlooking Monte Carlo.

Paul Harrington

FERRARI 308 GTSi

The i means injected—and improved

EMERSON WROTE THAT "Beauty is its own excuse for being." Some automotive enthusiasts feel that the same can be said for exoticars—such as Ferrari's latest 308GTSi. Here's a car that is expensive, not terribly space- or fuel-efficient, a car which demands that its driver pay it full heed while at the helm. Yet, the 308 is also a car whose beauty, craftsmanship and mechanical intricacies can overwhelm even a seasoned car buff such as one of our senior staff members who remarked, "Cars like this don't have to be justified." One of his co-workers concurred and added, "The 308 is an enthusiast's dream car and dreams don't have to have the least bit of practicality."

This is not to say that the Ferrari is impractical. It is simply less practical than, say, a Fiat 131. But then the 308 and cars of its ilk are designed primarily for personal and not group (family) transportation and as a means of pleasure and not just conveyance. They're not for everyone.

But we digress because the subject is the 308GTSi, Ferrari of North America's latest bauble. Notice the i suffix: That stands for *iniezione*, Italian for injection, in this case, Bosch K-Jetronic.

This continuous-flow system is used on a large number of imports because it is relatively simple to adapt to an existing powerplant. It's quite foolproof too, good news for those of us who have to live with it. The K-Jetronic represents the most significant change found in the 1980 (no, they're not 1981s yet) 308 models known as the GTBi and the GTSi. By the way, the B and the S stand for Berlinetta and Spyder and while one is a coupe and the other a targa, the two models are identical mechanically.

Other changes to the latest 308s include digital electronic ignition, as well as minor styling, mechanical and equipment revisions. For example, the interiors of both cars reflect subtle refinements that include extensive use of Connolly leather and cut-pile carpeting plus a redesign of the dashboard. It took our resident Ferrari expert to find this last change because to the casual observer the instrument pod looks the same except for its fascia which is flat black rather than brushed aluminum. "But look here," exclaimed the Ferrarista, "the clock and oil temperature gauge formerly located on the left side of the dash (and obscured by the steering wheel) are now on the console." Bravo. Our friend was not aware of the redesigned clutch pedal linkage which is said to reduce pedal effort by 20 percent. This simple fix involved changing the angle of attack of the pedal return spring.

The final change concerns wheels and tires. The wheels are Speedline cast alloys fitted with Michelin 220/55VR-390 TRX tires. They replace the Cromodora alloys and 205/70VR-14 Michelin XWXs fitted to the Spyder of our July 1978 road test. That car was one of the four best handling automobiles we've ever tested, at least insofar as skidpad and slalom times are concerned. One of the first Spyders (they had all been Berlinettas before that), it managed 0.852g and 60.8 mph in its evaluation runs and only the BMW M1, the Lamborghini Countach S and the 1978 Corvette could match or exceed its performance.

We expected as much (more, actually) from the latest TRX-shod 308, so you can imagine our surprise at its test scores. Although the GTSi's 60.6-mph slalom speed was only a tick slower than before, its 0.810g skidpad showing was markedly less than that of its predecessor. In fact, the new GTSi was suddenly relegated to the realm of such impressive but less sensational automobiles as the Pontiac Turbo Trans Am and most of the Porsche line, including the 924S Turbo, the 928 and the 930 Turbo. Admittedly, this is still fast company, but somehow you'd expect more from an exotic, especially one costing almost $53,000.

So we called Ferrari and were told that while the new tire/wheel combination is ½-in. taller than the old setup, it has a wider footprint and better all-weather (wet and dry) performance. You'd expect it to handle better than before but, in fact, this is not the case. Our tester found that what had formerly been a neutral-handling automobile now tended toward oversteer. "Its natural attitude is 20 degrees tail-out and this diminishes the excellent steering, suspension feel and tire grip," was the comment he made after repeated passes through the slalom. On the skidpad there was more of the same. The 308 would zip around with little drama, but with its tail hung out. Power-off brought the rear end out farther, power-on broke loose the rear wheels, causing the same reaction. Our test driver concluded, "This is not a car for the novice to drive on a winding, slippery road because it does not reward indecision."

On the bright side, other members of our staff find the 308's steering precise, except that at low speeds it seems heavy and a bit slow. But all of us admit that the steering's feel and responsiveness change when the car is driven at speeds more appropriate to a Ferrari.

There was never any question about the GTSi's brakes, which require only a modest touch of the large pedal to do their work. And excellent work it is, with virtually no fade and impressive stopping distances—154 and 254 ft from 60 and 80 mph respectively. That's better than the early Spyder's stops and one reason why the GTSi earned an excellent overall brake rating.

The 308GTSi's driveability is also excellent. Thank the Bosch K-Jetronic for that, even though its reason for being is more effective emissions control. With fuel injection, the 1980 308s get a clean bill of health from all 50 states, including finicky California. Their engines exhibit positive cold start and smooth cold running and have none of the former's flat spots and fuel starvation problems (when cornering). However, a somewhat frightening side effect of the engine's new emission controls is its 2500-rpm cold idle. Ferrari says it's high to meet smog control standards and assures us there's nothing to fear. Maybe so, but one can't help but cringe when a stone-cold engine lights and immediately revs to a rapid idle.

In addition to the new induction system, the latest 308 engines are also equipped with Marelli Digiplex electronic ignition. Among other things, the Digiplex automatically advances the spark along one of 90 different curves tailored to meet a variety of driving conditions and engine loads.

As effective as they are, all of these advances in induction and ignition contribute little to the car's net performance. Rather, they allow the 308 to maintain the performance level that existed before the heavy hand of the EPA was felt on the industry's shoulder. So even though the 308 is saddled with an air pump, thermal exhaust and catalytic converter, horsepower and torque (205 bhp, 181 lb-ft) remain unchanged and acceleration is only a trifle slower. The GTSi goes from 0 to 60 mph in 7.9 (versus 7.6) sec and reaches the quarter-mile in 16.1 (vs 15.8) sec.

The important news is that the engine is smoother and quieter

AT A GLANCE

	Ferrari 308 GTSi	BMW M1	Porsche 928
List price	$52,640	$115,000	$38,850
Curb weight, lb	3250	3325	3410
Engine	V-8	inline 6	V-8
Transmission	5-sp M	5-sp M	5-sp M
0-60 mph, sec	7.9	6.2	7.0
Standing ¼ mi, sec	16.1	14.5	15.6
Speed at end of ¼ mi, mph	88.0	97.0	93.0
Stopping distance from 60 mph, ft	154	156	138
Interior noise at 50 mph, dBA	76	79	71
Lateral acceleration, g	0.810	0.858	0.811
Slalom speed, mph	60.6	62.7	59.7
Fuel economy, mpg	11.5	13.0	16.0

than before and has a more flexible power curve. It can be lugged to as low as 1000 rpm without protestation, yet it will rev freely right up to the redline, even in 5th gear. That translates to almost 150 mph, mighty impressive in this day and age.

Driving the latest Spyder is not a unique experience. But it certainly is pleasurable. So much so that one's criticisms are easily forgotten, once the lightweight roof panel (secured by two clasps and stowable behind the seats) is removed, the engine started and the car set into motion. The visible shift gate is precise, although it can prove balky if rushed. The gear ratios are evenly graduated to exploit engine output to its fullest (no super-tall 5th gear here). As a result, the 308 is facile in traffic and an absolute flier on the highway. The lack of wind noise and buffeting, sans top, is a testimonial to the car's aerodynamics which are manifested in a design that looks as good as it works. We received nothing but compliments from other drivers who commented on the Ferrari's swoopiness. Because, personal gripes (limited seat adjustments, a ridiculous side-view mirror, a less than frugal 11.5 mpg in our admittedly enthusiastic driving) aside, the 308GTSi is certainly distinctive. It's exotic, even beautiful and, thus, is its own excuse for being.

PRICE
List price, all POE $52,640
Price as tested $55,040
Price as tested includes std equipment (air conditioning, elect. window lifts, leather interior), AM/FM stereo/cassette ($1800), pre-delivery insp ($350), black rocker-panel trim ($250)

GENERAL
Curb weight, lb/kg 3250 1476
Test weight 3415 1550
Weight dist (with driver), f/r, % 42/58
Wheelbase, in./mm 92.1 2340
Track, front/rear .... 57.8/57.8 ... 1468/1468
Length 174.2 4425
Width 67.7 1720
Height 44.1 1120
Trunk space, cu ft/liters 5.3 150
Fuel capacity, U.S. gal./liters .. 18.5 70

ENGINE
Type .... dohc V-8
Bore x stroke, in./mm .... 3.19 x 2.79 ..... 81.0 x 71.0
Displacement, cu in./cc 179 2926
Compression ratio 8.8:1
Bhp @ rpm, SAE net/kW .... 205/153 @ 6600
Torque @ rpm, lb-ft/Nm .... 181/245 @ 5000
Fuel injection .... Bosch K-Jetronic
Fuel requirement .... unleaded, 91-oct

DRIVETRAIN
Transmission .... 5-sp manual
Gear ratios: 5th (0.95) 3.52:1
    4th (1.24) 4.60:1
    3rd (1.69) 6.27:1
    2nd (2.35) 8.72:1
    1st (3.58) 13.28:1
Final drive ratio 3.71:1

CHASSIS & BODY
Layout .... mid engine/rear drive
Body/frame .... steel/tubular steel frame
Brake system.. 10.7-in. (272-mm) vented discs front, 10.9-in. (277-mm) vented discs rear; vacuum assisted
Wheels .... cast alloy, 390 x 190
Tires .... Michelin TRX, 220/55VR-390
Steering type .... rack & pinion
Turns, lock-to-lock 3.3
Suspension, front/rear: unequal-length A-arms, coil springs, tube shocks, anti-roll bar/unequal-length A-arms, coil springs, tube shocks, anti-roll bar

CALCULATED DATA
Lb/bhp (test weight) 16.7
Mph/1000 rpm (5th gear) 18.8
Engine revs/mi (60 mph) 3200
R&T steering index 1.30
Brake swept area, sq in./ton 263

ROAD TEST RESULTS

ACCELERATION
Time to distance, sec:
0-100 ft 3.5
0-500 ft 8.9
0-1320 ft (¼ mi) 16.1
Speed at end of ¼ mi, mph 88.0
Time to speed, sec:
0-30 mph 2.9
0-50 mph 6.1
0-60 mph 7.9
0-70 mph 10.6
0-80 mph 13.4
0-100 mph 22.1

SPEEDS IN GEARS
5th gear (7500 rpm) 147
4th (7700) 122
3rd (7700) 87
2nd (7700) 63
1st (7700) 41

FUEL ECONOMY
Normal driving, mpg 11.5

BRAKES
Minimum stopping distances, ft:
From 60 mph 154
From 80 mph 254
Control in panic stop .... excellent
Pedal effort for 0.5g stop, lb 20
Fade: percent increase in pedal effort to maintain 0.5g deceleration in 6 stops from 60 mph .... nil
Overall brake rating .... excellent

HANDLING
Lateral accel, 100-ft radius, g 0.810
Speed thru 700-ft slalom, mph 60.6

INTERIOR NOISE
Constant 30 mph, dBA 72
    50 mph 76
    70 mph 79

SPEEDOMETER ERROR
30 mph indicated is actually 27.0
60 mph 56.0

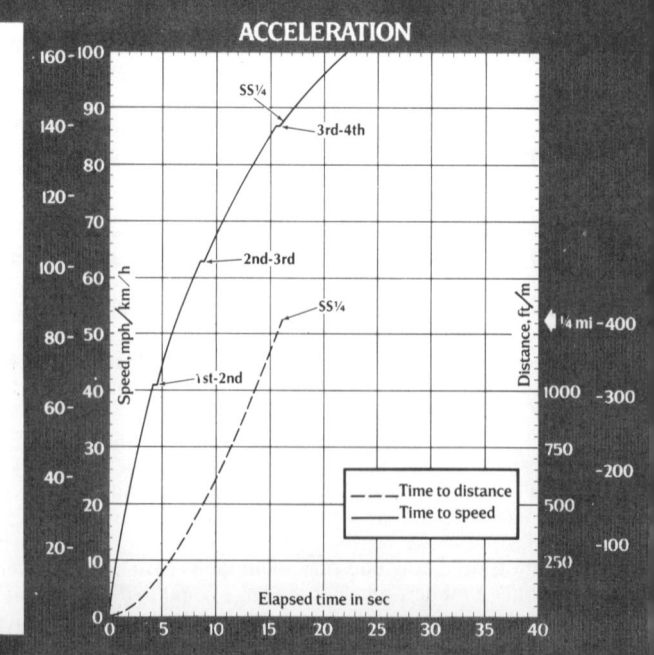

96

BUYER'S GUIDE

Ferrari Mondial 8

WHAT DO you get when you stretch a 308's wheelbase by 12.2 in. and add two occasional rear seats? Why, a Ferrari Mondial, of course. The Mondial is Ferrari's answer to those detractors who didn't consider the previous 2+2, the Bertone-bodied Dino GT4, to have enough pizzaz. So what we have here is a design that's a cross between the all-out swoopiness of a 308 and the more angular GT4. We don't feel the Mondial shape is as well integrated as the 308's and the GT4's because some of the details aren't as well worked out, bumpers and the side scoops for instance. But there's no mistaking the Mondial for anything but a Ferrari and for many people that's reason enough for wanting to own one.

The Mondial borrows many components from the 308, primarily the aluminum dohc V-8 and 5-speed transaxle, mounted mid-transversely, fueled by Bosch K-Jetronic injection and sparked by Marelli Digiplex electronic ignition. Transaxle ratios are unchanged and like the 308, the Mondial's clutch linkage is of an improved design offering reduced pedal pressure.

One of Pininfarina's design goals with the Mondial was an increase in head room and this has been accomplished—for front seaters. You'd be kidding yourself to think of the rear seats as serving anyone but small children. Except for the dash and armrests the interior is leather and a new feature is a warning check panel on the center console. However, it would be much more useful if the LEDs were more visible. But we're nitpicking and the Mondial is not a car to be picked apart but one to be enjoyed. It may not be the best car in its class but it's hard to argue with the Ferrari image and exclusivity.

SPECIFICATIONS

Basic price, base model	$63,939	Engine	dohc V-8
Country of origin	Italy	Bore x stroke, mm	81.0 x 71.0
Body/seats	cpe/2+2	Displacement cc/cu in.	2926/179
Layout	M/R	Compression ratio	8.8:1
Wheelbase, in.	104.3	Bhp @ rpm, net	205 @ 6600
Track, f/r	58.8/59.7	Torque @ rpm, lb-ft	181 @ 4600
Length	180.3	Transmission	5M
Width	70.5	Final drive ratio	3.71:1
Height	49.2	Suspension, f/r	ind/ind
Curb weight, lb	3640	Brakes, f/r	disc/disc
Fuel capacity, U.S. gal.	22.2	Tires	240/55VR-390
Fuel economy (EPA), mpg:		Steering type	rack & pinion
U.S.	10	Turning circle, ft	39.4
California & high altitudes	10	Turns, lock-to-lock	3.3

Ferrari 308GTBi & 308GTSi

TO ANYONE who loves cars, the Ferrari name is magical. And if a Ferrari is not the most practical of cars, that's understandable because practicality is not its main reason for being. Rather, a Ferrari should be thought of as a personal statement and as a means of pleasure, not just conveyance. Ferraris are expensive and not especially space or fuel efficient, but the world would be far worse were it not for unique and beautiful cars such as the Ferrari 308. As last year the 308 is offered in two versions, GTBi (B for Berlinetta, closed) and GTSi (S for Spyder, open or targa) with the i indicating injection in either case.

The Ferrari engine is a sophisticated piece of machinery that makes much use of aluminum and is fitted with double overhead camshafts for more efficient breathing. With the addition of K-Jetronic injection the engine is smoother and quieter and has a more flexible power curve than when it was fitted with Weber carburetors. The visible shift gate lets you know that 5 speeds are ready to do your bidding and if the shifter is a little balky when rushed, the ratios themselves are well spaced: The 308 is docile in traffic but a real highway flier topping out at 147 mph.

Both versions of the 308 have interiors trimmed in Connolly leather and including power assists for the windows and door locks. Ferrari installs a pair of door speakers and an electric antenna at the factory, but the choice of a stereo system to go with them is left to dealers and customers.

A 308 is not the most exotic car on the road, but you'll have to go a long way to name another marque that has as much mystique as the Ferrari prancing horse.

SPECIFICATIONS

Basic price, base model	$51,930	Engine	dohc V-8
Country of origin	Italy	Bore x stroke, mm	81.0 x 71.0
Body/seats	cpe*, targa/2	Displacement, cc/cu in.	2926/179
Layout	M/R	Compression ratio	8.8:1
Wheelbase, in.	92.1	Bhp @ rpm, net	205 @ 6600
Track, f/r	57.8/57.8	Torque @ rpm, lb-ft	181 @ 5000
Length	174.2	Transmission	5M
Width	67.7	Final drive ratio	3.71:1
Height	44.1	Suspension, f/r	ind/ind
Curb weight, lb	3250	Brakes, f/r	disc/disc
Fuel capacity, U.S. gal.	18.5	Tires	220/55VR-390
Fuel economy (EPA), mpg:		Steering type	rack & pinion
U.S.	11	Turning circle, ft	39.3
California & high altitudes	11	Turns, lock-to-lock	3.3

*indicates model described in specifications

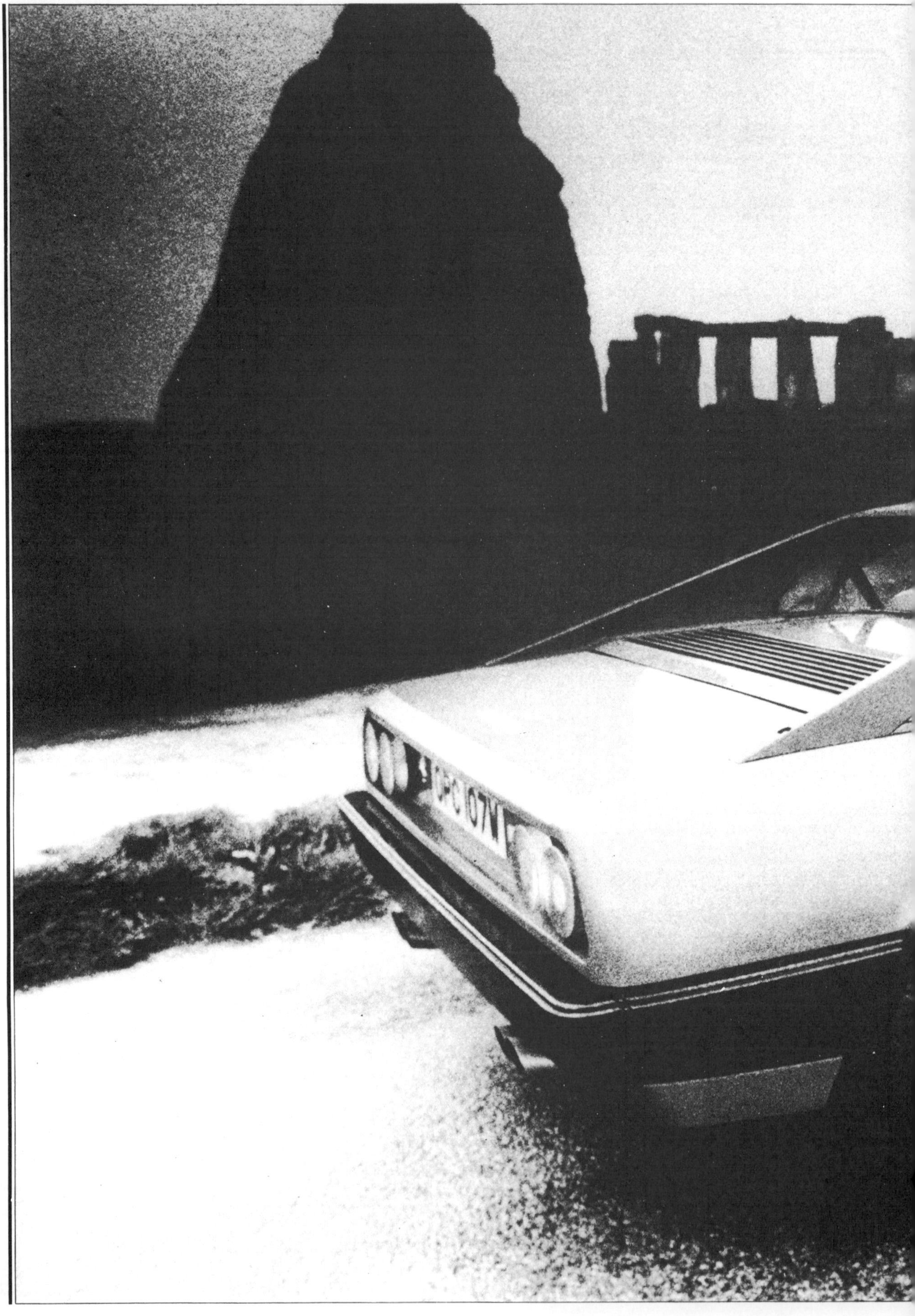

Mondial la magnifica!

*Ferrari's Mondial 8 – superbly blending dynamic ability
th masterful body design, clever electronics and Porsche-like
build quality – is proof positive that Maranello is reaching
bright new heights. Steve Cropley reports overleaf*

IT MAY NOT SEEM THE greatest of accolades to call the new four-place Mondial 8 the first decisively Fiat-influenced Ferrari. Yet that is the truth of the matter. And whatever it appears to be at first, it *is* an accolade – one of the biggest.

Any four-place Ferrari starts fairly well behind scratch. According to lore, Ferraris are built for their drivers, and for the one passenger each they might deign to choose. Any thought of Ferraris as people-carriers is scowled at by the cognoscenti, and that attitude is reflected by the fact that it is the two-seaters which become collector's items later in their lives. On top of that, it is the two-place cars which were always the best motor cars since they were invariably closer in design to the cars that really mattered at Ferrari, the racing cars.

But the Mondial 8 may be the car to change things for four-seaters. It has the inspired blend of compromises to be numbered among the best-designed Ferraris of them all. It has the build quality and durability to last longer than any that has gone before. And significantly, it is the first Ferrari to be wholly planned, designed, developed and built since Fiat took over the Ferrari road cars business at the end of 1969. Not that the Mondial is an aggressive, noticeable character of a car – it is not. On first acquaintance you might be excused for thinking it rather too conservative and easily missed. But in all of Ferrari's road car history it is the first to approach Porsche – even Mercedes-Benz coupe – standards of build and essential useability, and the achievement is the greater because it does not sacrifice a jot of the balance and sensitivity that divides the finest Italian cars from the rest. But these realisations do not come at once.

If you are like me, you will approach the Ferrari Mondial 8 with a measure of scepticism. You will be aware that the car's function is to supersede the wedge-shaped 308GT4 two-plus-two which, despite the fact that for years it was the marque's top-seller, was not an especially notable Ferrari. Many buyers liked its low price and had to have the extra room it offered over the pure two-seaters. But at the same time the car was rather scorned by some who felt that its Bertone body did not fit into the Ferrari-Pininfarina scheme of things, and who realised that its rear seats housed jackets and briefcases – or shopping bags – better than they housed people. Whatever the truth, the verdict has been reached: there are quite a lot of low-mileage GT4s on sale, and they don't cost much.

The Mondial starts much more strongly. It has a Pininfarina body that bears a relationship with the great cars that have gone before. But the roof is rather high – saloon-like, somebody said – and the lines seem a little straight-laced. And those rear-flank air scoop grilles are a little too 'pre-war Flash Gordon' in styling to be attractive. Certainly they aren't like the elegant ducts of the 308GTB and 246GT which make those cars look so good. But it's sleek, clean and, well, not ugly.

Broadly, the Mondial obeys Ferrari chassis principles by employing a tubular steel frame over which metal panels are mounted. The outer skins are substantially of steel but there are isolated alloy components (including the nose and engine covers) and glassfibre is used for some of the inner panels, particularly those inside the cockpit. But the designers say they have broken with tradition in two key ways. They have used more ribbed and boxed sheet in the structure in an attempt to improve its strength for no increase in weight, and they have made it possible, by removal of a sequence of bolts in the lower chassis, to drop out the engine/transaxle and rear suspension, complete with its subframe, to provide accessibility that will make the Mondial historically easy to service. In your Ferrari mechanic's opinion at least, the two-seater 308GTB is going to be a Mondial's poor relation.

Make no mistake, the two cars *are* closely related. The Mondial does have a unique chassis and body, but it uses much of the 308's suspension components and running gear, and the power packs of the two cars are identical, right down to their gear spreads and final drive ratios. That means both cars have coil-sprung unequal-length double wishbone systems at both ends, Koni-damped and balanced side-to-side by anti-roll bars. The brakes are powered discs all round, 11.6in diameter at the front and 11in at the rear. Steering is by manual rack and pinion. It takes 3.5turns to swing the Mondial's wheels from lock to lock, quite a lot when you consider that it can only just turn inside a 40ft diameter circle. Later, as you drive the car, the mild under-gearing of its steering is something you notice.

There is a considerable size difference between Mondial and GTB. At 104in the four-seater is a massive 12in longer in wheelbase, it is 13.8in longer overall at 180in, its tracks are an inch or so wider, it is a towering 5in higher than the 308 and nearly 3in wider – most of which seems to be reflected in increased shoulder width. All this bulk, and the extra accommodation, makes a Mondial around 350lb heavier than a GTB, though it is hard to be accurate about this because there is a considerable discrepancy between the 'official' figures quoted in the handbooks and those in the sales brochures. The car meets the road with Michelin TRX 240/55 VR390 tyres, bigger in section but similar in aspect ratio to the 225/55s recently fitted to the new, fuel injected 308GTBi. They're big, chunky tyres, noticeably fatter and taller than the GTB's. It's now clear that there were some problems in accommodating TRXs under the 308's wheel arches; one can't help thinking that the 240/55s might be the tyres for it as well.

The 3.0litre 90deg V8 has recently had its two banks of two twin-throat Weber carburettors replaced by Bosch K-Jetronic fuel injection, and has lost quoted power in the process. It has a bigger job to do in Mondial than 308 – it must propel a body with more frontal area, weighing 12percent more, on fatter and more air-resistant tyres that doubtless have a little more rolling resistance, too. The four cam engine, you will remember, was rated at 255bhp when things were easy back in the middle '70s. Now, according to the factory, the engine produces 214bhp at 6600rpm (though a direct conversion from their quoted output in kilowatts makes that 211bhp, and at one stage their spokesmen were calling it 210bhp). The torque rating is 176lb/ft at 4600rpm. What is important is that the 255bhp of 1975 probably wasn't the truth; the latest 214bhp has to be. The engine feels as though it has lost about 20bhp, not 45, and gained immeasurably in smoothness and crisp starting.

The Mondial's taller TRX tyres give it a higher overall gearing of 20.1mph/1000rpm against the GTB's 19.3mph/1000rpm. This might be thought to further handicap the Mondial against the 308, but it doesn't work out that way. Using the 7700rpm limit marked on the tachometer (once more there is a discrepancy: the handbook says 7700rpm is the maximum, the sales brochure says 7500) the bigger tyres extend the Mondial's gear maxima by between 2mph and 6mph in the lower four gears and that makes a surprising difference to the Mondial's feeling of effortlessness in eight-10ths touring over challenging terrain. Besides, as far as we could judge in a wet day's testing, there isn't anything like the discrepancy between the Mondial's and the GTB's performance that there was between GTB and the Lotus Esprit Turbo that it met in *CAR* last month.

Ferrari claim a 15.0sec standing 400m for the Mondial which seems around a second too fast to us. The top speed is claimed to be near-as-dammit to 150, though we'd be surprised whether you'd ever see much more than 140 in a standard-tune Mondial on open roads. The Mondial loses little, perhaps 0.3sec, on the GTB's 0-60mph time of 6.7sec because its second gear runs a little beyond 60, the GTB's wrung-out maximum. The Mondial would pass 100mph from standstill about 1.5 to 2.0sec behind the GTB – it would take around 21.0sec, which remains fast. And at that speed in the Mondial there's still the business end of fourth left, and fifth to come.

If you have reservations about the Mondial's outline (and you probably do, noting the classic beauty of the 308GTB) these are dispelled when you step into it to drive it away. The Mondial's cockpit is the most spacious of any of today's crop of mid-engined cars. You notice it immediately. You aren't at all concerned about headroom (as you are in 308GTB, Boxer, BMW M1, Lotus Esprit, Merak, Countach and all the rest) and for a change, the driver's door doesn't crowd you. There is actually a surfeit of elbow room, unheard of in a car like this. There is no question of a knees-high driving position – even for long legs there is plenty of room. That part of it is slightly spoiled by the fact that your legs must reach markedly left because the pedals are offset to clear the wheel arch, but you soon get used to it.

The gearlever (lengthened for easier operation, like the 308's) is a comfortable reach away and the wheel is set at that typical Maranello angle which allows the heel of the hands to be supported by the wheel rim, and the fingers to rest lightly on the leather – to sense its activity rather than to clutch at it for support. It's made many times better in this car because the column is adjustable for height and reach. Really, the Ferrari driver's cup runneth over. Visibility is excellent in all directions, even to the rear three-quarters. The windscreen pillars aren't especially thin, they just seem aptly placed to afford the driver the best view. The steering column

Mondial 8 *is less stunning than its exquisite 308GTBi sister, but considering difficulty of designing a mid-engined four-seater it's a masterpiece. Wraparound bumpers mean new level of protection for an Italian exotic; ground clearance good*

controls are now Lancia's latest type, nicer looking and acting than the old, cast-off Fiat kind the 308 use. The interior and exterior doorhandles are the up-market mechanisms from the Ferrari 400

The Mondial cockpit literally immerses you in English Connolly leather up to the armpit. It is used to cover the elegant bucket seats differently-styled but similar in dimension to the GTB's seats, and it also covers the doors to window height, and the facia. Deep, cut-pile carpet is used on the floor and footwells. The instruments are grouped in an oblong binnacle containing speedo and tacho, a double vertical row of electronic warning lights between them, and four minor gauges (oil pressure and temperature, fuel and water temperature) grouped nearer the centre of the car. The instrument graphics are fresh and new, but the whole binnacle tends to poin

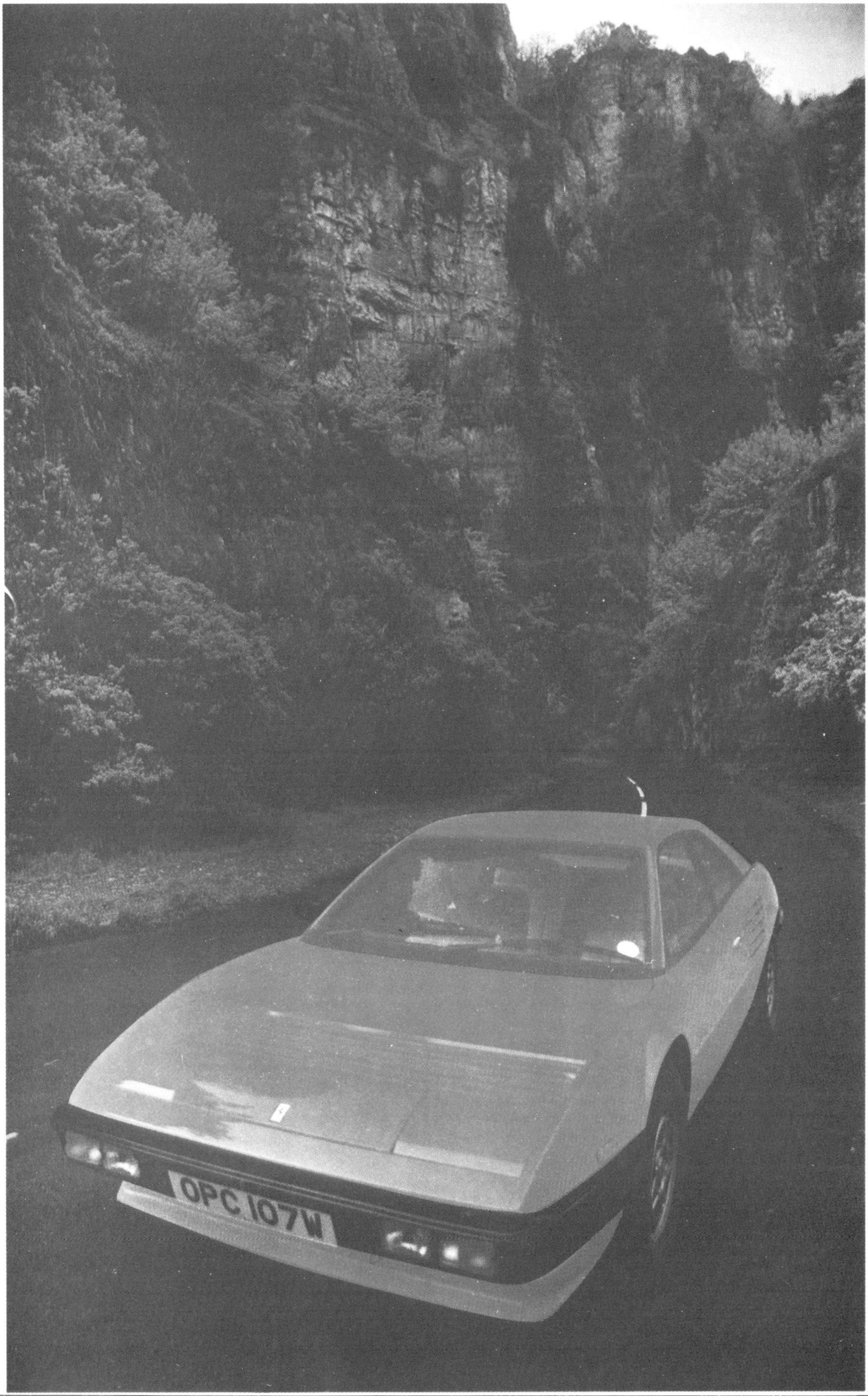
OPC 107W

at a tall driver's chest rather than his eyes. I yearned to be able to incline it upward about 10deg and bring the instruments' upper graduations out from under the binnacle eyebrow. Needless to say, the arrangment prevents the lighted instruments from reflecting on the screen at night.

One of the greatest surprises of the Mondial is the sophistication of its ventilation. You do not expect to find a decent, smoothly-flowing supply of fresh air in a car like this. That is why most of these cars have air conditioning plants crammed into their engine compartments – it is a strict necessity. The Mondial duly has air conditioning, too (with Fiat-familar buttons confusingly grouped just ahead of the gearlever) but the point is that you don't have to use it. Fresh air is available through a bank of three outlets in the middle of the car and through face level vents at each extremity of the facia. This, the car's visibility and its exceptionally roomy cockpit give it a degree of civilised comfort those in the Ferrari market could previously only dream about.

The Mondial's other major departure from tradition is its electronic monitoring system. There are sensors all over the car which feed information to a bank of warning lights, grouped between the seats (of all places) just ahead of the inner seat belt catches. The idea is that you check the lights as you buckle yourself in, presumably having started the engine, but I found that I always needed to remember to glance down there. Perhaps it's a habit you acquire. The list of monitored items reads like an aircraft checklist – engine coolant, brake lights, bonnets (are they closed?), transmission oil level, engine oil, screen washer reservoir, service due, stop lights, headlights, air conditioning. At one end of that row there are three 'general' warning lights. There's a yellow light that glows when there's a minor malfunction, a red one for when the problem is more serious, and a green that glows the all-clear. That's the one you're supposed to check for when buckling in. Indisputably, it's a good idea, but there's room for development of the layout.

The Mondial's reason for existence, its rear accommodation, is as impressive as it was disappointing in the GT4. The front passenger has lots of legroom – certainly enough to allow him to slide his seat well forward and accommodate another medium-height adult in the full-sized bucket seat behind. There is more headroom in the Mondial's rear than the 308GTB offers its driver. It isn't as easy to accommodate a passenger behind the Mondial's driver, if he is tall, but there is legroom enough for a child of up to 12 or so In sum, this mid-engined car's cockpit is much more roomy than plenty of front-engined coupes'

Dashboard *of Mondial is outstandingly complete for an Italian exotic. New emphasis on practicality means that wheel is adjustable for height and reach. Dial faces are new* **Gearlever** *still moves in classic gate but is now backed by console containing detailed readouts for electronic monitoring system complete enough to embarrass the best of the German makers* **Cabin** *is superbly trimmed and really can seat four. Visibility is excellent* cabins – and it beats the Porsche 911 (whose engine, you will remember, is outside the wheelbase to allow more people and luggage room) by a big, big margin. It even embarrasses the Porsche 928.

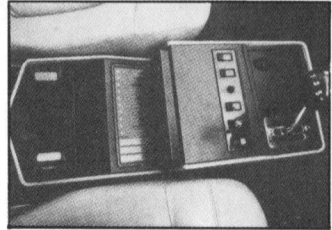

And so to the driving. After you become used to all of that extra Mondial cockpit room (and if you know how little space most exotics offer, that takes on the proportions of the New Deal) you'll find that a Mondial feels very similar to a 308GTBi. There are similarities in the seating, though the Mondial's buckets are a little higher off the floor and offer marginally better under-thigh support. The Mondial seats also seem to have better lumbar support but there is the same surprisingly low degree of side-support for hard cornering. Probably there has been a designer's trade-off between lateral support and ease of getting in and out. The Mondial's big doors make entry and exit very easy, except in confined spaces when you can't fully open the door.

Twist the Mondial key and there is the familiar delicate whine of the starter motor, followed a second later by the smooth rasp of a silken engine. There's no pumping of the throttle or initial spluttering built into this engine; those are things of the carburetted past. And for a V8, the engine is amazingly burble-free. Its sounds are inspiring, but they amount to a whine and a wail rather than the rumble most V8s emit. Mechan noise is of a low order in the cockpit anyway – even lower th the 308's discreet level. Mainly you hear the rasp of the exhaus whisper and a whistle from the mechanicals and some gear whine seems to bounce up off t road. You won't be disappointe it sounds just like a Ferrari.

Your initial impression is tha this Ferrari's steering *is* rather indirect. It is the 308's system, it must manoeuvre a car with a foot more wheelbase. Mind you the steering is light and entirely devoid of lost motion at the straight-ahead, so your objecti disappear as familiarity grows.

The longer wheelbase of the

r, allied to the quiet, stable chelin TRX tyres, becomes a al asset. The Mondial is around n longer than the 308, and actically all of that length is ide the wheels. This makes the erhangs, and their mass, a aller proportion of the whole d thus the car's polar moment of rtia exerts less influence in anouevres. The long wheelbase es the Mondial a decisive vantage over the 308 in straight- ead stability; the insignificant erhangs mean that it turns with e poise of a dancer but only en you turn the wheel.

n fact, this car's ability to hold a aight course, over bumps and even surfaces, is quite eerily od. Although the front suspension is largely from the GTB, there have been several areas of refinement, including the inclusion of some anti-dive, and modifications to the king-pin inclination, to improve its stability. Those meaty tyres give the Mondial the firmly-damped but level and quiet low speed ride of the 308, and add a wide-tracked 'gumball grip' quality. The balance of this chassis is sensational.

Essentially, the Mondial is a shallow understeerer, but as steering effort builds up towards the limit the characteristic changes to mild oversteer, perhaps a little more prominent than in the 308 because it is accomplished by a suggestion more body roll. But because of the effect of the longer wheelbase, the feeling that you get in the 308 – that it's safer to be on the power in really fast corners – is not nearly as noticeable. Throttling off in maximum effort 90mph bends causes no more than a dainty, stable tightening of your line. The 308 doesn't do much in those circumstances either, but there's a tendency for it to jig diagonally, inside rear to outside front, and this leads you to prefer power in bends. Our only (slight) criticism of the Mondial is a function of its body roll; in the tightest of s-bends, taken fast, there can be a suggestion of lurch as the side loads transfer from one direction to the other. But it only happens when you're driving nearer to the limit than you'd normally take passengers. Most of the time the Mondial balances out perfectly through the bend, feeling fast but secure and relaxed. If you play, seeing what happens when you take it all the way, it can be held in huge oversteer incredibly easily for a mid-engined car. You soon know that it's very much a friend – a striking contrast to the old 308GTB4, which broke into oversteer like sudden death when you overcooked things (see *CAR* Oct 1975).

As far as power goes, there is a negligible difference apparent between Mondial and 308, even though it's there on paper. The four-place car feels fast but not brutally powerful. Overtaking manoeuvres of the 'confined' kind must be advisedly made, because a Mondial is certainly not an any-gear-open-the-throttle car like the Boxer. But if there is one thing about the power that does distinguish it from the 308 it is the extra length of its gears. It might not seem much for a car to have an extra 2mph available in first and 6-7mph in fourth, yet it matters. There is a relaxed quality about the Mondial's progress and when the going is really fast, the tacho needle doesn't so often seem to be running towards the 7500s.

The rest of the controls work in a familiar Ferrari way (it's always a pleasure to discover that in a new one). The Mondial has the 308's sensationally direct and crisp clutch, the skinny gearlever snaps around its gate with the same panache, sometimes with a whisker of notchiness, but always precisely. The brake pedal is the same low-travel but beautifully progressive control, now weighted to match the clutch.

This car, like all Ferraris, is for consuming distance. The 120mph cruise, if you can find the road, is a doddle. The car can stop exceptionally quickly from that, and if you are unlucky enough not to see a kink coming at that speed, it has the chassis to cope with it, as long as it also has the driver. At 120mph the mechanical noise level is low as is the wind noise, generally speaking, although our test car had some air leaks around the lower edges of its windows.

This Mondial, in every sense, is the sensible Ferrari. It has a boot as big as many small saloons, it has a big cabin, it has superb assembly and rustproofing standards, it starts easily and has maintenance-minimising items like fuel injection and contactless ignition (not to mention the easily-removed rear subframe). These things cut ownership costs. Yet to belie the pallid, characterless component of that word 'sensible', the Mondial has poise, speed, panache and beauty. For £24,500 (with the air conditioning installed) it is very fine value against a lot of expensive cars – and let's be candid about it, the Mondial's own very beautiful sister, the 308GTBi is among them.

Engine is paragon of alloy and black crackle in visual appeal but has new accent on ease of maintenance and repair: entire drivetrain and suspension package drops easily on its own subframe from main chassis
Headlights rise smartly and lock solidly in position; electric powered
Boot is very roomy indeed for the type
Body shape is nicest in profile, displaying balanced proportions and typically graceful Pininfarina lines

ROAD TEST

MARANELLO'S MONDIAL 8

A Ferrari for all reasons

IT is almost twelve years since Enzo Ferrari's road car building concern passed into the control of the giant national Fiat company, and there can be little doubt that, Gianni Agnelli fully appreciated just how special and valuable this acquisition was. In consequence, although ownership of the most exclusive, talked-about and fascinating car manufacturer in the World may have changed, Fiat in no way forced the pace of major change and only altered the character of the company by pursuing the same policy of unhurried evolution which characterised Ferrari progress ever since the end of the Second World War. It therefore comes as no surprise that it has taken some years for Fiat's obvious influence to make itself felt in the character of the machines which roll off the Maranello production line. We have lately been spending some time driving the Ferrari Mondial 8, a recently-introduced two-plus-two central-engined coupé with such versatility and docility coupled to the performance side of its character as to demolish much of the criticism and apprehension which has surrounded the prospect of Ferrari ownership in the past. Temperamental? Intractable in traffic? Inflexible? Not at all. The Ferrari Mondial 8 is truly a Ferrari for all reasons, whether they be simply A-to-B transport, open road motoring, or even heavy-traffic commuting. We employed our test model for all these purposes during a hectic five days' motoring, five days in which many of our own pre-conceptions were laid to rest as well!

Motoring writers risk becoming either totally blasé about the varied cross-section of excellent cars they are lucky enough to test — or, alternatively, they tend to heap praise unreservedly onto anything which is expensive, exclusive and out-of-the-ordinary. Consequently, when faced with a car like the Mondial 8, it's of paramount importance to be honest with oneself. A new Ferrari model doesn't come our way every month of the year. But let's say here and now that the Mondial 8 is *not* the perfect sports coupé to match all others. It has got some shortcomings, which we'll deal with later, although they are not of enormous consequence. But what the Mondial 8 *has* got is a tremendous all-round character, combining the magic of the Ferrari legend with

CONTACT WITH HOME! Our road test Ferrari Mondial 8 poses alongside the Ferrari F1 transporter at a recent Silverstone test session.

COCKPIT: fine layout and superb leather trim characterise the Mondial's interior.

HEART: the 90-degree V8 develops around 210 b.h.p. and is situated just ahead of the rear axle line.

THE clean lines of the Pininfarina styled Mondial 8 are well illustrated in these two shots on this page. Centre, the spare Michelin TRX nestles under the Ferrari's nose section.

Ferrari Mondial

impressively high standards of finish and ease of driving. It's interesting to watch the Mondial's effect on bystanders in comparison with their reaction to, say, a Porsche. With so many Porsches coming into England each year, you might be forgiven for feeling that these German cars are a little commonplace. But if Porsche's philosophy tends towards a wider market for their products, in effect educating Mr. Everyman to appreciate fine machinery, Ferrari's approach is still unashamedly elitist. Heads turned in respectful acknowledgement when we tried Porsche's 924 Carrera turbo and 928S, in a manner which suggested they had a quiet appreciation of the car's quality. But with the Mondial 8, heads *spun* round, joyous smiles abounded and little clusters of people appeared from nowhere when we pulled up at petrol pumps. A Ferrari is intended for the select few like no other car on earth!

At a tax paid price of £24,488.25 the Ferrari Mondial falls between the 308GTB two-seater and the big V12 front-engined 400i in the Maranello range, and is intended to replace the two-plus-two 308GT4. Obviously, the Mondial 8 is only what might be described as a "nominal" two-plus-two, for although there are two very positively styled rear seats beneath the gently sloping Pininfarina roofline, there's no question of getting any full-size human being in behind the driver if he chooses to use the full range of the seat adjustment. Three people can be squeezed in with a moderate degree of comfort, but it's really for two and the shopping/dog/kids — something which might, in itself, be regarded as heresy to those Maranello addicts who believe that any Ferrari with more than two seats is something of an imposter!

The Mondial's splendid body shape is rooted in the traditional Ferrari practice of overlaying a tubular frame base with steel panels, although alloy is employed for the engine and front bonnet cover and there are some glass fibre panels within the structure itself. Viewed from any angle, the Mondial 8 is really a most aggressive and imposing machine, and although some frowned slightly at the sight of the heavily finned air cooling ducts behind each door, preferring the more subtle vents on the 308GTB models, we felt that they enhanced the car's sleek appearance. Power comes from a 90-degree transverse mounted, five bearing, light alloy V8, sited just ahead of the rear axle line, with a bore and stroke of 81 x 71 mm. for a total capacity of 2,926 c.c. Its four overhead camshafts driven by toothed belts, this Ferrari V8 develops a maximum of around 212 b.h.p. at 6,600 r.p.m., delivering maximum torque at 4,600 r.p.m. A single dry plate clutch transmits the power through a delightful five-speed gearbox which is controlled from the cockpit via the "traditional", notchy, but satisfying open Ferrari "gate". Bosch K-Jetronic fuel injection has recently replaced Weber carburation on this V8 power unit.

On first acquaintance, the interior of the Modial 8 makes quite an impact. If you've been educated to suspect that all Italian super-cars subjugate every aspect of their make-up to sheer performance, you'll be surprised with the first-rate level of finish in this Ferrari. The driver sinks into a superbly upholstered and trimmed leather covered seat — a trifle too narrow for the writer, perhaps — and the smell of that top quality leather is all-pervading. The floor is covered with high quality carpeting and leather trim extends to the fascia and door panels. It's all executed to a very high standard. Nowhere could be found any trace of shoddy workmanship; everything fitted and, with the exception of a broken rear ashtray, worked impeccably. This was proof of Fiat's influence; the sheer quality of the car's production engineering and assembly standards. It really was approaching the level we've been used to expecting from Porsche....

The driving position was excellent, we discovered, although the writer had to make full

FERRARI MONDIAL —

use of the adjustable steering column, pulling the wheel as far as it would come towards him in order to make use of the rearward adjustment of his seat. That arrangement guaranteed that his six foot-plus British frame could be accommodated in comfort without any of the concessions to the Mediterranean physique which usually has to be made when driving Italian cars. There was adequate headroom, more than sufficient elbow room and, although the pedals are slightly off-set owing to the obvious intrusion of the front wheel arch, this proved absolutely no bother at all. The pedals, incidentally, are superbly positioned in relation to each other and allowed my sloppy interpretation of "heel and toeing", whereby I simply roll the side of my right foot from brake to throttle, to be carried out with no problems whatsoever!

The Mondial's V8 fires up instantly, with no fluffing or spitting, a pleasant mixture of deep rasping and mechanical whirring. By no stretch of anybody's charitable imagination could the Mondial's cockpit be described as quiet once its engine burst into life, but it's not obtrusively worrying. First gear is dog-legged to the left and back towards the driver and it takes total depression of the long-travel clutch pedal to engage it without a gentle crunch. From that point onwards the clutch takes up smoothly and progressively, although, again, it needs to be fully depressed before second gear goes in without mild protest. At first, second baulks slightly until the box is fully warmed up and we found ourselves double-declutching "just to be sure". But from that moment on, it's a dream — not in the Ford-fashion, but in the way it rewards precise and conscientious use. In fact, the Mondial is a car which brings out the best in a driver. Drive it sloppily and it will look after you, but it won't be fun. Drive it as it's meant to be driven and you suddenly become aware that you're paying more attention to your own driving style, concentrating more and taking a pride in your progress.

Faced by the tempting 180 m.p.h. speedometer and matching rev. counter which is yellow-lined at 7,000 r.p.m., there's a choice of the Mondial's temperament to explore. Either you marvel at its flexibility, running down to as little as 20 m.p.h. in fifth with matching behaviour in the other ratios, or you get that rev. counter needle swinging its way round the dial. That's the great joy of this engine, it simply revs. and revs. And it's quick, with 60 m.p.h. coming up in fractionally more than seven seconds and just over 20 seconds to 100 m.p.h., figures which may mean the Mondial is lagging marginally in the split-second contest to be the quickest super-car, but by no means making it a sluggard. Fourth gear is good for 115 m.p.h., by which time you're ready for the slog up towards its claimed limit of 150 m.p.h. Judging by the way in which it started to run out of breath at around 125 m.p.h., we must agree with a rival magazine which suggests that a Mondial straight off the production line probably wouldn't be good for much more than 140 m.p.h. But, again, it's not always what things do in this world, it's the way that they do it!

Contact with the ground is made through Michelin TRX 240/55 VR390 radials mounted on distinctive alloy rims. There is a perceptible degree of roll from the coil spring/unequal length wishbone suspension when hurrying in this Ferrari and if you make the mistake of running into a sweeping corner on a trailing throttle, you may emerge a little disappointed at the ensuring understeer. Thus caught out, it takes a little time to become acquainted with the reassuring way the Mondial behaves under hard acceleration, within a corner. With the power hard on, the Michelins really do their stuff, absorbing most of the bumps and ripples without any wavering from their prescribed line. The ride is excellent, although some adjudged it a little too firm for their taste during sustained motorway cruising. Certainly, although there's not too much kick-back through the steering wheel from minor road undulations, any large pothole or distortion doesn't receive the same favourable reaction. The brakes are superb by any standards and never gave us the slightest worry, even in the wet.

Visibility forward is fine, although one tends to guess when it comes to knowing precisely where the nearside front corner is at any particular moment. Rearward visibility is awful when it comes to parking and I was slightly surprised to find that there is no passenger door mirror on the Mondial, something I would have liked a great deal. These two points, allied to the heavy clutch and second gear's reluctance to engage, made a crawl round the North Circular Road in the rain a bit of a misery although the luxury of the efficient air conditioning, the first-class stereo and the effective windscreen wiper/washer system must be mentioned as mitigating factors!

To the right of the fascia, three buttons electrically activate the rear luggage compartment lid, the engine cover and the front boot (where the spare Michelin TRX resides). There is a bank of warning lights on the central console in a position where *nobody* is ever going to see them until too late, unless they look down to slip another tape into the stereo system. On the central console, just behind the gear lever, there are controls for the electric windows, the fuel filler cap and the retractable aerial on the right rear flank. The headlights fold upwards from the front of the nose section and proved more than adequate for our needs.

During its spell in our hands, the Ferrari Mondial 8 returned an average fuel consumption of 16.6 m.p.g. It is obviously not a cheap car to run, nor is it meant to be, but it is a very positive step in the direction of persuading Jaguar, BMW, Mercedes and Porsche owners that here is a car to match their current machines. . . and it carries a Ferrari badge. It is a very realistic *alternative*, rather than an extravagant *indulgence*. For that reason, the Ferrari Mondial 8 should be regarded as a significant step forward for the Italian marque. It's not the quickest car available for that money, but it's about as well finished as you could wish and offers the indefinable magic of Ferrari motoring as an everyday experience rather that as an occasional treat. — A.H.

FERRARI 308 GTB
continued from page 52

with a grey-black felt-like material that had zero reflective quotient. At no time did either the instruments or the trim reflect against the steeply canted, curved windshield. In the case of the 308, the main instruments are still in a pod in front of the driver, but not all of them. The oil temperature gauge and clock are stuck, as though an afterthought, under the dash to the left of the steering column. What was most annoying, however, in fact downright dangerous, is that the dash and instrument pod are covered with a shiny vinyl material that reflects afternoon sun and overhead lights at night directly on the windshield. The worst reflection comes from the pod, placing the glare directly in front of the driver. It is serious enough to actually interfere with vision under all too many circumstances. The controls for heat, defrost, fan and air are all grouped on a console between the seats. They are handy enough but, unlike those of the Dino 246, they require that you take your eyes off the road to operate them. A small but annoying difference. The lesson is simple: Turin should take lessons from Maranello rather than the other way around, forcing Ferrari into the Fiat mold.

Caviling aside, the 308 is definitely a successor to the beloved 246. It has the all of the good road manners, the fantastic roadholding and cornering that make it feel as though it was an extension of the driver's body that is to be expected from a modern Ferrari. There are the brakes that, when applied, make the car feel like it is being halted by a rubber wall, and all of the performance the reasonable enthusiast can ever use. It is, in short, a true Ferrari.

To sum it all up, the Dino 246 was a car for retired racers who still desired the feeling without the noise and harshness of the real thing. The 308 GTB, not to put too fine a point on it, is a car for the would-be or early-retired racer who still wants some of the sound and fury of such devices as the 275 LM.

Now, let's see; where does one look for a duke's daughter caught up in a sticky situation? ■

Finally. See the USA in your new
Ferrari Mondial

by Tony Swan
PHOTOGRAPHY BY TONY SWAN AND BOB D'OLIVO

CROSS-COUNTRY TEST

"What the hayull izzat, anyway?" The Arizona Highway Patrol officer didn't know it, but he was asking the first question in what I had come to think of as the Mondial Catechism. Minutes earlier, in true Western peace officer style, he'd beaten me to the draw, his hand-held K-band radar gun against my Escort radar detector, the first lawman to get the drop on the Escort in over 3,200 miles. And without putting too fine a point on it, many of those miles blazed by at speeds the NHTSA regards as Certain Death.

It seemed like a good idea to keep my firsthand knowledge of the car's speed potential to myself, particularly since I already knew what the officer's next question would be.

"How fast'll this thang go?"

He didn't read me my rights—I'd managed to get in a good lick of brakes when the Escort beeped, so with only 64 mph on the radar gun I wasn't looking like Public Enemy Number One—but it seemed like a good idea to avoid self-incrimination.

"The European factory literature says top speed is over 140, but the European magazines say 140 is absolute tops. I'd be surprised if U.S. models will do that much."

In fact, the most I'd seen was just a tad over 130, achieved on a long, straight stretch that I shared only with an armadillo or two. This speed represented about 6,500 rpm on the tach, and it took a good long run to get there. Although the engine seemed willing to do more, I ran out of straight before I could find out.

The officer digested the top speed info with raised eyebrows, then went on to question No. 3.

"Well, whut's it cost?"

"About $70,000," I said, which ended the conversation.

In the course of a three-and-a-half day, 3,600-mile trip that started in New Jersey,

107

Ferrari Mondial

swung down through the Gulf Coast states, across Texas and along the Rio Grande, then back up through New Mexico and Arizona to L.A., those three questions were repeated virtually every time I stopped.

Whut issit?
Whut'll it do?
Whut's it cost?

There was something charming about this guileless innocence, especially in contrast to the jaded denizens of Sunset Boulevard; about the only double takes we drew in Hollywood were from drivers of other Ferraris, who weren't quite sure what they were looking at, but knew it didn't look to be a 308. In every other port of call on this cross-country voyage, however, I became an instant one-man parade.

The Car

Debuted at the 1980 Geneva show, this is the first Ferrari conceived and executed under the Fiat aegis, designed in the Pininfarina studios as a replacement for the seven-year-old 308 GT4. The latter, a wedge-shaped creation from Bertone, was the only production Ferrari designed outside Pininfarina in almost 30 years. Even though the GT4 sold well and got good marks as a driver's car, it was never really accepted by true cloth *Ferraristi* inside the organization or out.

At first glance, the Mondial appears to be a 308 GT with a few appearance changes and enough extra interior volume to accommodate a couple of extra seats. However, while it's true the car shares the general look that's distinguished this series since the advent of the Dino back in 1966, and there is extensive mechanical commonality, the Mondial does represent a fresh sheet of paper. It's got about a foot more wheelbase and around 3 inches more width, and is about 5 inches taller—not to mention about 150 pounds heavier. It's also bigger than the GT4—3.9 inches in the wheelbase, 3.2 inches across the beam and 1.6 inches from top to ground, which adds up to 55 pounds heavier on the scales.

The most readily visible evidence of all this expansion is the rather tall greenhouse, although Pininfarina has done an excellent job of keeping the proportions right. And virtually all the dimensional increases have been passed along directly to the consumer, which makes for outstanding front seat legroom and a back seat area that could actually accommodate a couple of smallish adult people (provided they don't mind traveling in full fetal position), or maybe two full-size inflatable dolls at half-pressure. The 2+2 concept is, after all, always a compromise. But we can say that this particular execution of the concept is a substantial improvement over the preceding effort.

Pininfarina's good work is also evident in the cockpit furnishings. The Connolly-clad leather buckets are remarkably comfortable for extended voyaging, although their side bolsters don't offer much in the way of lateral support. The big leather-wrapped steering wheel is adjustable for rake, and the instrumentation has also been given some attention—new graphics, cowling and layout as compared to the 308 GT or the GT4. About the only complaints in the area of interior design, fit and finish are the awkward radio location (Ferrari continues its practice of leaving radio installation up to its dealers), and the door handles. The latter are of poor design, affording insufficient leverage, plenty of confusion and far too many rattles. The left-side mirror on our car was inoperative, but I suspect this was due to the hasty radio installation.

Aside from an LED clock/stopwatch combination that seemed to defy most efforts at adjustment and/or general use, the dashboard is mercifully free of the digital dementia that seems to be sweeping much of autodom. However, this Ferrari does possess an on-board self-diagnostics system that looks to be rich with potential for electronic hysteria, its oil warning provisions in particular. More on this later.

Below the skin there's a tube steel frame, with fully independent double A-arm suspension and vented disc brakes at all four corners—very much like the 308 GT, and all of it mainstream Ferrari chassis practice. There is, however, a major distinction, one that's sure to score with mechanics. The rear portion of the frame, complete with engine and drivetrain, is designed for easy removal as a complete unit, which will cut major maintenance time in half.

Aesthetically, the car has drawn some flak for its cheese-slicer side scoops, particularly in Europe where the scoop grille-work is painted black. The going-away view has an unfinished feeling to it, created in part by the utilitarian look of the muffler cowling. But even if the overall shape does give something away to its 2-seat counterpart, there's absolutely no way you could justify calling it homely. And it *definitely* looks like a Ferrari.

The Powertrain

The Mondial's aluminum 3-liter 4-cam V-8 has been around since 1975, and is a proven, reliable performer. Ferrari replaced the original quartet of Weber 40DCNF 2-throat carburetors with Bosch K-Jetronic injection last year, a change that whittled a little bit off peak horsepower but made overall response much smoother and easier to live with. Ignition is now Marelli's electronic Digiplex unit, which is capable of handling eight different spark curves and is maintenance-free, contributing to the 7,500-mile service interval.

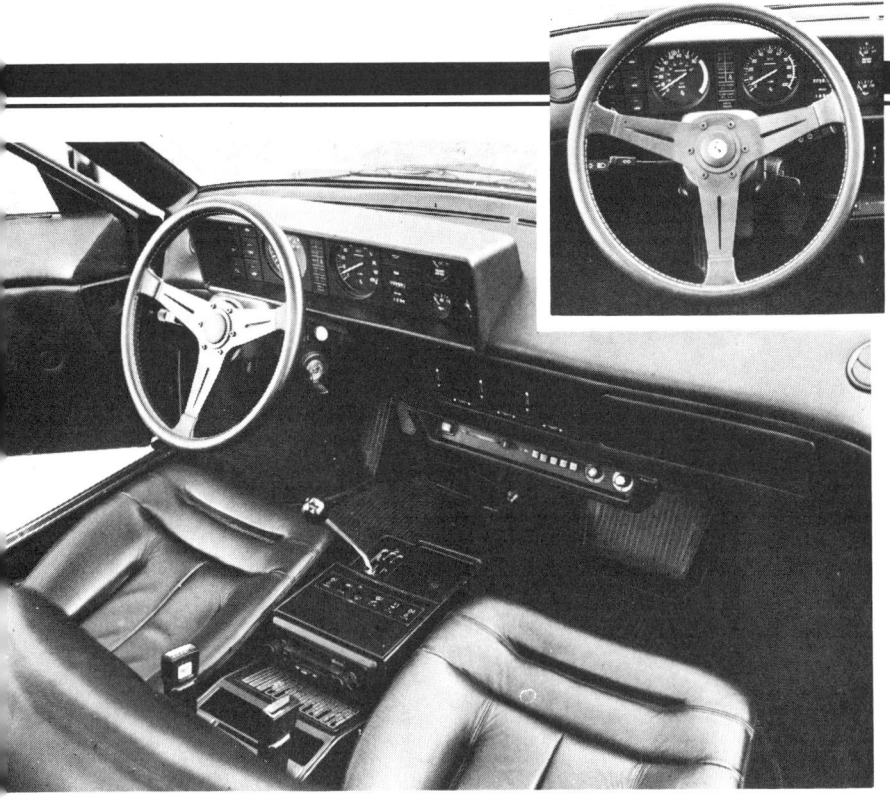

Like the 308 GT, the Mondial's engine is mounted athwartships, just behind the passenger compartment. It has the typical mid-engine problem of noise, but not as much of it as you'd expect, at least at legal highway speeds. The louvers that punctuate the rear deck help keep the engine compartment relatively cool, although it still generates enough heat into the coach work to give your luggage—stowed in the surprisingly generous trunk—a thorough baking. Keep your cameras and film in the main cabin.

The engine's cooling system holds a whopping 25.4 quarts, and this, coupled with the oil radiator, keeps operating temperatures well within normal limits even in prolonged stop-and-go city combat.

As noted, the oil warning system gets a unanimous raspberry from the *MT* staff. The sending unit is attached to the oil dipstick, which means you've got to unhook it to check the oil level. The idea is to drive along until the warning light comes on, then add oil. This may be fine for those drivers whose idea of maintenance is filling the gas tank, but leaves much to be desired for anyone who really cares about his car. When it come to the presence of oil in the sump, we want to know more than yes or no; we want to know how much.

Like the engine, the Mondial's drivetrain is lifted directly from the 308, including the internal ratios of the 5-speed gearbox. The gearbox has its own little oil pump, to make sure everything inside stays nice and slurpy. Shift action is precise but stiff, although the process can be augmented by heel-to-toe. The foot controls are nicely designed for this activity, as they should be.

Performance

With more mass, more frontal area, slightly fatter and taller tires (effectively raising final drive and rolling resistance), it's not surprising that the Mondial is a wink or two slower than the 308. The 2-seater will turn sub-16-second quarter miles, the Mondial low 16s; terminal speeds for that distance favor the 308 by about 2 mph. It's not an easy car to get out of the blocks in a hurry, and there are obviously cars around that'll out-quick it. Thanks to its downhill gearing, the Mondial will generate an impressive top speed, but this, too, requires a certain amount of patience to achieve. Then again, how many cars today can top 125 mph in *any* distance? The Mondial is no rival to the Boxer, but it is nevertheless a member of a select performance fraternity.

When it comes to getting around corners, the Mondial inspires confidence. The extra foot of wheelbase makes the car feel a good deal more stable than the shorter 308 in practically any operating situation—straight ahead at high speed, caning it around fast sweepers or scrambling into decreasing-radius sphincter-tighteners. Once you've established a rapport, the car lets you get away with all kinds of bad habits, like lifting in mid-turn. The nose tucks in a little (but not enough to be called oversteer), and its general behavior could be called neutral—surprising, considering the rearward weight bias. While there isn't enough power to indulge in a lot of wild, tail-out power slides, the car seems well suited to making a fast entry and holding a respectable speed all the way to the exit with little more body roll than the 308.

The Mondial's rack-and-pinion steering requires healthy effort at low speed but is very responsive once you're rolling, and naturally contributes a lot to the feeling of stability. And the popular TRX Michelins do even more. They're remarkably sticky on dry pavement, provide better-than-average stability in the wet, don't make noise and undoubtedly offer a better ride than low-profile numbers like the Pirelli P7.

For a machine sporting a moderately firm racing-style suspension system, the Mondial performs very well in the ride department. The Koni shocks have a habit of simply ignoring small, sharp bumps, which is irritating, especially when the bumps begin contributing to the steering. However, aside from this harshness, the car deals well with other surface irregularities. Again, the extra wheelbase helps out here.

Braking is superb. There's plenty of power in the system, and excellent control to go with it. Like the car's forgiving cornering qualities, the brakes are designed to inspire confidence.

Fuel economy isn't much of a concern to the people who will buy these cars (about 60% of them are in North America), but the Mondial does reasonably well on this score. The car has EPA ratings of 10 city, 18 highway, and as usual we were able to exceed the highway number for legal cruising, scoring as high as 21 mpg. For the whole 3,622-mile trip the average was 18.7 mpg, a thoroughly impressive number considering some of the cruising speeds involved. On the *MT* 73-mile gas loop, the Mondial recorded 17.1 mpg.

The car went through just a tad under 2 quarts of oil over the cross-continent haul, which, according to the Ferrari folks, is also remarkable.

The Experience

While attending the University of Minnesota a couple of lifetimes ago, I managed to scrape together enough money to buy a sweet little Porsche 356 coupe, already well into middle age by the time it came to me, but in excellent condition. I found that it immediately altered the structure of my days, because I'd wake up as soon as there

Ferrari Mondial

was enough light to see more than 10 feet and look at it from my bedroom window. I found that, even during perilous times like final exams, the Porsche was able to make each day's prospect considerably brighter.

A great quantity of machinery, two-wheeled and four, has come and gone since those days. But I found that this Ferrari was one of the few cars able to rekindle those old passions. As the crew at Fiat headquarters in New Jersey put the finishing touches on the car's prep—including installation of the Pioneer AM/FM/cassette unit—I caught myself looking at the car from the office window and experiencing *deja vu* hot flashes.

Basically, the prospect of settling into a new Ferrari for a trans-continent shuttle makes the whole country look like one gigantic dessert. The world's best highway system, even though it's wrapped in the chains of the lowest common denominator, suddenly gets a massive jolt of magic. This, of course, has been the Ferrari stock in trade since the beginning, in 1940. And it seems quite reasonable to me to assume that anyone who takes delivery of a new Ferrari can feel the élan of all those races, on all those tracks, over all those years. As you settle into the leather command seat, you're sharing a bond with men like Alberto Ascari, Phil Hill, Wolfgang von Trips, Fangio, Jody Scheckter—the list is long and distinguised.

Your hand closes on the key, pausing at the portal of this fresh adventure, then brings the beautifully orchestrated machinery to life (no throttle required). The famous Weber induction noise is gone, but there's still a sound that conjures up visions of hurtling down the Mulsanne Straight in some misty dawn, of rocketing half-blind into the tunnel at Monaco, of attacking the giddy, off-camber downhill drift that makes up Turn 1 at Brands Hatch. Illusions, to be sure, and not the sorts of things you'd care to offer up to explain to the patrolman why you were doing 120 down that dead-end street. But how many of today's motor vehicles include on-board fantasy-fulfillment generators at no extra cost?

There is one down side to the experience. This car, and presumably any Ferrari, is an absolute top-drawer, double throwdown jerk magnet. Just as the sight of a jogger will send every dog in a neighborhood into instant frenzy, the sight of a Ferrari sorting out the traffic will inspire certain drivers to strange rites of manhood. The Ferrari driver regularly finds himself being pursued and/or overtaken by cars that moments before had been going 10 or 20 mph slower. Better yet, the Ferrari will

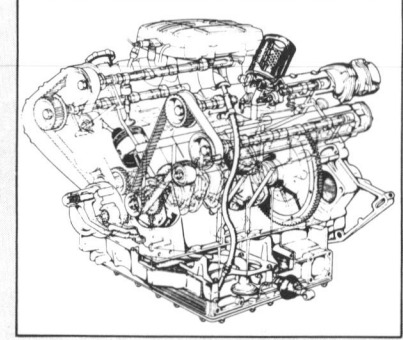

Ferrari's 2,926cc 90° aluminum V-8 entered production in 1975 with four 2-throat Weber carburetors and a rating of 255 horsepower. Bosch K-Jetronic injection replaced the Webers last year, and tightened emissions regulations have reduced horsepower in current U.S. models to 205. (European editions are rated at 214.) Fired by Marelli electronic Digiplex ignition, the engine employs toothed belts to drive its quartet of overhead camshafts. The crank turns in five main bearings, with a conventional 8-quart oil system plus an oil cooler. Ferrari uses the 3-liter V-8 in both the new Mondial and 308 GT.

The original Ferrari Mondial appeared in 1954. Inside the sleek Pininfarina-designed, Scaglietti-built bodywork was a streetified version of the 2-liter 4-cylinder engine that had been virtually unbeatable in the two previous seasons of World Championship racing. This roadster, known as the 500 Mondial, also saw some racing service, mostly in the hands of privateers. Although it was not conspicuously successful in competition, it did finish second overall and first in class in a Mille Miglia. Generally overshadowed in racing by contemporary 2-liter Maseratis, the 500 Mondial soon gave way to the famous Testa Rossa.

pull out to pass on a two-lane highway and find the car he's passing suddenly leaping forward under full acceleration. Interestingly enough, most of this sociopathic driving isn't displayed by other sports or GT car drivers. It seems to be much more prevalent among drivers of older American cars. And the bottom line here is that you occasionally find yourself taking chances to get past some guy in a '65 Ford who's kicked it up from 60 to 90 or so when he saw you coming, his car shedding anonymous little pieces as it wallows down the highway. The 55-mph limit is clearly too *fast* for these people.

I encountered plenty of these cretins going across America—an experience that I'm certain is in no way unique—but they couldn't really scotch the joy of the going. The Mondial is not the greatest Ferrari ever issued by Maranello, but it is by no means the least. It is a long-legged, high-gloss profile piece designed for the buyer who wants to pack as much class into a cross-continent trip as possible, and it achieves this goal in ways the 308 cannot.

But more than that, it is a Ferrari. It looks like a Ferrari, it smells like a Ferrari, and it sounds like a Ferrari. And knowing there's a Ferrari waiting in the carport, with an open road just beyond, makes any day seem too short.

/MT/

ROAD TEST DATA

Ferrari Mondial

SPECIFICATIONS

GENERAL
Vehicle type	Mid-engine, rear-drive, 2 plus 2 coupe
Base price	$70,000 (est.)
Options on test car	Electric sunroof, remote-control right-hand mirror, AM/FM/cassette

ENGINE
Type	V-8, water cooled, aluminum alloy block and heads, 5 main bearings
Bore & stroke	3.18 x 2.80 in. (81 x 71 mm)
Displacement	178.5 cu. in. (2,926 cc)
Compression ratio	8.8:1
Fuel system	Bosch K-Jetronic injection
Recommended fuel	91 unleaded
Emission control	3-way catalyst, EGR, air injection
Valve gear	Dual overhead cams, belt-driven
Horsepower (SAE. net)	205 at 6,600 rpm
Torque (lb.-ft., SAE net)	180 at 5,000 rpm
Power-to-weight ratio	17.75 lb./hp

DRIVETRAIN
Transmission	5-speed manual
Final drive ratio	4.06:1

DIMENSIONS
Wheelbase	104.3 in.
Track, F/R	58.9/59.7 in.
Length	180.3 in.
Width	70.5 in.
Height	49.2 in.
Ground clearance	4.9 in.

CAPACITIES
Curb weight	3,460 lb.
Weight distribution, F/R	41.5/58.5%
Fuel	22.2 gals., incl. 4.75-gal. reserve
Crankcase	9.5 qts.
Cooling system	25.4 qts.
Trunk	10.6 cu. ft.

SUSPENSION
Front	Independent, unequal-length A-arms, coil springs, telescopic shocks, anti-roll bar
Rear	Independent, unequal-length A-arms, coil springs, telescopic shocks, anti-roll bar

STEERING
Type	Rack and pinion
Turns lock-to-lock	3.5
Turning circle, curb-to-curb	39.4 ft.

BRAKES
Front	11.1-in. vented discs
Rear	11.7-in. vented discs

WHEELS AND TIRES
Wheel size	7.08 x 15.35 in. (180 x 390 mm)
Wheel type	Cast alloy
Tire make and size	Michelin TRX 240/55 VR 390
Tire type	Steel-belted radial
Recommended pressure (psi), F/R	33/33

TEST RESULTS

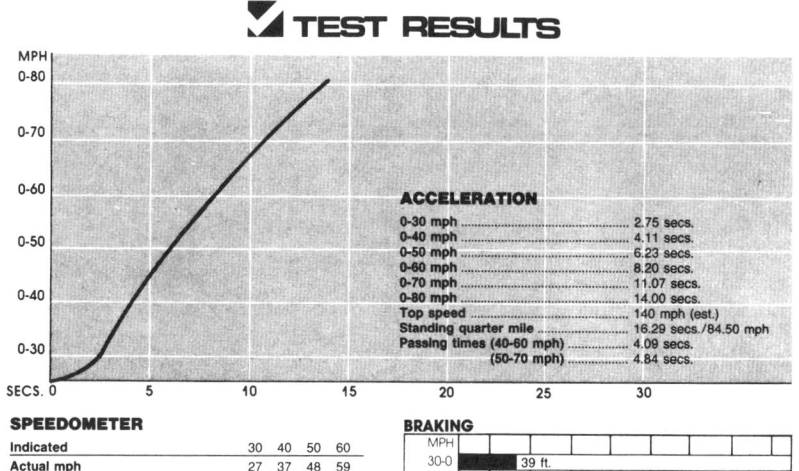

ACCELERATION
0-30 mph	2.75 secs.
0-40 mph	4.11 secs.
0-50 mph	6.23 secs.
0-60 mph	8.20 secs.
0-70 mph	11.07 secs.
0-80 mph	14.00 secs.
Top speed	140 mph (est.)
Standing quarter mile	16.29 secs./84.50 mph
Passing times (40-60 mph)	4.09 secs.
(50-70 mph)	4.84 secs.

SPEEDOMETER
Indicated	30	40	50	60
Actual mph	27	37	48	59

BRAKING
30-0	39 ft.
60-0	164 ft.

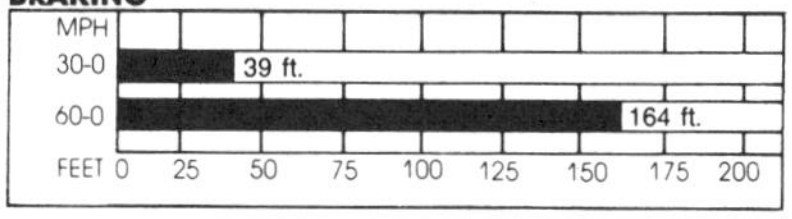

FERRARI FEAST

Ferrari Mondial 8

Is there life after rip and snort?

BY PATRICK BEDARD

• It must be hell trying to design new Ferraris. Over the years the legend has thickened, becoming a Jell-O inhibiting every stroke of the pen, every flight of the imagination. How can a new model live up to expectations? Too many twelve-cylinders have shrieked down the autostrada pumping too much adrenalin along the way. Too many road testers have fired too many salvos of hyperbole. For years we car critics have reviewed the world's finest sporting cars, pronounced them nice, even exciting, but not Ferraris. Ferrari was always atop the pedestal, and that pedestal was always being jacked up a bit each year. Now the altitude is such that even new Ferraris can't measure up. A Ferrari owner of our acquaintance drove the Mondial 8 a few miles and judged it nice, but definitely not a Ferrari.

So how are we to decide the truth of this latest product of Maranello—by measuring its stifled snorts and screams against the legend (in which case it will inevitably fall short) or by holding it up to the sporting requirements of the Eighties? Were it of any other brand, we would unhesitatingly do the latter. But with a Ferrari, the legend always lurks. Perhaps, by starting with the basics, some appropriate yardstick may evolve.

The Mondial 8 is a transverse V-8 mid-engined coupe with an upholstered section in back that appears to be a rear seat. So you would expect this to be a two-plus-two. And you would be wrong. Ferrari has built models in the past officially designated two-plus-twos, and they always had sufficient leg and head room back there to accommodate the occasional occupant of adult dimensions, but only Venus de Milos need apply for the rear compartment of this car.

Such a configuration has precedent in the 308GT4 that was discontinued two years ago, so the Mondial 8 must be accepted as consistent with past Ferrari practice. Still, it *is* a dumb way to build a car—okay in a hatchback where the trunk space can be extended forward by folding the seats, but essentially useless in a mid-engined design.

The Mondial 8 is also a rather unattractive lump. Pininfarina is known for soft shapes that approach the zaftig, but this one just came out vague. Except in the side view, that is, where air-intake grilles the size of storm sewers ruin even the fundamental blandness. Apart from the Lusso Berlinetta, the 1964 GTO, and the current 308s, Ferraris have always looked sort of *ehhhh*—and the Mondial 8 continues the tradition. We therefore cannot deny its Ferrarihood on this count either.

MONDIAL

But what about the way it drives? Objectively, Ferraris have always been a pain in the butt in this regard, a quality appreciated only by those who thought the very definition of a man's car was that nobody else could even get it out of the driveway. Ferraris have been uniformly balky of shift, stiff of clutch, and hard of steering for as long as anybody can remember. Here the Mondial 8 may not be a true-red Ferrari. Your sister could drive it. The steering is not bad, the brakes no sweat, the clutch so gradual in action that nobody would ever kill the engine and so moderate in effort that there should be no complaints. The five-speed shifter is still genuine Ferrari, however—maybe not quite as hard to stir as some past models, but a purebred for notchiness. No other brand has so many traps in the pattern waiting to catch the lever.

But maybe it's time to stop beating around the bush. People buy Ferraris neither for the mazelike qualities of the shifter nor for the hospitality of the back seat. Instead, they seek the essential prancing-horse rip and snort, and if the Mondial 8 can deliver that, no question, it's a Ferrari.

Here we may be in trouble. The rip is subdued—a velvet purr, more song than shriek, that sweetly changes pitch as the engine climbs through its broad rev range. It's a splendid sound, but it soothes rather than incites to riot. That's not very Ferrari.

And, sadly, there is no snort whatsoever. The Mondial 8 will barely get out of its own way, or, more correctly, out of the way of other Ferraris. It's the slowest one in memory. Weight is largely the cause. The Mondial 8 shares the same Bosch K-Jetronic–injected three-liter V-8 with the GTBi and GTSi, but the car weighs in at 3560 pounds, 280 more than the GTBi that we tested in October 1980. This extra mass burdens it down to the point of being dog meat for the turbocharged Porsche 924 and

Datsun 280-ZX. A Ferrari that slow is certainly an enigma and maybe even a contradiction in terms.

It's not much fun to drive either. The Mondial 8 doesn't make you giggle, doesn't goad you into trying some foolish feats of antigravity. Instead, it suggests serious grand-touring transportation. It whispers, "C'mon, let's head for the coast." And it's not kidding. You could go anywhere in this car; it wouldn't fry your nerves in the manner of past Ferraris. No zingy noises, no jouncy ride, no hang-onto-the-wheel-with-both-hands-lest-it-get-away-from-you feeling. Just get in and go. How much more un-Ferrari could it be?

You may think, since the Mondial's back seat is worthless, that it ends up merely a slower and uglier GTBi. Actually, the two are much different. The **GTBi** is a full-time sportster. It'll never

114

MONDIAL

let you forget. Its roof presses down against your forehead, the door against your elbow, the console against your thigh—it's tight. And noisy. And demanding. The Mondial, in contrast, is relatively roomy. The front wheel arch takes a bite out of the spot the driver would like to have for his left leg, but that's the only encroachment.

The Mondial is also more relaxing. It doesn't have the low cowl of the GTBi (or the old GT4), so you can't see the road streaming directly under the nose. You are forced to take a longer view, and that's less dynamic, less stimulating. The suspension doesn't batter you either. The Michelin TRX tires are notably resilient, and the shocks have been calibrated to merely damp ride motions rather than prevent them. Except for some expansion-joint *kawop*, the ride is very pleasant. Nor does the Mondial make you keep your guard up. It doesn't kick back through the steering like the 308s. You may have noticed that the Mondial's wheels have an uncommon amount of "inset" to reduce the scrub radius. This is a new idea at Ferrari, and it takes much of the twitch out of the steering.

You add all of this up—the twitchless steering, the elbowroom, the friendly (if not quiet) acoustics, the low-effort controls, the civilized ride—and you find a pretty nice sports car, not a Ferrari in the traditional sense, but not bad either. The real question at this point is, Are there enough drivers in the Eighties who would buy a real Ferrari if it were available? It's easy for all of us sitting around wondering how we're going to cover the next Visa-card bill to say yes. But those who make their livings in the car business have noticed that by far the majority of those with resources to buy a Ferrari go for a Mercedes 450SL or SLC instead. Fiat, which now pulls the strings at Ferrari, is in the business of making money (or at least of trying to), and it's very tempting to dilute the Ferrari rip and snort in favor of some proven-in-the-market M-B civilization. We even hear oblique references in that direction from Fiat of North America, which imports Ferrari.

It sounds sacrilegious. Be assured that the Mondial is still a whole lot more Ferrari than it is Mercedes, but at the same time there is a conspicuous drift away from the joyfully mechanical persona that made up the traditional Ferrari. It shows up particularly in the use of electronics and electrics—maybe we

Technical Highlights

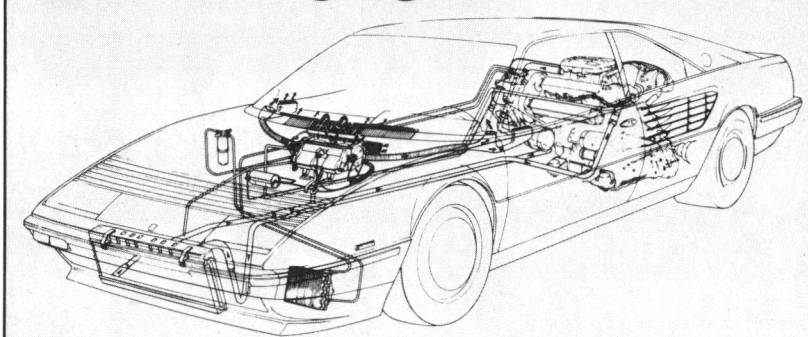

• Fiat's influence on Ferrari hasn't produced a wealth of horsepower as yet, but the relationship has at least borne a few practical improvements. Every Ferrari body is now dipped in an electrophoretic rustproofing solution. Zincrometal is used in particularly vulnerable areas of the coachwork for corrosion resistance. For a final layer of protection, every car is undercoated with abrasion-resistant PVC.

Access to Fiat's engineering resources has encouraged the use of ribbed and boxed steel stampings in the body structure, although the basic construction is still a relatively crude steel-tubing skeleton fleshed out with panels of steel, aluminum, and fiberglass. Electronics, at long last, have also penetrated the Ferrari works: Marelli Digiplex ignition is standard equipment, providing eight different spark-advance curves to better suit the needs of the double-overhead-cam V-8 engine, with the added advantage of practically no maintenance. The Mondial also introduces an electronic "check control monitor" that scrutinizes various liquid levels and lighting systems for proper operation every time the engine is fired up. An inductive pickup in the transaxle generates an electrical signal that drives the Mondial's speedometer. Finally, the instrument panel boasts a combination electronic stopwatch and clock.

The Mondial's gift to the world's highest-paid mechanics is a new system whereby the whole engine, transaxle, and rear suspension can be removed as a unit for more convenient (major) service or repair. A subframe assembly carries the whole power pack, and after approximately 40 hoses, linkages, shields, and electrical cables are disconnected, the removal of six more nuts and bolts will permit the subframe to be detached from the Mondial and wheeled off on a dolly for ready access to problem areas. This of course assumes your local mechanic sees fit to purchase the dolly through the Fiat/Ferrari service system. —*Don Sherman*

MONDIAL

should call it electricks—in the interior. Somebody decided that remote trunk releases are nifty, so the Mondial has an array of solenoid buttons on the dash to open the front hood, the engine cover, the trunk, and even the gas-filler door. This is harmless fun, but it gets a bit silly when applied to the glove box, the door of which is neatly devoid of any latch—the effect is spoiled by a big black button poking out below the dash that remotely releases the door from a full six inches away. The "computer" early-warning system, which tells of trouble with *liquido raffredd.* or *lavacristallo* or any of eight other possible menaces, is similarly misguided, because the signal lights are on the tunnel down by your hip, where you'd never see them until the problem became apparent anyway. These gimmicks are certainly typical of cars of the Eighties. Maybe we should even be reassured that Ferrari is less adept at them than other automakers; maybe this is proof that Ferrari has not wholeheartedly embraced electricks.

In any case, Ferrari spokesmen anticipate some redesign of the interior before full-scale production begins for American models. The console will be less conspicuously plastic, the air-conditioning controls below the dash will be relocated, and the brow over the instrument cluster will be reangled. This latter will be a mixed blessing. Right now it is both flat and level, the perfect place to clamp your radar detector. But the semigloss vinyl surface also reflects a shiny spot onto the windshield right where you're supposed to be looking at the road, so some alteration would be appreciated.

But enough of this minutia. Returning to the original question, is the Mondial 8 truly a Ferrari? We say yes, albeit the most democratic one ever built. Anyone with the price of admission can drive it. Maybe that's not the way Ferrari cars were built in the past, but this is the Eighties and things are different. For one thing, people aren't buying anachronisms. ●

COUNTERPOINT

● This Mondial has plunged me into a dilemma from which I see no escape. All at once, I think it's terrific *and* that I can't stand it. Every time it grows on me I discover something about it that's unforgivable. Almost.

Everything about the Mondial, from its sleek sheetmetal to the magnificent yowl of its V-8, sets you up, tells you it's going to be some kinda ride. But the most barbaric, proletarian Trans Am will suck the headlights out of this chic flyer at any stoplight you choose. "Emasculated" is the word that comes to mind.

But even so, it's a ball to barrel around in—the kind of car that talks to you, eggs you on. I went too fast too often and loved every minute of it.

Then again, here is a megabuck driver's car that you have to sit sidesaddle in, and that you can't even heel-and-toe. But, God, every time you turn the Mondial down a new street, it's an ego-stroking grand entrance.

I could go back and forth like this forever. It's probably just as well that most of the people who can actually afford a Mondial won't care much about how it runs, just that it makes them look so good. —*Rich Ceppos*

There are but two things wrong with the Mondial: somebody in Italy slipped on the wrong nameplate and the wrong price tag. If the truth were known, the Mondial is a Fiat. It runs like a Fiat, handles like a Fiat, it even has an interior full of Fiat parts. There is no denying that it is the product of Fiat thinking. The Mondial would make a terrific replacement for the Fiat 124 Sport Coupe that left our midst several years ago. The 124SC was never replaced, and I've always wondered why. Now I know. In Ferrari's defense, the Mondial does reflect progress. The engine sounds as if it's in the car behind you instead of in the right front seat. Conventionally built human beings now come close to a proper match-up with the steering wheel and the pedals. The ride is perfect. And the air conditioning works. With a little work on the torque curve, the throttle linkage, the shift linkage, the rear passenger compartment, the seats, the interior decor, and the console, Fiat—or Ferrari, if you prefer—could make a real machine of the Mondial. Right now, it's a 10-grand car hiding behind a 68-grand nameplate. —*Don Sherman*

Ferrari should include a lifetime supply of hoi-polloi repellent with this new Mondial. Although initially it's ambrosia for one's ego to encounter a crowd of gawkers every time you park on the street, you soon just want to get into the car and drive off without answering questions about what it is, how much it is, and how fast it will go.

Particularly that last question, because after hearing that it's a 68-grand Ferrari, people are unmoved by a 138-mph top speed and acceleration figures that can be bettered by a good-running RX-7. In other aspects the Mondial is very civilized for a car wearing the prancing horse. Its road manners are reassuring; its driving position, although it appears to be Classic Italian, is really livable; the rear seat can accommodate a pair of supple adults for short periods; and the general level of quality and solidity marks a new high for Ferrari. But Ferraris are supposed to be fast, first and foremost. When they're not, like this Mondial, the patina of civilization is a very inadequate substitute, and their whole reason for existence falls into doubt. —*Csaba Csere*

MONDIAL

Vehicle type: mid-engine, rear-wheel-drive, 4-passenger, 2-door sedan

Price as tested: $68,000 (estimated)

Options on test car: metallic silver paint, $780.

ENGINE
Type .... V-8, aluminum block and heads
Bore x stroke .... 3.19 x 2.80 in, 81 x 71mm
Displacement .... 179 cu in, 2927cc
Compression ratio 8.8:1
Fuel system .... Bosch K-Jetronic fuel injection
Emissions controls .two 3-way catalytic converters, feedback fuel-air-ratio control, EGR, auxiliary air pump
Valve gear .... belt-driven double overhead cams
Power (SAE net) .... 205 bhp @ 6600 rpm
Torque (SAE net) .... 181 lbs-ft @ 5000 rpm
Redline .... 7700 rpm

DRIVETRAIN
Transmission 5-speed
Final-drive ratio 3.71:1

Gear	Ratio	Mph/1000 rpm	Max. test speed
I	3.59	5.7	44 mph (7700 rpm)
II	2.35	8.7	67 mph (7700 rpm)
III	1.69	12.1	93 mph (7700 rpm)
IV	1.24	16.4	126 mph (7700 rpm)
V	0.95	21.5	138 mph (6400 rpm)

DIMENSIONS AND CAPACITIES
Wheelbase .... 104.3 in
Track, F/R .... 58.9/59.7 in
Length .... 180.3 in
Width .... 70.5 in
Height .... 49.2 in
Ground clearance .... 4.9 in
Curb weight .... 3560 lbs
Weight distribution, F/R .... 42.1/57.9 %
Fuel capacity .... 22.2 gal

CHASSIS/BODY
Type .... unit construction
Body material .... welded steel and aluminum stampings

INTERIOR
SAE volume, front seat .... 48 cu ft
           rear seat .... 27 cu ft
           trunk space .... 3 cu ft
Front seats .... bucket
Recliner type .... infinitely adjustable
General comfort .... poor fair **good** excellent
Fore-and-aft support .... poor fair **good** excellent
Lateral support .... poor fair good **excellent**

SUSPENSION
F: ...unequal-length control arms, coil springs, anti-sway bar
R: ...unequal-length control arms, coil springs, anti-sway bar

STEERING
Type .... rack-and-pinion
Turns lock-to-lock 3.4
Turning circle curb-to-curb .... 39.4 ft

BRAKES
F: .... 11.4 x 0.8-in vented disc
R: .... 11.7 x 0.8-in vented disc
Power assist .... vacuum

WHEELS AND TIRES
Wheel size .... 390 x 180mm, 15.4 x 7.1 in
Wheel type .... cast aluminum
Tire make and size .... Michelin TRX, 240/55VR-390
Test inflation pressures, F/R .... 35/39 psi

Car and Driver Test Results

ACCELERATION	Seconds
Zero to 30 mph	3.1
40 mph	4.6
50 mph	7.1
60 mph	9.3
70 mph	12.9
80 mph	16.0
90 mph	21.7
100 mph	27.8
110 mph	37.7
Top-gear passing time, 30–50 mph	13.1
50–70 mph	11.8
Standing ¼-mile	16.9 sec @ 83 mph
Top speed	138 mph

BRAKING
70–0 mph @ impending lockup .... 195 ft
Modulation .... poor fair good **excellent**
Fade .... **none** moderate heavy
Front-rear balance .... poor fair **good**

HANDLING
Roadholding, 282-ft-dia skidpad .... 0.79 g
Understeer .... **minimal** moderate excessive

COAST-DOWN MEASUREMENTS
Road horsepower @ 50 mph .... 15.5 hp
Friction and tire losses @ 50 mph ..... 8.0 hp
Aerodynamic drag @ 50 mph .... 7.5 hp

FUEL ECONOMY
EPA city driving .... **10 mpg**
EPA highway driving .... 18 mpg
EPA combined driving .... 13 mpg
C/D observed fuel economy .... **14 mpg**

INTERIOR SOUND LEVEL
Idle .... 71 dBA
Full-throttle acceleration .... 94 dBA
70-mph cruising .... 77 dBA
70-mph coasting .... 75 dBA

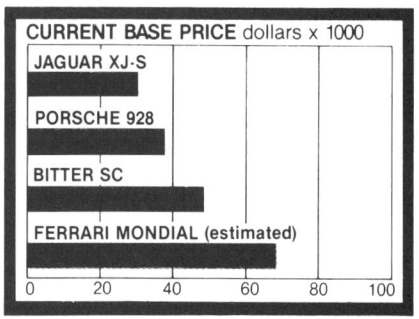

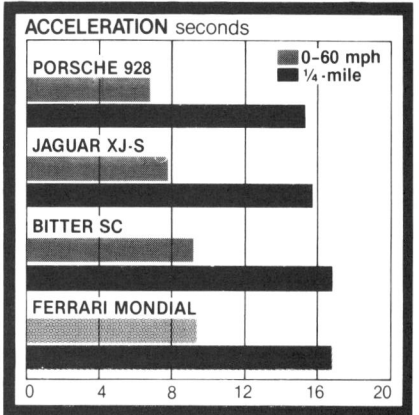

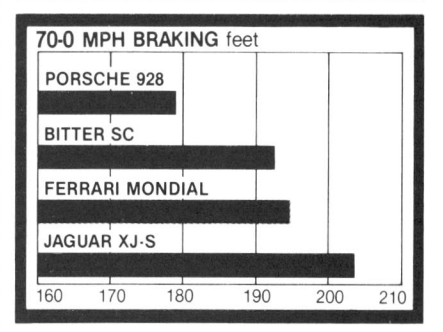

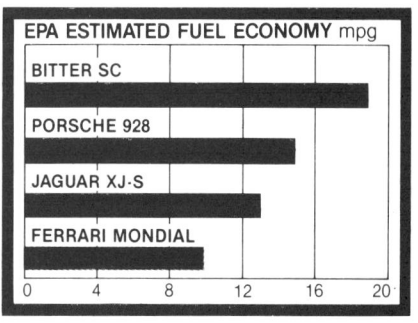

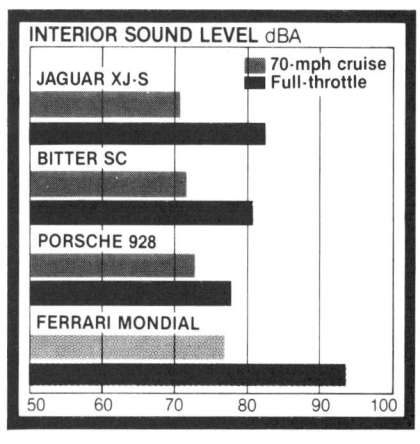

CURRENT BASE PRICE dollars x 1000
- JAGUAR XJ-S
- PORSCHE 928
- BITTER SC
- FERRARI MONDIAL (estimated)

ACCELERATION seconds (0–60 mph / ¼-mile)
- PORSCHE 928
- JAGUAR XJ-S
- BITTER SC
- FERRARI MONDIAL

70-0 MPH BRAKING feet
- PORSCHE 928
- BITTER SC
- FERRARI MONDIAL
- JAGUAR XJ-S

EPA ESTIMATED FUEL ECONOMY mpg
- BITTER SC
- PORSCHE 928
- JAGUAR XJ-S
- FERRARI MONDIAL

INTERIOR SOUND LEVEL dBA (70-mph cruise / Full-throttle)
- JAGUAR XJ-S
- BITTER SC
- PORSCHE 928
- FERRARI MONDIAL

FOUR SEATS AND THE BEST HANDLING

Ferrari's 2+2 Mondial 8 is meant to be the most practical of Ferrari's exquisite machines. But it also manages to have their best chassis. Rex Greenslade reports

TO THE committed Ferrari enthusiast, a "true" Ferrari has only two seats. It is an uncompromising and uncompromised machine, offering exhilarating and shatteringly fast transport, fully reflecting the sporting involvement that's always been such a large part of life at Maranello. But the reality is that most of the people who can *afford* Ferraris, and even in these dismal times there still seem to be plenty of those, buy the 2+2 variants.

Thus the latest Ferrari, the Mondial 8, which was launched at Geneva in March 1980, may not cut the most ice with the *cognoscenti*, but it certainly will with Ferrari customers. In Britain, for instance, Maranello Concessionaires expect that more than half of their 1982 target of 190 to 200 cars will be Mondials, each sold for the asking price of £24,488.25. The traditional (and gorgeous) front-engined V12 400i — again a 2+2 — has amazed many, including Ferrari, with its continuing sales, though as it costs a whopping £35,299.66, there aren't quite so many people falling over themselves to sign for it on the dotted line.

From the start, then, the Mondial 8 was meant to be a major bread-winner for Maranello, to take over where the Bertone-designed 308 GT4 left off. *Motor*, like many publications, was never a GT4 fan, though in retrospect that may have been because it had to follow the legendary Dino 246 and never managed to capture the smaller car's charisma. In fact, the GT4 was never meant to replace the 246 — the 308 GTB eventually did that — and the GT4 went on to become one of Ferrari's most successful cars of the late '70s. Yet the wedge-shaped GT4 lacked the sensual curves that one had come to expect of the normally-Pininfarina-styled Ferraris, as well as the sharpness of handling that so seduced a 246 driver.

Most of all, it was the severe lack of space in the so-called rear seats that let the GT4 down. So a high priority with the Mondial design — by Pininfarina — was to increase the wheelbase and interior headroom. Accordingly, the Mondial has a wheelbase that's four inches longer than the GT4 and it stands nearly two inches higher. The result is an interior that's remarkably roomy for a car of this ilk, the headroom in the front being exceptional, and in the rear acceptable for even an above-average sized adult. Likewise, front seat legroom is more than enough to please even a 6ft 5in giant, and while an adult would find it impossible to sit behind such a driver, two above-average sized adults can sit in tandem on the left side of the car without undue negotiations on the front seat position. To balance the increased length (the Mondial now measures 15ft 0in from stem to stern — about 10 inches *longer* than a Ford Cortina), the Mondial has grown more than three inches wider; that helps the elbow room and general impression of roominess enormously, but it also means that the Mondial is a bulky car indeed, as witnessed by the quoted (dry) kerb weight of 3188lb — 28.5 cwt. With oil, water and fuel for 50 miles (*Motor*'s specification for measuring unladen weight) the figure would rise to about 29.3 cwt — or about 4 cwt more than the GT4.

The Mondial shape has, to our eyes, an elegance and flow to it wholly lacking in the old GT4. Some people don't like the Garth Vader-style plastic embellishers for the side air intakes (and certainly their finish doesn't stand close examination too well), but overall the shape is a superb embodiment of Ferrari style, Ferrari movement. It's particularly difficult to make a 2+2 mid-engined car look right, as the engine's right where the rear passengers would sit in a front-engined car. But Pininfarina has managed to disguise the forward siting of the passenger cell, by cleverly extending the rear-threequarter glass rearwards far behind the rear seat (one side-effect being excellent rear threequarter visibility) and by extending the roofline almost as far back as the tail (a more usual mid-engined designer's trick).

As with the Boxer and the 308 GTSi and GTBi, the body is built by Scaglietti. It follows traditional Ferrari practice in that it clothes a tubular steel frame; Ferrari designers have incorporated more folded and boxed members in the chassis itself, though, in the interests of enhanced strength with little increase in weight. Another departure is that the Mondial is the first Ferrari ever to have a subframe at the rear which carries the engine, transmission and suspension complete; this subframe assembly can be unbolted from the car to aid servicing. The body panels are mostly of steel, with some aluminium (bonnets/engine cover) and some glassfibre ones (out of sight).

Mechanically, the Mondial is pure 308 GTBi/GTSi, with the all-alloy four-cam 90 deg V8 sitting transversely midships in the chassis, driving the five-speed close ratio gearbox below and slightly behind via drop-gears, Mini-style. A limited slip differential is standard. Double-wishbone suspension (what else?) with coil springs and Koni dampers is used front and rear, in both cases in conjunction with an anti-roll bar. The steering is by rack and pinion. In the Mondial, the wheels, and hubs have been modified to reduce kingpin offset (and hence steering kickback over bumps) while the front suspension's upper wishbone mountings have been angled to give a degree of anti-dive.

Ferrari say that the Mondial even has the same ratios as the GTBi/GTSi but the situation is complicated by the fact that the Mondial has new, fat Michelin TRX 240/55 VR 390 tyres. Ferrari quote a mph/1000 rpm figure of 20.06 in fifth in their sales brochure, stated 19.88 over the telephone, while our calcula-

notice this right on start-up from cold when you don't need to floor the throttle a couple of times in the time-honoured fashion; you simply climb into the car, turn the key and drive off — no hesitations, no stutter, no stumbles. The most sensitive of mechanical minds may detect a smoothening of the engine during warm-up as all the mechanical tolerances achieve their correct dimensions, but no more than that.

Warmed-up, the engine must have one of the widest power bands of any power unit in production today, for it revs cleanly on full throttle right from the idle speed of just over 1000 rpm right through to the red line of 7,700 rpm. You can make respectable progress using just 3,000-3,500 rpm, when the engine noise is surprisingly well suppressed, with much of the low speed intake gobble of the old unit missing. Cunning exhaust tuning means that there is little of the classic V8 woofle, the engine sounding every bit the competition-sired motor it is. Above 5,000 rpm, the note deepens noticeably, a unique mixture of exhaust rasp with cam and valve scream, and whirring gear whine. At 7,000 rpm it's almost like having a whole GP grid lined up behind you. Yet at no stage does the engine become excessively noisy — from the front seats at least; and if you've got a full complement of passengers would you thrash the car anyway? Mechanical noise and thrash is commendably well suppressed up to 100 mph but the low overall gearing (at 100 mph the engine's already spinning

tions show it to be 19.7. Unfortunately we couldn't clear up these anomalies before we went to press, but there is no doubt that the Ferrari is low-geared by any standards, a situation highlighted when you remember that even relatively mundane family cars like VW's new Passat have long-legged gearing approaching 25 mph/1000 rpm.

The engine differs from the 308's in two important ways. Firstly, it has Marelli Digiplex ignition (electronic of course, and developed in conjunction with Ferrari) and Bosch K-Jetronic fuel injection in place of four Weber 40 DCNF carburetters. Neither of these changes was made in the traditional Ferrari search for power and performance; even Ferraris must comply with international exhaust emission regulations and comply with these (in the USA anyway) over 50,000 miles. Both the ignition and injection changes enable Ferrari to meet these goals, the ignition because it has no points or centrifugal masses or springs, a selection of advance curves being stored and called upon electronically, the injection because it allows the adoption of efficient feedback control and catalysts for the American market.

But what has been the USA's gain has been Europe's loss. Because the quoted power outputs of the USA and Europe versions of the Mondial engine are so close (205 and 214 bhp), we can only assume that they are identical apart from the former's catalyst and air pumps. And "our" 214 bhp (at 6600 rpm) is a long way short of the carburetted GT4's 255 bhp (at 7,600 rpm). Still, it is a reasonable output for an engine of 2926 cc (unchanged, as are the bore and stroke of 81 and 71 mm, and the compression ratio of 8.8:1) and it may well be that in 1975 — when quoted figures weren't watched so closely by the authorities — that that 255 was a little more generous than it should have been. Maximum torque, too, has decreased from 209.8 lb ft at 5,000 rpm to 179 lb ft at 4,600 rpm.

With about 16 per cent more weight, more frontal area (wider and taller body, fatter tyres) and less power, there is obviously no way that a Mondial is going to match the superb performance of the GT4 — we called it "shatteringly quick" in our January 1975 road test: as that car had a maximum of 152 mph, a standing quarter mile of 14.7 sec and 0-60 mph and 0-100 mph times of 6.4 and 16.7 sec respectively this was not an overstatement. The Mondial is fast — particularly over winding country roads where full advantage can be taken over its simply stunning cornering and high speed stability — but in absolute terms in a straight line it must be regarded as a modest performer. The factory claims of a top speed of 143 mph and a standing quarter mile of 15 sec look considerably optimistic to us, our car's acceleration tailing off so markedly above 120 mph that we doubt whether it would exceed 135 mph. Certainly that's the figure that the American magazine Road and Track achieved when testing a US-spec Mondial; in turn, their measured 0-60 mph time of 9.4 sec and 0-100 mph time of 28.1 sec look way too slow — our car appeared capable of about 8.5 sec and 25 sec for the corresponding accelerations.

While the Mondial might not have the stunning punch of its predecessors, it delivers its power in a wholly new and untemperamental way. You at over 5,000 rpm) means that at 120 mph the engine's more insistent than it need be.

In our car, the overall gearing's shortcomings were exaggerated by a pessimistic speedometer (2.5 mph slow at 100 mph) and an optimistic rev counter (400 rpm fast at 100 mph/5000 rpm), two characteristics which didn't help to enhance the way the modest performance felt either. But there was nothing subjective about our full-tank-to-full-tank fuel consumption figure of 14.6 mpg, obtained over 470 miles of some hard to very hard driving. Perhaps that's not too surprising in view of the Government urban figure of 10.6 mpg but is hardly commendable when quite a few 140 mph coupes can approach 20 mpg in our hands nowadays. Perhaps it won't be of too much importance to the Ferrari customer though....

As always, the Mondial has the chromed, slotted external Ferrari gear-lever gate, with the top four ratios arranged in a normal H and first across to the left and back. Again typically, the change is awful when cold — second being well nigh unobtainable — but it loosens delightfully within a very few miles. Our Mondial's change was one of the best yet, being superbly fast and positive provided that the full (and rather long) travel of the clutch was used. There are few more pleasant experiences than zapping a Ferrari through the box, hearing the clonk of the lever as it hits the end of each gear slot and the surge of power as the clutch bites home. One quirk of our car — hopefully untypical — was that occasionally if you rushed it into reverse — on a three point turn, for instance — second was selected in error; perhaps the linkage needed adjustment.

Where the Mondial really shines is in its chassis. Like the GTB and the GT4 before it, the steering is initially unendearing, managing to feel low-geared yet excessively heavy at parking speeds. Through tight, low speed corners taken gently, the Mondial seems to understeer too, calling for a degree of wheel twirling that hardly augurs well for when you start trying. How misleading those initial impressions are! When you corner quickly, particularly on fast curves, when the car is developing some real weight transfer, the feeling of understeer all but disappears, the chassis becomes alive with feel — it almost seems as if the car can defy the laws of motion, so great is the lateral acceleration that can be developed. The revisions to the front suspension certainly have reduced steering kickback from the thumb-cracking level though there's still a little too much on really bumpy surfaces. But the kickback does help endow — and the TRX tyres too, no doubt — the Mondial with a feel of a quality (dare I say Porsche-like quality?) hitherto lacking in Ferraris. On a twisting country road, the steering writhes gently in your hands, letting there be no doubt of the state of the road under the front wheels.

Enter a corner too quickly and lifting the throttle produces a mild tightening of the line — enough to scrub speed off without requiring a specific steering correction — and even if you're forced to brake in mid-corner, hard, the Mondial slows without an excessive change of attitude. This stability is one of the Mondial's fortes and the Ferrari engineers deserve the greatest compliment for managing to blend such good high speed stability (even at 120 mph on a bumpy country road with the wheels pounding up and down like pistons the Mondial feels rock-solid on line) with a lack of understeer and neutrality in strong cornering.

The scale of the Maranello chassis team's achievement becomes even more evident when the excellent ride is taken into account, and the unusually good (for this class of car) suppression of road noise. The Mondial does feel firm and jiggly at low speed, though never uncomfortable as any vertical jarring has been cunningly removed by subtle tuning of the dampers. At speed, over all surfaces, the ride smooths out to become more than acceptable — on motorway and smooth A-roads it almost qualifies for the magic carpet class.

The only time where caution is needed is on slippery surfaces such as damp leaves (it didn't rain during our test), where excessive throttle can make the tail step out of line very smartly indeed. You have to be very quick and accurate applying opposite lock, though just the right amount of castor action is a considerable help.

Matching this superb road behaviour is a brake system that must be as good as that of any road car in the world today. Massive ventilated disc brakes larger at the rear (11.78 in) than at the front (11.0 in) in deference to the car's rearward weight bias and a vacuum servo provide a progressive and positive pedal action, whether the brakes are hot or cold. The handbrake, on the right of the driver's seat, is of the fly-off type and is sited too close to the stereo speaker in the door: we lost count of the times we barked our knuckles pulling it on.

With a steering wheel adjustable for both reach and tilt, the Mondial is one of the first Ferraris where even tall drivers can become perfectly comfortable. Ferraris are made for driving and the care with which the pedals and footrest have been arranged to make heel and toe changes second nature is obvious. A full three-stalk (Lancia-based) stalk system is offered, which is an improvement though we'd like a slightly longer indicator and a less stiff wiper stalk.

Only one extra, a UK-fitted sunroof, is available for the Mondial, a high level of equipment, being Ferrari's aim. So the car has electric windows, tinted glass, central locking, air conditioning and a Pioneer stereo radio/cassette player as standard. There's lots of deep-cut pile carpet and a veritable expanse of leather trim and upholstery, which certainly endows the car with a feel of luxury. We're generally not fans of leather seat coverings as they tend to let you slide around on corners, but this wasn't noticeable in the Mondial, partly due to seats with good lateral support, partly due to excellent seat belts that help hold you in place.

Incorporated in a rectangular nacelle directly in front of the driver is the instrument display containing all the normal circular dials, plus a separate odometer and a digital clock/stopwatch; the deep top shroud for this nacelle completely eliminates stray reflections but prevents tall drivers seeing the tops of the dials, the clock display or, most importantly, the indicator tell-tales. To the right of the speedo are touch-sensitive switches for the heated rear window, front/rear foglights and for the electric catches of the bonnet/engine/boot lids.

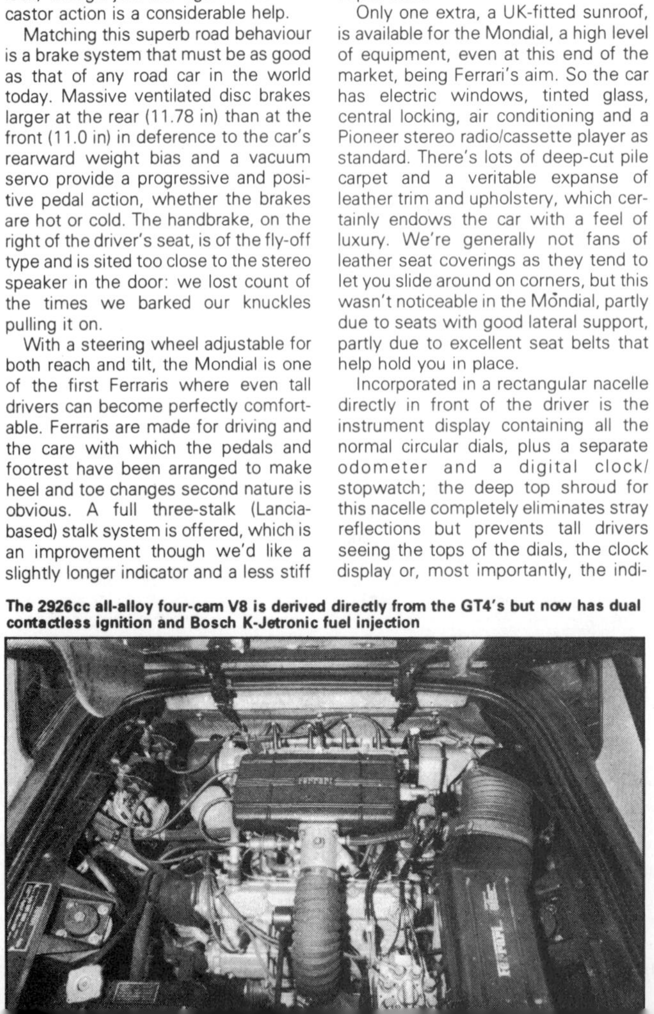

Incorporated in the centre of the facia (down low) is a row of three individually adjustable air vents together with a pair of face-level vents, one at each end of the facia, which, in conjunction with a powerful and controllable air conditioner, mean that the Mondial is by far the best ventilated Ferrari yet for hot weather. But in cooler times, the arrangement is less satisfactory, for the heater — though powerful — is slow to warm and it isn't possible to get cold air from the vents with hot air to the footwells; some, barely noticeable, temperature stratification does appear to take place when the heater is switched down low, but it's not enough. Moreover, the air temperature drops at engine idle.

Only the Mondial designer knows why it was necessary to place the heating and air conditioning controls so far down and away from the driver, and to site the radio so far back on the centre console that you can't see its face. The siting of the electronic check control display is also strange — it's even further back than the radio — but the system itself is admirable in theory, if over-complex in practice. Each time the ignition is switched on the system monitors levels of engine water, engine oil, transmission oil, screen-washer fluid and air conditioning fluid, as well as tracking down any failures in the electrical circuits of the brake warning lights (on the facia), the lights and the stop lights and informing you if any of the bonnets haven't been closed properly. There's even a "service due" light which comes on 3000 miles after the last check. At the end of the display there are three major warning lamps to indicate (and draw your attention to) a major failure, a minor failure or to give you the all systems "go".

Ask a Ferrari salesman who buys the Mondial and he'll tell you that it's the man who wants the traditional Ferrari virtues but can't live with just two seats. It has to look like a coupé, it has to be a sports car — but it has to have four seats, even if most of the time they are occupied by briefcases or coats rather than people. To these customers the Mondial must fit the bill well, being spacious, practical and luxurious — and beautiful. It should be faster and it must surely become more economical. But it must be icing on the cake for the Mondial owner to know that he has bought not just the most rational Ferrari design, but certainly the best handling one. And if that's a surprise to you, it certainly was to us.

Above left and right: luxury interior, as reflected in the large expanses of leather upholstery. Space is notably better than in the old GT4. Below: full, but shrouded instrumentation and the radio and check control (left) are too far back

The 2926cc all-alloy four-cam V8 is derived directly from the GT4's but now has dual contactless ignition and Bosch K-Jetronic fuel injection

For Ferrari owners who believe the 308GTB is underpowered, Britain's Maranello Concessionaires team with Janspeed of Salisbury to come up with a potent turbocharging package. Steve Cropley reports

Ferrari with fight

SACRILEGE AND DESECRATION! How else would you describe the act of sticking proprietary hot bits onto a Ferrari? We make no bones about it; when we heard that Britain's Ferrari importers, Maranello Concessionaires, were at the centre of a move to fit a turbocharger kit to a 308GTB, our scepticism rose high.

A Ferrari is one of the most completely developed and lovingly crafted of all cars. There was no reason, we felt, to think that its delicate fabric would be improved by the same single Garrett TO4 turbocharger that is used so commonly in kits to make Cortinas and Avengers go a bit faster. Rather, the reverse was likely. The traditional turbo problems of coping with massive engine bay heat, keeping the engine cool, lubricating the sorely-tried turbocharger bearing and handling extra internal stresses in the engine, would overtake any benefit of top end power. Ferrari V8s already have top-end power: the whole of the Turbo version would be far less than a sum of its parts, we doggedly believed.

To be fair, there were some tempering factors. First, the job was being handled by Janspeed Engineering, Salisbury-based turbocharging experts whose reputation is impressive. Second, while in Modena recently we learned of Ferrari's own work on turbocharged V8s. Theirs are to be twin turbocharger systems but we couldn't help feeling that the single blower, low-boost, British forerunner (wastegated at 5.25lb/in^2) would show how factory Turbos might feel and sound. And besides, the factory's interest in turbocharging indicated that the 308's expensive internals – transmission, engine bottom end, axles and clutch – could all stand some extra power.

And so it proved on test. We drove the Janspeeded car for several days, at times right to the limit of its performance, but nothing broke or showed a sign of undue strain. In our admittedly-short 700mile test, the Turbo was utterly reliable. Perhaps even more importantly, it was *fast!* If you were to match this car against the same Lotus Turbo Esprit which decisively outran the latest 308 version, the fuel injected GTBi (*Prize Fight, CAR* June) in an old-fashioned tyre shredding drag race, the tussle would be close and momentous.

According to our watches, the cars would be locked together as they rocketed through 100mph and would fight wheel-to-wheel right up to their top speeds, both decently beyond 150mph.

Of course, it remains a tribute to the super-efficiency of the Lotus Turbo Esprit that it takes a 3.0litre Ferrari V6 with turbocharger even to *match* its 2.2litre, four-cylinder with blower – yet the message to the Ferrari faithful is clear. This 308GTB Turbo of Maranello Concessionaires is the car to put a determined driver right back into the game. Depending on how well he copes with the heavy clutch (lightened in the 308GTBi) he can make the Hethel car's driver work very hard – something the Lotus's breathtaking ability usually spares him.

Janspeed have done their work on a one-year-old carburettor 308GTB, developing it over several months so that its reliability matches its speed. And quite a lot of the work has entailed finding a specification that can be repeated in customers' cars without production problems.

The 308 Turbo uses a single Garrett TO4 turbocharger that sits high behind the transversely-mounted engine on specially-made exhaust manifolding which feeds the front cylinder bank exhaust gas in low down, out of a casual observer's sight. Induction air is drawn in through a filtered box on the right rear of the engine bay, put through the turbocharger and then into an induction pipe which curls upward and to the right over the rear cylinder bank to a big flat rectangular chamber above the four twin-throat downdraught Weber carburettors. These are in their usual position on a standard inlet manifold. This Ferrari's engine forfeits its intricately-crafted stainless steel exhaust pipe and silencer system in favour of one straight-through box mounted across the rear under-body, with one no-nonsense outlet.

The standard compression ratio of 8.8 to one is maintained, the standard cylinder head gaskets are used and there are no modifications to the engine's ignition or internals at all. According to Janspeed, they just don't need it.

The project's principal engineer, Howard Askey, argued immediately with our opening assertion that this was rather a cursory turbocharging job, done to justify the badge on the 308's rear body (it comes from a Renault 18 Turbo) rather than to improve the car. He made it clear that the installation is effective and unusually sophisticated.

The maintenance of the standard compression ratio allows the engine to retain its 'sharpness' at low revs. It does not, in the usual manner of those with lowered compression, lack sparkle off boost. Pre-ignition problems, often threatened by the combination of turbo-heated air and a high compression ratio, are countered with a unique form of inter-cooling, which has the more often mentioned bonus of making denser the air about to be inducted. The flat, rectangular box over the eight carburettor throats contains a radiator matrix through which the freon of the standard air conditioning system is circulated. Freon circulation starts when a pressure switch senses turbocharger boost and pulls in a magnetic clutch on the air conditioning pump. Both Ferrari and Janspeed say the cabin air conditioning is little affected by the antics of the intercooler, and our test backed up the claim. Cabin cooling was rather erratic, but always there.

Though the 308 Turbo's carburettors sit in their usual place, theyre not of standard specification. It is a characteristic of blow-through turbocharging layouts that they tend to 'wear out' their carburettors by causing leaks through the holes for the carburettor spindles. The fuel-air mixture removes lubricant from the spindles and the resulting metal-on-metal contact wears and loosens them in their mountings. A crude turbo installation of this kind can 'chew up' carburettors in 10,000miles, according to a concensus of turbocharging experts we contacted.

Howard Askey used special versions of the usual Weber 40DCNF carburettors, with seals fitted to the spindles to beat the problem. The carbs, designated WX, were specially made and imported for this Ferrari. Lotus, incidentally, also use carburettors with spindle seals on their Turbo Esprits.

Janspeed considered a twin turbocharger installation for the 308, after the style of those under development in Maranello (a 2.0litre V8 Turbo will be released in Italy next year, the 3.0litre is merely 'planned'). Howard Askey says such an installation would have been neater from a 'plumbing' point of view and there was room in the engine bay, low down ahead of the block, for a second blower. But cost and the unavailability of a suitable smaller-capacity turbocharger (such as the Garrett TO3 which has more recently become available) put paid to the idea.

Besides, even more elaborate lubrication arrangements would have been needed for another blower. Janspeed lubricate their single Garrett TO4 by providing it with its own pump (a BL A Series unit driven from the end of the rear bank's inlet camshaft), a sump, a cooling radiator, a pressure light and oil temperature gauge. First quality Aeroquip hose carries the oil. Although it sounds complicated, it is impressively neat and functional. The boost governed at 5.25psi by a wastegate on the induction pipe. Janspeed tried higher boost but found that it brought pre-ignition problems for a surprisingly small performance improvement.

What, then about the performance? The engine starts in a similar manner to the fuel-injected 308 – no choke, no pumping of the accelerator. On test, we often found the car hard to start from cold; it would fire instantly, but it wouldn't catch. If you use choke, you flood the carburettors. Even when warm, the engine won't burst into life with the exultancy of an all-Ferrari 308. But once it's running evenly, it is very, very smooth and there are none of the hiccups of the usual cold engine. It's silken.

And it's powerful. There is such push from the engine – realistically speaking, probably 60-70bhp stronger than that of a 308GTBi, that the driver is hard-worked just to manipulate the clutch and gearchange fast enough. The injected car has a longer, more easily wielded gearlever and a lighter clutch. A 308 Turbo driver driver could often do with both.

The 308 Turbo, even with its Michelin XWX tyres on optional 7.5in wide wheels (the standard wheels are 6.5in) will spin its wheels with determination if the clutch is popped in the turbocharger's operating band. Then it arrows down the road, front wheels light and nose high, in comparative silence but propelled by power enough to match a Turbo Esprit. In the longer gears, at full throttle, the boost gauge starts to register positively at about 2900rpm. It provides its maximum boost of 5.25psi at about 4400rpm. From there up – a further band of 3300rpm – the power is massive, smooth, effortless, long-legged and quiet in a way no 308 has been before.

We never reached this Turbo's maximum speed. We merely discovered that at a genuine 150 there was more to come, so much more that we felt that the engine might even spin right to its 7700rpm redline and 165mph! But all this performance comes at the expense of high fuel consumption. We ran several checks, some when using the potential and others when treading carefully. Our best figure, over 170miles on open roads, was 14.1mpg. This fell to only 11.8 when the turbocharger was regularly doing its work. That's no better than a Boxer returns and certainly somewhat worse than either an injected or carburettor 308 would return. It's about double a Lotus Turbo Esprit's fuel consumption.

There are no dyno figures for the turbocharged engine yet, but an extra 40bhp is claimed from it. We've already asserted, in other stories, that the 255bhp claimed for the carburettor 3.0litre V8 seems too much; that 230 is nearer the mark. If that is so, the Turbo is probably producing 260-270bhp which is only 70bhp less than the Boxer – much bigger, heavier and probably less aerodynamic – puts under its driver's right foot. If you pay any attention to power-to-weight ratios, the 308 Turbo's 11.1lb/bhp is now closer to the Boxer's 9.8 than to the Lotus Turbo Esprit's 12.8. No wonder it's fast.

It might seem – it seemed to us at first – that to give a proven 150mph car a lot more potential is a patent waste of fuel and enterprise. Yet the extra power can be used in a surprising number of places. You use it uphill for consuming lines of crawling cars (where you can't *have* too much power). You use it to boot the car out of bends on the limit of its rear adhesion (not nearly so easily reached with the standard power). And the extra potential suits the more relaxed, high gearing of the carburettor 308s; it was lowered from 21.7mph/1000rpm in top to 19.3mph for the 308GTBi to compensate for the GTBi's lower, 210bhp power output.

So great is the extra cornering kick out of bends that Maranello Concessionaires have found it advisable to fit stiffer shock absorbers to this pilot car. It does stay rigidly, ruthlessly flat, now, even in S-bends which might otherwise set up a modest body lurch. The ride is still good; flat and completely lacking in untoward bounce, but there is more bump-thump than we remember of recent 308s, part of it doubtlessly due to this firmer damping and part to the inferior bump absorption properties of the older XWX tyres. The new-fit GTBi TRXs are in another league.

In fact, this older GTB's chassis served to show up starkly the extent of the improvements in 3.0litre Ferraris of current manufacture. Its steering, very heavy at parking speeds, felt quite dead at times, though it was always precise. The XWXs just didn't have the low-speed pliability of the new low-profile Michelin TRXs which 308GTBi and Mondial now wear. At times the XWXs are very noisy and feel quite clumsy. The clutch is 50percent too heavy, too.

So far, you have heard only good things about this Ferrari's turbocharger conversion, but there are some bad ones. First, it is no longer a *complete* Ferrari. If

Engine is impressively modified by add-on turbo experts Janspeed to gain an extra 30 to 40bhp. Garrett blower (alongside engine stay, lower centre of picture) has its own oil cooler and main bearing lubrication. Compressed air is pushed into airbox covering the four Webers. **Roadholding** and **handling** is stronger and better than ever thanks to firmer dampers and wider wheels/tyres

part of the appeal of a 308 is that an overwhelming majority of its components were specifically designed and crafted at Maranello only for *that* car, then you must regard the 308 Turbo as being smaller in stature for sporting after-market hot bits of little pedigree, whatever their efficiency. Second, with the Turbo conversion, the car's appearance and sound suffers.

The engine's low-speed rasp and its howl-wail as the revs rise have been replaced by a typical turbo'd mixture of whistles and hissing of compressed air, together with a reduced, smoothed exhaust note. 'It's only a vacuum cleaner car, now', said one kerbside critic. In a way, he was right. This Ferrari sometimes sounded like a Renault 18 Turbo. Not only that, the GTB tail is the poorer for its new, ugly silencer and some undisguised, rather rusty piping. Perhaps these things, along with the 'Turbo' badge so recognisably Renault, do not deserve harsh criticism on the first car, but out on the road they do underline the fact that

122

ere is something amiss with the ar's pedigree. On the other hand, e differences draw onlookers nd, 'experts' like little else.

There are some other echanical shortcomings. First is e unhappy cold starting. If you joice in the way fit Ferraris abitually burst into life as if ey'd been waiting for you to ask, e catch-and-die performance of e test car is disappointing. The ar also has a tendency to get hot nder its engine cover in traffic hough there is never a problem ith high radiator temperatures) nd for the car to stall when quired to idle. This is particularly ad when it stands for a few inutes after a hard run; it just on't keep going – or blip cleanly the throttle – when you require

The cabin heat-soak seemed eater than standard to each of r testers though this may have een because we drove the Turbo hot weather, and because the st Ferrari we drove was a ondial whose ventilation is stly better than other models.

The question of the car's absolute reliability remains unanswered – and with it, the question of its warranty. There are quantities of heat in that engine bay with which it was never designed to cope. Janspeed have fitted an asbestos shield between turbocharger and engine cover to protect the panel and its paint, but there are times when the heat haze actually disturbs rear vision. On test, the paint on the asbestos cover quickly blistered on the turbocharger side. Those things make us anxious. Janspeed guarantee the reliability of the components they fit, as they do their own workmanship, but you couldn't blame them for not wanting to know about engine damage to a secondhand Ferrari which might or might not have been caused by its 20percent power transfusion. The same attitude is likely to be displayed by dealers in year-old carburettor 308s.

Interestingly, Howard Askey sees 'no great problem' in turbocharging a current, fuel injected Ferrari 308GTBi. Janspeed have done Bosch K-Jetronic cars quite often before. Maranello's men say that if a buyer wanted to turbocharge his new 308GTBi, they'd be happy for him to have it. We formed the impression that the warranty details would be a matter for negotiation, though the cover on the parts not directly affected by a turbocharger ought to be clear.

Janspeed's price for their job is 'from £4000' which means that clients who want refinements like a turbocharger bearing oil supply linked to the engine's own – or safety-first cylinder head gasket 'tweaks' – might pay as much as £4500. That premium on top of the already-hefty Ferrari price means they'd be paying a great deal more than the price of the £17,000 basic Esprit Turbo to match the Hethel car's straight-line performance (though as we've said before, there are uniquely intangible features of a Ferrari – heritage, 'feel', sound – which go towards compensating for any price disadvantage). But the plain fact is that the Ferrari after-market turbocharging job is not nearly as well integrated into the overall design and performance. There's no way it could be, given that Janspeed have had the hardest of raw material to work with – a fully developed, built and working car.

Maranello Concessionaires must have known full well that the Turbo would appeal only to a minority of owners and new buyers. For this reason, and because of the project's potential to generate purist criticism, they were brave to begin it. In a sense they have achieved what they sought: there is now a Ferrari 308 which will match the Lotus – on sheer performance, king of the mid-size exotics – in a sprint and then shade it for top speed.

But desirability? That's another question. It's quite easy for us to decide whether we'd prefer a Ferrari Turbo or a standard 308GTB with £4000 still in the bank – we'd take the money. But apt or not, the conversion makes it clear that there will be some shattering cars rolling out of the green gates of Maranello inside the next couple of years.

123

SILVER DREAM MACHINE

With four camshafts, four valves per cylinder (hence the "Quattrovalvole" tag), and fuel injection, the Ferrari Mondial now produces a claimed 240 bhp. It is a beautifully made and practical 2-plus-2, but it is as quick as you'd expect for a price of nearly £26,000? Jeremy Sinek investigates

Photographs by Peter Burn

Far left, top: the Mondial is surprisingly commodious, with a good driving position, though space in the rear is limited. Far left, below: access to the engine is better than in most mid-engined cars. Left: all the lids are opened by buttons on the facia, activated only with the ignition on

Ferrari 308 GTB

Silver Dream Machine
continued from colour

FERRARI ARE fighting back. After a debilitating few years when ever-stiffening emissions regulations sapped the strength of Maranello's three-litre V8 screamer, the progress chart has turned upwards again. A magic ingredient code-named *Quattrovalvole* has effected a miracle cure, and Ferrari's mid-engined trio are back running strongly in the supercar league where they belong.

Ferrari fans could be excused if for a time their loyalty *had* wavered. A Ferrari without real performance is about as unthinkable to followers of the prancing horse as alcohol-free lager is to a real ale connoisseur, and it was following the launch of the Mondial 8 in 1980 — a 2 + 2 replacement for the unlovely yet popular 308 GT4 — that their loyalty was put to the toughest test. The Mondial's debut coincided with developments to the classic four-cam V8 that may well have cleaned up its exhaust, but which also slashed its power from a one-time 255 to a comparatively puny 214 brake horse power. The adoption of Bosch K-Jetronic injection and Marelli Digiplex fully electronic ignition did wonders for the engine's tractability and social acceptability, but softer cam timing knocked hell out of the engine's top end bite, and to compound the felony the Mondial was substantially heavier and bulkier than the car it replaced. The result was a supercar hard pressed to exceed 135 mph, and which could be matched on sprinting power by quite a few full sized luxury saloons. That the new car had a superbly well sorted chassis served merely to sharpen the awareness of its comparative lack of performance. And although when fitted with the same engine, the lighter 308s GTBi and GTSi were quicker than the Mondial, they were still only shadows of their former selves.

Appropriately, it was the Mondial that benefited first from the development that has now remedied the situation, making its appearance this summer with a discreet *Quattrovalvole* badge on its tail denoting a new four-valves per cylinder layout which, according to Ferrari, owed much to the company's Formula 1 programme. In conjunction with a compression ratio increase from 8.8 to 9.2:1, power was boosted from 214 bhp (DIN) at 6,600 rpm to 240 bhp at 7,000 rpm, and torque from 179 lb ft at 4,600 rpm to 192 lb ft at 5,000 rpm — not, on paper, quite back to square one but, as we shall see, it seems a fair bet that 1982's horses are bigger than those of the mid-seventies when trades descriptions legislation wasn't as inhibiting as it is today.

Otherwise the Mondial is unchanged from the car we reported on in December last year. The 2,927cc engine is an all-alloy 90 deg V8 and is installed transversely, just ahead of the rear wheels which it drives via a five-speed transmission and limited slip differential. The suspension is by a classic double-wishbone layout front and rear with coil springs and Koni dampers, an anti-roll bar at each end, and anti-dive geometry at the front. The steering is by rack and pinion, and there are massive ventilated disc brakes at both ends. All this is installed on a tubular steel chassis frame, which is in turn clothed in a mostly steel body (the boot lids and engine cover are of aluminium), styled by Pininfarina and built by Scaglietti. For the first time in a Ferrari, the power train and rear suspension are all assembled on a separate, detachable subframe.

Like most supercars, the Mondial is a deceptively large machine — one inch longer than an Austin Ambassador, for example, and a full two inches wider. It's a heavy car, too: at 28.5 cwt, it tips the scales at about 150 lbs more than a 2.8 litre Ford Granada. In that context, an engine displacing just over 2.9 litres doesn't look over-generous when you're aiming at performance in the supercar league, and to achieve it, something has to give. In the Mondial, the price Ferrari's engineers have elected to pay is refinement and economy on the motorway. At just 19.7 mph per 1,000 rpm in top the Mondial is by modern standards absurdly low geared — most rivals have a longer stride in fourth than the Mondial does in fifth. In an era when many car manufacturers are relying on high torque, low weight and minimal aerodynamic drag to be able to pull ultra long-legged gearing, the Ferrari's rev-forever engine and sprint gearing are like a throwback to another era.

The fact is, however, that it works. On a blustery day with the barometer hanging low we lapped Vauxhall's Millbrook high speed bowl at 146.1 mph, making the manufacturer's claim of 149 mph seem if anything on the conservative side given a good day and a long straight where there's no power absorbed in cornering forces. Our car felt capable of pulling the red line in fifth, which corresponds to 151.3 mph.

Conditions on the acceleration straights were far from ideal too, and the need for a second-to-third gearchange at 59 mph wasn't going to do the 0-60 mph time any favours. Even so, the times are impressive: 0-60 mph in 6.4 sec and 0-100 mph in 16.2 sec are virtually identical to those of the lighter 308GT4, and firmly re-establish the Mondial in the supercar league. So too do such fourth gear pick-up times as 30-50 mph in 5.6 sec, 50-70 mph in 5.2, and 70-90 mph in 5.7 sec. Only when left to accelerate in top gear does the Mondial lag behind the standards of its class, though even then the corresponding times for the same increments — 8.5, 8.6 and 9.3 sec — say that it's far from sluggish.

What the bald figures cannot convey is the *way* this performance is delivered. The Ferrari formula of a relatively small, high-revving screamer of an engine may be an anachronism, but what a magnificent one. The breadth of its power band is simply breathtaking. In town you can stuff it in top and the engine will accelerate whistle-clean from below 1,000 rpm. In your moments of forgetfullness, it'll pull away from rest in third gear. It's as tractable as a steam engine when conditions demand it.

Yet show it an open road and it's transformed. From 5,000 rpm, when the power comes in hard and the engine's note becomes an exultant howl, it's a classic thoroughbred of the kind only the Italians know how to build. The tachometer red line is at 7,700 rpm and the engine will spin to it so eagerly, so utterly smoothly, that it's just as well that Ferrari also fit an electronic rev limiter that cuts in at 7,800 rpm . . .

Needless to say a Mondial engine when given its head is far from quiet — who in their right minds would want it to be? But driven with moderation — kept below about 3,500 rpm, say — the engine is nicely muted. And at higher speeds its smoothness, its lack of strain, does much to mitigate the effect of low gearing. Even so, 100 mph corresponds to a very busy 5,100 rpm in top gear, and an effortless stride on the autobahn is what you sacrifice to achieve stout-hearted performance from a small engine.

The price in economy, however, is not as great as you'd expect. Admittedly the 14.8 mpg returned by the Mondial over its first few hundred miles with us is poor, but we suspect

127

that at that stage the engine was suffering from a mystery power loss that was also affecting its performance, if our early subjective impressions are anything to go by. By the time we took our figures, however, this had evidently cleared itself, and it's significant that over the second part of our test, which included performance testing, the Mondial returned a very respectable 18.6 mpg. Achieved without the benefits of low weight and high gearing, this speaks volumes for the engine's inherent efficiency. Just think what might be possible in a car 20 per cent lighter and correspondingly higher geared . . . Nonetheless, it's a major step in the right direction, and suggests that more than 20 mpg should be attainable given only a modicum of moderation.

You could also say that moderation is a part of the Mondial's appeal in a broader sense. As a Ferrari should, the Mondial has the power to thrill in the short term; but it's also, to a greater degree than any previous Ferrari, a *sensible* car — an undemanding, even gentle car to drive, and an easy one to live with in the long-term. For a start there's that flexible and utterly untemperamental engine. Then there's the clutch, which is moderately weighted pendulous oversteer that affects some supercars if you inadvertently enter a corner too fast and react by lifting sharply off the throttle.

For the serious driver, the Mondial is at its least endearing in the wet, when the grip of its 240/55 VR 390 Michelin TRXs is modest, and the consequent lack of loading on the suspension leaves the steering short on feel just when you need it the most.

On a dry road it all comes together. As you start to explore the high outer limits of its roadholding and get some load onto the suspension, the steering gains in both weight and feel. There is scarcely any body lean, and the chassis is marvelously well balanced. Mild initial understeer can be balanced out on the throttle to give a virtually neutral attitude, or even, through tighter curves, it can be powered out into an easily held tail slide. Lifting off or braking deep into a bend tightens the car's line without upsetting its poise or its balance and the suspension takes mid-corner bumps in its stride, though steering kickback can become tiresome on lumpy secondary roads and makes it tricky to hold the car accurately on line at speed.

One of our testers also found the Mondial a handful under hard braking in

Above: redesigned central console houses the stereo, electronic check panel, and air conditioning controls, and electric controls for windows, glove box, aerial fuel filler. Rear seats, right, are comfortable but short on legroom

and as gentle in its engagement as any small saloon's. And its steering is easy on the arms.

Only the Ferrari-typical metal gated gearchange still demands conscious inputs from the driver. Almost impossibly obstructive when cold — especially the dog-leg shift from first to second — it retains a notchy, heavy-metal quality when warm that won't tolerate limp-wristed treatment. Handle it with the appropriate authority, however, and it rewards you with tremendously fast and positive upshifts. There's something enormously satisfying about the metal on metal clickety-clack — like a train's wheels passing over points — as you slam the lever through from slot to slot.

Conversely, while the Mondial's handling also responds satisfyingly to firm and skilled treatment, it is also forgiving and manageable for the less skilled or the lazy driver. The steering is beautifully precise, and if it doesn't transmit a great deal of feel in gentler motoring it does compensate with its direct gearing and moderate weighting. There is little sign of the heavy, understeery cumbersomeness that afflicts many mid-engined cars through tight, low speed corners. Equally, there's little likelihood of provoking the the wet, complaining that it locked up its front wheels prematurely and then twitched sideways. Perhaps it was a freakishly greasy surface, though, since we were unable to reproduce the problem at the test track even under deliberate provocation. In the dry the brakes are simply fabulous, delivering immensely powerful and fade-free retardation through a firm and progressive pedal action.

As a driving machine, then, the Mondial is rarely found wanting. Just as important, however, is that with the Mondial Ferrari have achieved a measure of practicality and useability that makes it as pleasant to drive in the daily urban grind as it is thrilling to drive on a fast and winding country road.

Although always firm, the suspension's efficient damping and the low levels — by exoticar standards — of tyre rumble and thump ensure that the ride is never harsh or uncomfortable, while at motorway speeds it smooths out beautifully. With a steering wheel adjustable for both reach and height, the driving position can be tailored to suit most sizes and shapes of driver — even our resident 6ft 4in beanpole could find little to complain about. Although offset to the left, the pedals are well sited in relation to each other — heel and toe gearchanges come easily — with plenty of space around them and a rest for your left foot. Sensible shaping of the seats holds you firmly in place despite their slippery leather upholstery. The minor switchgear is logical and conveniently located, and includes touch-sensitive electric switches to open the bonnet, engine and boot lids. The instrumentation is comprehensive and clearly marked, though the individual glass covers do catch stray reflections and our tallest tester found that part of the tachometer was shrouded from view.

For a mid-engined car the Mondial is exceptionally easy to see out of. With plenty of glass there are no serious blind spots, the wipers and lights are excellent, and the two electrically adjustable door mirrors give a panoramic view to the rear. Only in tight spaces, when you cannot easily position the nose of the car, is special care required.

As a two-seater the Mondial is very spacious, with ample front legroom and headroom, a fair sized boot, and the option of carrying further luggage on the rear seats. Oddments space is not over-generous however (though the provision of a small cassette rack is a nice touch) and for longer journeys the rear seats are only suitable for children. It *is* possible to accommodate a medium sized adult behind a similarly proportioned driver, but only if he is prepared to splay his knees either side of the front seat backrest. Access to the rear is facilitated by a tilt and slide mechanism for the front seat.

Perhaps it would be too much to ask any Italian car to have an efficient and versatile heating and ventilation system. The Mondial has air conditioning as standard, which is fine when the weather outside is hot enough to warrant it. Like most such systems, however, it is unable to supply stratified air — cool above and warm below — and is only designed to maintain a steady overall temperature. In practice, even that function proved beyond the capabilities of the Mondial's system: its output varied between too hot and too cold in the cool autumnal weather of our test, and it proved almost impossible to achieve a happy medium.

Perhaps the system was in need of adjustment. If so, it was the only thing we could find wrong with the test car's finish. We couldn't fault the fit or quality of the interior trim, the external paintwork was superb, and the car structure was solid and squeak-free.

That much, of course, is only as it

Instrumentation is comprehensive and clearly marked. Touch-switches to the right electrically open the engine cover and boot lids. Chrome lever on steering column is for wheel adjustment

should be when you are paying £25,851.20 for a motor car. Yet the plain fact is that too many high-priced exotica are fragile, temperamental, scrappily finished, difficult to drive in traffic, hard to see out of, cramped and impractical. Fine for short-term thrills, a pain in the proverbial in the long run.

Not so the Mondial Quattrovalvole. When you're in the mood, it's as exciting to drive as any Ferrari should be — as Ferraris always have been. The difference with *this* Ferrari, though, is that if your mood changes, if traffic or road conditions force a change of tempo, the Mondial is still a friendly car in which to drive and to ride. When your weekend in the country is over and you're back in the workaday routine, that matters.

PERFORMANCE

MAX SPEEDS

	Ferrari Mondial 2.9	Porsche 928S 4.7	Maserati Merak 3.0
	mph	mph	mph
Mean in top	146.1*	155e	150e
in 4th	112	119	124
in 3rd	82	91	88
in 2nd	59	65	60
in 1st	41	44	40

* See text

ACCELERATION

mph	sec	sec	sec
0-30	2.2	2.4	3.1
0-40	3.3	3.3	4.3
0-50	4.7	4.9	6.1
0-60	6.4	6.2	7.8
0-70	8.2	7.8	10.3
0-80	10.2	9.7	12.4
0-90	13.0	12.0	16.0
0-100	16.2	14.8	19.8
0-110	20.7	17.9	24.9
0-120	—	22.1	32.9
St'g ¼ mile	14.5	14.2	16.0
St'g km	26.7	25.8	28.9

IN TOP

mph	sec	sec	sec
20-40	9.3	7.6	—
30-50	8.5	7.7	9.5
40-60	8.6	7.6	10.7
50-70	8.6	7.3	10.8
60-80	8.7	7.6	10.4
70-90	9.3	8.4	11.0
80-100	10.6	8.7	11.8
90-100	12.5	10.2	13.6
100-110	—	10.6	—

IN FOURTH

mph	sec	sec	sec
20-40	5.9	5.2	7.0
30-50	5.6	5.2	7.2
40-60	5.5	5.3	7.3
50-70	5.2	5.2	7.1
60-80	5.4	5.0	7.2
70-90	5.7	4.9	6.7
80-100	6.1	5.1	7.3
90-110	—	5.7	9.0
100-120	—	7.2	—

FUEL CONSUMPTION

Overall	18.6	16.0	13.3

GENERAL SPECIFICATION

ENGINE
Cylinders — Vee 8
Capacity — 2,927cc (178.5 cu in)
Bore/stroke — 81/71mm (3.19/2.80 in)
Cooling — Water
Block — Aluminium alloy
Head — Aluminium alloy
Valves — Dohc, 4 per cylinder
Cam drive — Toothed belts
Compression — 9.2:1
Fuel system — Bosch K-Jetronic injection
Bearings — 5 main
Max power — 240 bhp (DIN) at 7,000 rpm
Max torque — 192 lb ft (DIN) at 5,000 rpm

TRANSMISSION
Type — 5-speed manual
Clutch dia — N/A
Actuation — Hydraulic
Internal ratios and mph/1000 rpm
Top — 0.920:1/19.7
4th — 1.244:1/14.5
3rd — 1.693:1/10.7
2nd — 2.353:1/7.7
1st — 3.419:1/5.3
Rev — 3.248:1
Final drive — 4.063:1

BODY CHASSIS
Construction — Tubular steel frame with steel, aluminium and glass-fibre body panels

SUSPENSION
Front — Independent by double wishbones and coil springs; anti-roll bar
Rear — Independent by double wishbones and coil springs; anti-roll bar

STEERING
Type — Rack and pinion
Assistance — No

BRAKES
Front — Ventilated discs, 11.0 in dia
Rear — Ventilated discs, 11.8 in dia
Park — On rear
Servo — Yes
Circuit — Split front/rear
Rear valve — No
Adjustment — Automatic

WHEELS/TYRES
Type — Alloy, 180 TR 390
Tyres — Michelin TRX, 240/55 VR 390
Pressures — 33/35 psi F/R

ELECTRICAL
Battery — 12V, 66 Ah
Earth — Negative
Generator — Alternator, 80 Amp
Fuses — 23
Headlights
 type — Rectangular retractable
 dip — 110W total
 main — 220W total

FERRARI 308GTSi
A thing of beauty is a Ferrari forever

IT'S CERTAINLY NOT the latest thing anymore, this open mid-engine Ferrari. The 308GTB coupe from which it sprang first appeared in 1975, followed by the open GTS in 1977. But it continues to be one of the best-looking, most exciting cars available today—simply because Pininfarina gave it timeless lines that say "speed" and "elegance" with every swing and curve of its sheet metal.

Ferrari still builds 12-cylinder cars, but for some years now the 308 GTs and Mondial 8, all V-8s, have been the only models officially exported to North America. The 308 pair take the lion's share of Ferrari production at Maranello, accounting for about 60 percent of the 2565 Ferraris built in 1981. After the current supply of coupes is exhausted, Ferrari North America will import only the roadster, with the expectation of selling around 600 of them in 1983. The Mondial 8 will be the other Ferrari model for America in 1983.

The 308GTSi, as it has been called since Bosch K-Jetronic fuel injection replaced four Weber carburetors in 1980, is largely handbuilt at the Ferrari-owned Scaglietti works in Modena. Its all-steel body is carried by a tubular steel frame—a traditional Ferrari construction technique that results in greater weight than if the car had a modern unit body designed with the aid of computers. To its credit, however, it is very well put together: rigid for an open car, and with minor exceptions extremely well finished. Its panel finish and paint quality deserve special mention, being fully in line with the $60,000 price.

Generally, the same can be said of its cockpit. The leather seats are beautifully made and well shaped for holding driver and passenger in place even in spirited driving; the door panels are handsomely designed and nicely finished. Only an ill-fitting right-hand carpet detracted from the air of quality materials well assembled.

Ergonomics is another matter. One doesn't expect a low-built, mid-engine car to be especially comfortable, and the

AT A GLANCE	Ferrari 308GTSi	Lamborghini Jalpa	Porsche 911SC
Curb weight, lb	3250	3305	2805
Engine	V-8	V-8	flat-6
Transmission	5-sp M	5-sp M[1]	5-sp M
0–60 mph, sec	7.9	7.3[1]	6.7
Standing ¼ mi, sec	16.1	15.4[1]	15.3
Speed at end of ¼ mi, mph	88.0	92.0[1]	91.0
Stopping distance from 60 mph, ft	154	144	140
Interior noise at 50 mph, dBA	76	80	72
Lateral acceleration, g	0.810	0.846	0.798
Slalom speed, mph	60.6	56.6	59.7
Fuel economy, mpg	11.5	est 12.0	18.5

[1] European version

308GTSi isn't. Its seats have a sharply limited longitudinal adjustment range, so long-legged drivers (and maybe even passengers) may not fit, as was the case with one six-feet-fiver who tried to find a driving position in the 308. The Clarion Spec II electronic push-button stereo system Ferrari was promoting as a dealer-installed option at the time of our test is impressive-looking, but you need a magnifying glass to read its controls and its reception and tone are mediocre.

Even in mild weather the cockpit gets hot, partly because of the large, rakish windshield and partly because the engine is very close by. The Ferrari's rudimentary ventilation system cannot begin to cope with the heat, and the equally primitive air conditioning is hard pressed to keep things comfortable on an 85-degree day.

Which brings us to the solution: the removable roof. It fits right, keeping wind noise to a minimum when in place. But it can also be removed very easily, merely by loosening two fasteners at its rear edge, and stored neatly in a vinyl envelope behind the seats. Even with it off, wind noise is moderate as long as the windows are rolled up—right up to speeds well over 100 mph. So the 308GTSi is an open car that probably will be driven open most of the time, as we did while we had it.

A long journey with the roof off, in fact, seems a wonderful thing to do with such a car. But if the Ferrari's owner and a companion decided to undertake same, they might well decide to send their luggage on ahead. The compartment behind the engine encloses a mere 5.3 cubic feet of space and is kept very hot by the engine.

With the advent of fuel injection, the 3.0-liter, 4-cam Ferrari V-8 lost some of its performance and nearly all its exotic sound while gaining improved tractability. It's anything but quiet; nor is it particularly smooth for a V-8, feeling as if it's bolted directly to the frame without any rubber mounts (it isn't). One disconcerting characteristic is that upon being started from cold, the engine speeds up to 2800 rpm—an emission-control measure—where it idles for about the first minute of operation. Ferrari assures us that this will do no harm, but one cannot help but think of dry bearings at 2800 rpm . . .

The Ferrari's performance is clearly more than adequate for American driving, but its acceleration (0–60 mph in 7.9 seconds, 0–100 in 22.1) can be matched or approached by quite a few much less exotic cars. Despite the V-8's willingness to run at low revs, most drivers will feel compelled to use the gears to keep the revs up somewhere around the 5000-rpm torque peak. Then, and only then, does the 308 show much élan. From there up to its 7700-rpm redline, it pulls strongly and freely, as you expect a Ferrari to do.

Using the gears is both pleasurable and annoying. On the one hand, the driver gets the satisfying feeling of manipulating a lot of good machinery in shifting the visually gated lever, once he or she has mastered it. On the other, it takes some practice to master it as well as some patience even after mastering it. For this is not the slick, precise sort of shifting one gets from the best transmissions in front-engine cars, whether front- or rear-drive. Linkage to the transverse mid-engine powertrain is long and moving the gearbox innards takes considerable force, so one has to shift rather slowly and deliberately or the innards don't quite cooperate.

Actually, comments like this are typical for mid-engine race cars too, and it would be fair to say that the 308GTSi drives much like a road-tamed racing machine. The same goes for its handling.

Though not particularly quick, its steering is precise and (thanks to the wide tires and considerable weight) rather heavy. A large turning circle detracts from in-town maneuverability. But get the Ferrari out on a winding road and the fun begins.

There, one gets a reinforcement of the race-car feeling. In moderate-to-brisk driving the 308 understeers mildly and imparts a sense of almost unlimited cornering power; in short, it inspires a lot of confidence. But when we went to the race track and skidpad to explore its limits we found these initial impressions contradicted.

This year's test car, like the one tested for the 1982 *Guide*, had Michelin TRX tires, size 220/55VR-390, and apparently they cost the Ferrari both sticking ability and the at-the-limit security that used to be typical of this model when it had the same tiremaker's XWX rubber. Now the Ferrari manages "only" 0.810g on the skidpad, a figure rapidly becoming middle-of-the-road for sporting cars. And at the limit, its tail decidedly wants to hang out. It is not a car whose limits the non-professional driver should explore on public roads.

Brakes are an aspect of the 308GTSi that earn less qualified praise. Though with only 20 pounds of effort for 0.5g deceleration they are a bit too light for our tastes, they are certainly capable. Here the TRX tires help them haul down the Ferrari from 60 mph in a respectable 154 feet and from 80 in 254. As you can well expect from a car of this type, the 4-wheel vented discs resist fade staunchly.

For all the excitement of looking at it and driving it, the Ferrari 308GTSi demonstrates that even the world's leading maker of exotic cars has to keep developing new models lest the mass producers catch up in the objective sense. To be sure, there is no inexpensive, quantity-produced car that can match its esthetic and semi-race-car feel. But many of them are packing equal or better performance and handling into far less exotic, less costly and much more fuel-efficient automobiles. Too, it doesn't seem too much to expect more effective ventilation and air conditioning at nearly $60,000.

At least the Ferrari's performance is in for a boost in summer 1983, when a new 32-valve version of the 3.0-liter V-8 replaces the present 16-valver in both the 308GTSi and Mondial 8. With its four valves per cylinder, the revised engine is expected to deliver about 240 bhp—which should put the GTSi's 0–60 mph time down in the 6-to-7-sec range. The new engine is also said to be a little more economical of fuel, though probably not enough to stretch its 200-mile tank range much.

R&T will publish a test of the new 308GTSi version as soon as it is available.

PRICE
List price, all POE	$59,295
Price as tested	$60,780

Price as tested includes std equipment (air conditioning, elect. window lifts, central locking, leather interior), AM/FM stereo/cassette ($1485)

GENERAL
Curb weight, lb / kg	3250	1476
Test weight	3415	1550
Weight dist (with driver), f/r, %		42 / 58
Wheelbase, in. / mm	92.1	2340
Track, front / rear	57.8 / 57.8	1468 / 1468
Length	174.2	4425
Width	67.7	1720
Height	44.1	1120
Trunk space, cu ft / liters	5.3	150
Fuel capacity, U.S. gal. / liters	18.5	70

ENGINE
Type	dohc V-8
Bore x stroke, in./mm	3.19 x 2.79....81.0 x 71.0
Displacement, cu in./cc	179....2926
Compression ratio	8.8:1
Bhp @ rpm, SAE net/kW	205/153 @ 6600
Torque @ rpm, lb-ft/Nm	181/245 @ 5000
Fuel injection	Bosch K-Jetronic
Fuel requirement	unleaded, 91-oct

DRIVETRAIN
Transmission	5-sp manual
Gear ratios: 5th (0.95)	3.52:1
4th (1.24)	4.60:1
3rd (1.69)	6.27:1
2nd (2.35)	8.72:1
1st (3.58)	13.28:1
Final drive ratio	3.71:1

CHASSIS & BODY
Layout	mid engine/rear drive
Body/frame	steel/tubular steel frame
Brake system	10.7-in. (272-mm) vented discs front, 10.9-in. (277-mm) vented discs rear; vacuum assisted
Wheels	cast alloy, 390 x 190
Tires	Michelin TRX, 220/55VR-390
Steering type	rack & pinion
Turns, lock-to-lock	3.3
Suspension, front/rear: unequal-length A-arms, coil springs, tube shocks, anti-roll bar/unequal-length A-arms, coil springs, tube shocks, anti-roll bar

CALCULATED DATA
Lb/bhp (test weight)	16.7
Mph/1000 rpm (5th gear)	18.8
Engine revs/mi (60 mph)	3200
R&T steering index	1.30
Brake swept area, sq in./ton	263

ROAD TEST RESULTS

ACCELERATION
Time to distance, sec:
0–100 ft	3.5
0–500 ft	8.9
0–1320 ft (¼ mi)	16.1
Speed at end of ¼ mi, mph	88.0

Time to speed, sec:
0–30 mph	2.9
0–50 mph	6.1
0–60 mph	7.9
0–70 mph	10.6
0–80 mph	13.4
0–100 mph	22.1

SPEEDS IN GEARS
5th gear (7300 rpm)	140
4th (7700)	122
3rd (7700)	87
2nd (7700)	63
1st (7700)	41

FUEL ECONOMY
Normal driving, mpg	11.5

BRAKES
Minimum stopping distances, ft:
From 60 mph	154
From 80 mph	254
Control in panic stop	excellent
Pedal effort for 0.5g stop, lb	20
Fade: percent increase in pedal effort to maintain 0.5g deceleration in 6 stops from 60 mph	nil
Overall brake rating	excellent

HANDLING
Lateral accel, 100-ft radius, g	0.810
Speed thru 700-ft slalom, mph	60.6

INTERIOR NOISE
Constant 30 mph, dBA	72
50 mph	76
70 mph	79

SPEEDOMETER ERROR
30 mph indicated is actually	27.0
60 mph	56.0

SHORT TAKE

Ferrari 208GTBi Turbo

Prancing right around the tax man.

• Italians change governments the way most people change socks, but the one thing all Italian governments never seem to lose track of is that it's a good idea to sock it to the rich guys with the fast cars. The reasoning runs that, while the Italian rich have as many ways to tuck their fortunes away safe from taxes as the next country's well-to-do, it's good to tax the bejeezus out of their big-engined toys because this at least pays lip service to that highly vocal portion of the population which can never hope for more than a loaf of bread, a bottle of wine, and a 1913 Vespa scooter with half a cup of gas in the tank.

All new cars are taxed in Italy, but engine size determines the amount of the tax—20 percent of a car's purchase price if the engine is 2.0 liters or less, or a whopping 35 percent for the more exciting end of the scale, where swoop and top speed count for more than social responsibility. The rich will never miss the tax difference, and the poor will be satisfied that this week's government is no less diligent than last week's government when it comes to cashing in on conspicuous consumers and their consumingly conspicuous motorcars.

Enter Enzo Ferrari and turbocharging, which none of the governments foresaw. Ferrari reasoned that applying a turbocharger to a two-liter engine could make it perform just about as well as a three-liter engine, while saving fifteen percent in taxes.

Sold only in Italy, the Ferrari 208GTBi and GTSi are proof positive that the tax man doesn't always win. Starting with Ferrari's existing 3.0-liter dual-overhead-cam V-8, the engineering boffins in the company's horsepower department first reduced the engine's displacement by one-third (with a smaller bore), then applied their ever-growing fund of turbocharging knowledge. This turbo know-how is swelling by leaps and bounds with the progress of the horrendously powerful twin-turbo Formula 1 cars, which issue forth from the same factory gate in Maranello as the road cars.

If there is one thing the men of Maranello know how to do, it is design engines. Turbocharging for more horsepower was no problem, merely a matter of obtaining the right combination of fuel flow, turbo size, combustion

Vehicle type: mid-engine, rear-wheel-drive, 2-passenger, 2-door coupe
Price as tested: $31,570 (Italy)
Engine type: turbocharged V-8, aluminum block and heads, Bosch K-Jetronic fuel injection
Displacement .... 121 cu in, 1989cc
Power (SAE net) .... 211 bhp @ 7000 rpm
Transmission 5-speed
Wheelbase .... 92.1 in
Length .... 171.0 in
Manufacturer's performance ratings:
Fuel economy, European city cycle .... 12 mpg
 steady 56 mph .... 26 mpg
 steady 75 mph .... 21 mpg
Top speed .... 147 mph

efficiency, and exhaust configuration. The problem was that all these factors had to be squeezed into the same tight, mid-engined space allowed for the normally aspirated 3.0-liter, and this was what took the real doing. The result is a spaghettilike tangle of runners and components that fits the allotted space like plump fingers in a snug glove. The disadvantage of any turbo system, and one magnified by the closeness of quarters in the original 308 body shell, designed by Pininfarina, is that the turbo system generates withering amounts of heat, which is the last thing a good engine man wants building up around his pride and joy. The solution on the 208 is a pair of NACA ducts, one low on each side just in front of the rear wheels. Subtle in appearance, the ducts supply the necessary breath of fresh air beneath the rear deck.

The engine itself is a breath of fresh air, too. Frankly, we hadn't expected it to be too exciting, somehow dreading that it would prove overworked in the heavyish package. Not true. While the 308 is perhaps a bit happier around town, where its torque is an advantage, the 208 Turbo is an extremely healthy companion on the road, where its high-rev output (7800-rpm redline) can be put to best use. Claimed top speed is 147 mph, and proper use of the gated five-speed shifter keeps the revs in the perfect power band. Our rosy view of the 208 admittedly came on Ferrari's legendary Fiorano test track (where it was photographed in front of Ferrari's summer home).

Like any track and unlike the less predictable environment of street driving, Fiorano allowed the free-revving little V-8 to be kept exactly in its happiest power band. The influence of throttle position on boost level was particularly hardy, perfect for skimming between the right and wrong sides of control at midcorner, never sagging into the off-boost doldrums and instantly pumping up for beautifully controllable power-slides onto the straights. What's more, given the fact that in Italy, most Ferraris are usually garaged, taken out only for fast-lane forays to Roma or burbling cruises to Portofino, low-end horsepower doesn't really count for all that much.

All the other Ferrari virtues still apply—sexy bodywork, a highly roadworthy chassis, and the best levels of fit and finish in the exotic-car business. The driving position remains best suited to those of Italian proportions (i.e., long arms and short legs), but with this high level of performance at hand, one tends to forget minor misfits. And even rich, impatient Italians who fit perfectly can adapt nicely to the 208 Turbo's power characteristics, a task no doubt eased by the tax break. —*Larry Griffin*

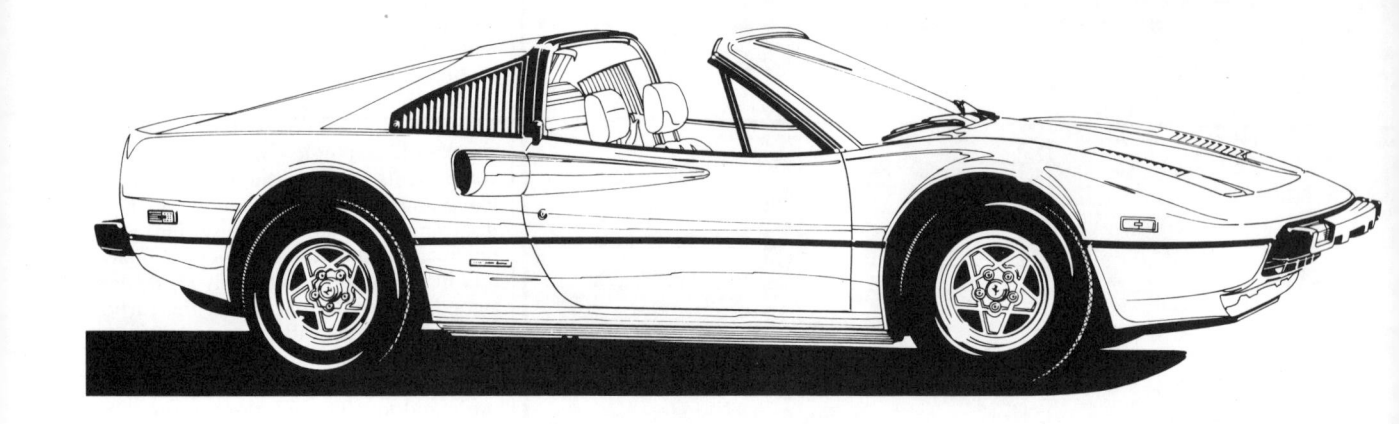

Road & Track Owner Survey
FERRARI 308

Is it, as one owner suggests, merely the best all-around Ferrari ever built?

ILLUSTRATION BY BILL DOBSON

WE'RE GOING TO open this Owner Survey report with a mild disclaimer. To wit: The number of cars included in the tabulation is too small to be statistically valid. In our sample we had only 58 cars that met our criteria for inclusion—10,000 or more miles on the odometer (so defects, if any, would have a chance to make themselves known) and purchased new by the person supplying the information (so the respondent knows the whole story of the car). Fifty-eight isn't enough to assure that the results are truly representative.

So why are we printing this? Two reasons. First, the history of 58 cars, while of little significance for a mass-produced automobile like a Chevette or Corolla, does represent a worthwhile percentage when the car is something as low-volume and exotic as a Ferrari. Our second reason is that the information extracted from these 58 surveys is simply too interesting *not* to report. Which is a good enough excuse for printing any story.

The Owners

TO START OFF, the owners themselves are interesting—largely professional men, obviously automotive sophisticates and impressively affluent as well.

As a group, they have attained a certain maturity of years, 60% being over 35. They are well educated (81% college graduates), almost half (45%) own four or more cars, doctors make up the largest single owner group (24%) and the median annual income of all owners is more than $75,000. Obviously not many poor folks are driving Ferrari 308s these days.

Reasons for Buying

WE NOTED earlier that these Ferrari owners seemed to be rather sophisticated, automotively speaking. We base this on the fact that they seemed to have known what to expect when they made their selection. And then got what they expected.

For example: The five most common reasons for having chosen the 308 were: handling, fun to drive, styling, performance and engineering. And when we asked, "What do you consider the five best features of this car?" the answers, in order, were: handling, performance, fun to drive, styling and engineering.

Which, if you compare the two lists, suggests that they got what they expected from their 308s. The only significant difference in the two lists is that "Performance" moved from 4th among the reasons for their selection to 2nd among the car's best features.

Mileage & Use

THE FERRARIS in our survey were not especially high-mileage cars although they ranged from 1977 to 1981 models; the highest odometer reading reported was 68,000 and the median for the whole group was 20,000.

Half of the owners say they use their 308s for daily transportation, 38% use them for vacations and long trips and roughly 14% enter them in rallies and slaloms. A number of others race their cars and several noted that they drove theirs only on weekends.

On the whole, they do consider themselves vigorous drivers, however, and 71% say they drive either "hard" or "very hard."

Maintenance & Service

OUR FERRARI owners take good care of their cars, more than 80% reporting that they either follow the manufacturer's recommended maintenance schedule (57%) or do more than is recommended (26%).

They are also not greatly displeased with the service they have received from their dealers, less than 30% rating it as either fair or poor. At the other end of the spectrum, 45% say they've had "excellent" service from their dealers. This is uncommonly high, the average for all cars covered by these surveys during the past eight years being 25% "excellent."

The availability of parts is always one of the uncertainties with an exotic and here the 308 seems better than average. Yes, roughly 35% of the owners did report having their 308s laid up for a day or longer while waiting for parts. But there seemed to be no particular part that was hard to obtain, causing owners to wait days or weeks for delivery.

Problems

IN OUR questionnaires we ask for information about the problems that have been encountered. In general, we'd have to say that the Ferrari 308 has more problems than most of the cars we've surveyed. We classify a problem area as one reported by 5% or more of the owners. With the 308, the owners reported a total of 14 problem areas, nine of which involved 10% or more of the cars. The average number of problem areas for all the cars we've surveyed since 1975 has been 11 and in the only other Ferrari survey we've ever done (November 1971), which included an assortment of models, the owners reported a total of seven problem areas.

It was also interesting to note that there were very few major engine problems. One owner did report having gotten a new engine on warranty because of oil leaks but the most serious problem other than this seemed to be with a camshaft. Even here, however, only 5% of the owners had camshaft trouble and there wasn't a single incident involving the failure of a cylinder block, piston, connecting rod or crankshaft.

It is also notable that most of the owners were perfectly happy with their cars. One owner from Boston whose problems seemed to be typical had this to say: "These cars are special. Every time I drive mine, it's an ear-to-ear grin. It drives well, looks great, smells great, sounds great and does all the things a recreational exotic car should do."

Best & Worst Features

IN SOME of the cars we have surveyed, there has been little agreement about the best features of the car, probably because the cars themselves were fairly bland and homogenous. Not so with the Ferrari 308.

With them, there's a surprising unanimity that handling (agreed upon by 88%), performance (86%), fun to drive (79%), styling (78%) and engineering (55%) are the best features.

There is less agreement among the owners about the 308's worst features. This list starts with economy of operation (noted by 69%), and is followed by fuel economy (55%), price (52%), use of interior space (36%) and ventilation (36%).

It would seem from this that even these sophisticated, affluent owners have been surprised how much it costs to operate the 308—or perhaps this is merely a common attitude they masochistically enjoy sharing, as if they were saying, "Sure, it's expensive but it's worth it."

Because some people think of exotic cars as being essentially temperamental and undependable, we think it is worthwhile to point out that only 5% of the owners said reliability was a worst feature.

All things considered, perhaps the owner from Burlington, Ontario, Canada, summed it up best when he said, "Anyone purchasing such a car should be prepared to live with some of the worst features. In my estimation, the best features far outweigh the worst ones. Besides, they are not all that bad and it's a Ferrari!"

Buy Another?

WHEN ASKED if, after their experience with this car, they'd buy another of the same make, 90% of our 308 owners said yes, they would. This is an impressive percentage. The average "buy another" percentage for all the cars we've surveyed since 1975 has been 71% and only two cars, the Mazda RX-7 (September 1982) and Honda Civic (July 1975) have scored a higher percentage (95% and 94%, respectively) than the 308. Buy-again percentages on some of the other enthusiast cars we've surveyed include BMW 530i (80%), Mercedes 450 (77%), Volvo 240 Series (67%) and Triumph TR7 (42%).

So what's the final word from the Ferrari 308 owners? An owner from Burlingame, California, who uses his 308 for everyday transportation, races it on occasion and in general uses it the way it should be used, put it this way: "Overall . . . this is merely the best all-around Ferrari ever built."

He's probably right.

SUMMARY: FERRARI 308

	Ferrari 308	Avg[1]
How Owners Rate Dealer Service		
Excellent	45%	25%
Good	19%	32%
Fair	14%	21%
Poor	15%	20%
Do own work	7%	2%
Maintained by the Book?		
Yes	57%	48%
No	7%	6%
Mostly	10%	15%
More than recommended	26%	31%
Buy Another of Same Make?		
Yes	90%	71%
No	5%	16%
Undecided	5%	13%

[1]Average for all cars surveyed since 1975

	Ferrari 308	Avg[1]
Problems		
Common to 10% or more	9	7
Clutch*		
Electric window lifts		
Upholstery		
A/C compressor		
Instruments		
Master cylinder*		
Shift linkage		
Paint		
Distributor*		

Five Best Features
Handling
Performance
Fun to drive
Styling
Engineering

	Ferrari 308	Avg[1]
Common to 5–10%	5	5
Head gasket*		
Transmission*		
Exhaust system		
Fuel injection*		
Camshaft*		
Affecting reliability	7	4

*Represents a reliability area that could make the car unsafe or impossible to drive.

Five Worst Features
Economy of operation
Fuel economy
Price
Use of interior space
Ventilation

Putting a punch into the Prancing Horse

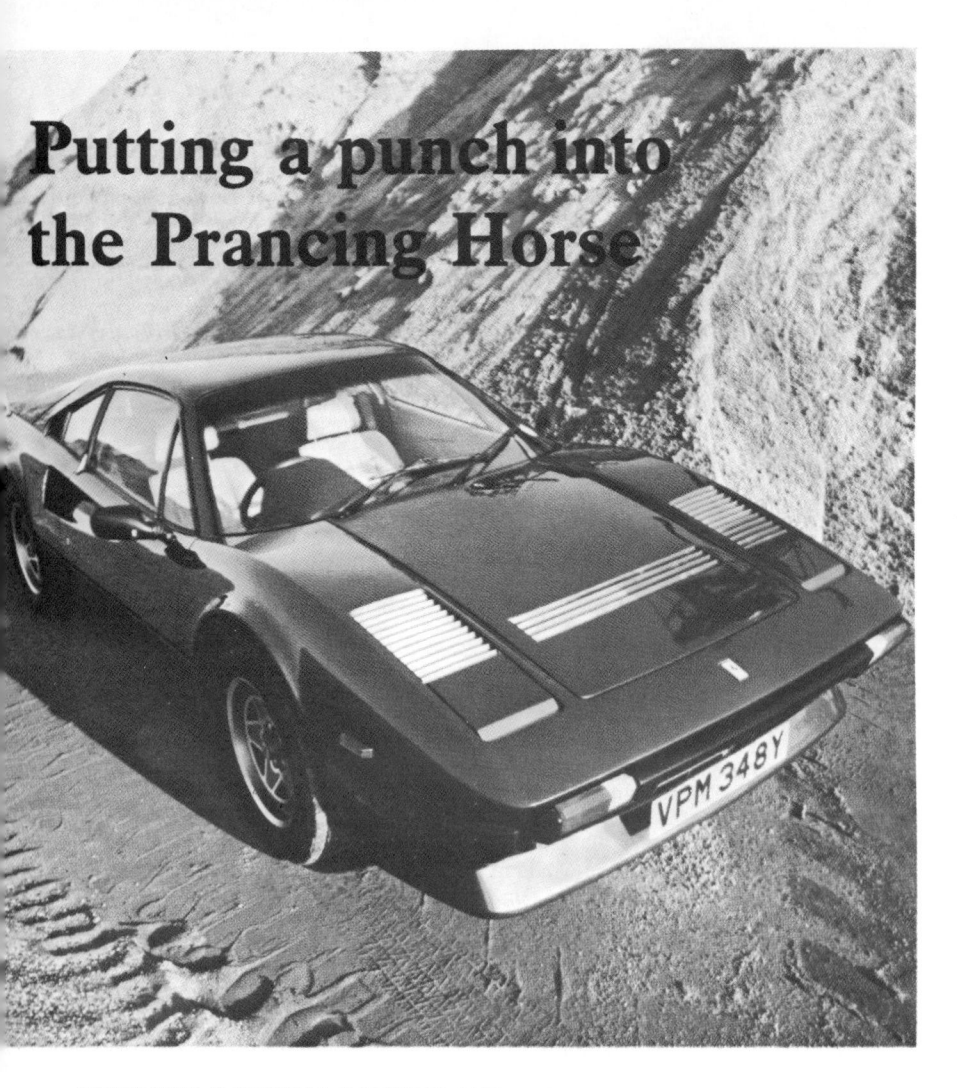

AGGRESSIVE and functional: the 308GTBi is an impressively finished, well-equipped two-seater coupé which more than adequately upholds the Maranello image for high performance road machines.

FERRARI 308GTBi QUATTROVALVOLE

"QUATTROVALVOLE" proclaims a discreet little legend on the rear of the latest Ferrari 308GTBi, one of the few outward visual signs that the smallest of Maranello's delectable road coupés is any different to its forbears which have been a regular sight on the roads of Europe for more than six years. When D.S.J. assessed the original, carburated 308GTB within the pages of the December, 1976, issue of MOTOR SPORT, he didn't quite damn the car with faint praise, but he did make it very clear that he felt that the marque's endearing idiosyncracies "seem to have been discreetly swept away by the dead hand of Fiat." Those who'd enjoyed the delightful little Dino 246GT, which seemed to become happier and happier the more it was revved, initially professed themselves to be quite happy with the 308GTB, even if the standard of workmanship applied to this two-seater coupé failed to reach the standards expected from a £12,000 motorcar. But with a reputed 255 b.h.p. available at 6,600 r.p.m. and 209 ft./lb. torque at 5,000 r.p.m., there was no doubt that it was an exciting car to drive. Unfortunately, the necessity to fit fuel injection, partly to comply with emission control requirements for the United States market, later saw the power output stifled to 214 b.h.p. (with a corresponding drop in torque), slicing the best part of 10 m.p.h. off the car's top speed and dampening its appeal to those who'd experienced the original machine.

Slightly under two years ago MOTOR SPORT tested the Mondial 8, two-plus-two central-engined coupé which was then quite new on the British market. It was the writer's first experience of the Ferrari V8 engine, only the 4.4-litre Daytona V12 and Dino 246GT V6 having previously come his way for road test purposes. We were impressed by the Mondial, of that there is no question, but once over 5,000 r.p.m. one was definitely conscious that the acceleration was tailing off and by the time 6,000 r.p.m. had been reached (just under 125 m.p.h.) it really was beginning to be a bit of a slog to ease any extra speed out of it. At the time the Mondial's claimed top·speed was in the order of 150 m.p.h. and we remarked that we didn't really feel that an example straight from the production line would be good for more than 140 m.p.h. Only when we tried the comparable 308GTBi, with its newly fitted cylinder heads, did we realise that we had been missing out quite badly and that there was rather more to the latest V8 Ferrari than simply the addition of the extra inlet and exhaust valves.

The bare facts of the matter are that the "quattrovalvole" now produces 240 b.h.p. at 7,000 r.p.m. and 191 lb./ft. torque at 5,000 r.p.m. from its 81 x 71 m.m., 2,927 c.c., 90-degree V8 engine. The Ferrari factory's latest brochure makes the point that it has applied a lot of its F1 "know-how" to the development of these cylinder heads (which are also added to the Mondial's specification, incidentally) and since the design and engineering work was carried out at Maranello it's not difficult to imagine the bespectacled Mauro Forghieri taking an interested glance over the road car engineers' shoulders from time to time as work progressed on these alterations. It's also interesting that, as a parallel to the four valve head development, a turbocharged 2-litre V8 version has also been developed, the capacity reduction achieved by a smaller bore while retaining the same stroke. This machine unfortunately is only marketed in Italy, where peculiar tax laws heavily penalise any car over 2-litres: a shame, because we would like to have compared it against the four-valve 3-litre V8.

As we said, it wasn't simply a case of slapping on a four valve head and leaving it at that. A great deal of thought went into such things as the shape of the combustion chambers, the volumetric efficiency and revisions to the water-cooling passages. The compression ratio has been increased from 8.8:1 to 9.2:1 and the heads are fashioned from an aluminium / silicon alloy. Valve seats are of a purpose-made special cast steel, valve guides are of tellurium copper and the exhaust valves of nimonic alloy, already well tested by the marque's F1 engines but now employed for the first time by Ferrari in production road car engines. Bosch K-Jetronic mechanical fuel injection ensures a crisp response from the transverse-mounted V8 engine which, as before, drives through a notchy, adequately precise five-speed manual gearbox mated to a dry single plate mechanically operated clutch.

Since we tested that original 308GTB in 1976, other aspects of Ferrari's production engineering have also improved. The finish, both of the body and the interior, are dramatically improved and the paint finish is of a lustrous quality, even though our test car was an appealing dark blue rather than the blazing red true Ferrari buffs would more fully approve of. The tasteful blend of light tan and black leather in the snug two-seater cockpit complemented the external high standards and there was no trace of the irritating rattles or water leaks which have been referred to by 308 testers in the past. As we noted at the time of the Mondial 8 test, a tremendous amount of attention has been lavished on the Ferrari's finish and it is this factor, allied to the V8 engine's delightful blend of performance and flexibility which makes the 308GTBi and its similarly equipped stablemate such a realistic alternative to BMW, Porsche and Mercedes-Benz for use in everyday motoring.

Any driver over six feet tall will find that headroom is pretty limited inside the 308GTBi and the writer reckoned he could do with about half an inch more on the rearward adjustment of the comfortable seat which offers just enough support without being excessively firm. But this is a minor complaint. Far more irritating was the fact that the tinted anti-dazzle strip blending into the upper edge of the gently sloping windscreen tends to "chop-off" a tall driver's forward vision. There is also an annoying reflection from the instruments into the screen which becomes much worse at night when all the warning lights on the fascia are projected upwards into one's line of sight. We and others commented on this back in 1976, so it's a shame that this rather elementary problem hasn't been rectified.

All such ergonomic grumbles fade away, however, once you turn the key and fire up the V8 engine into willing action. Starting from cold in

INTERIOR: a tasteful blend of light tan and black leather helps enhance the snug cockpit (above). Below, the rear view of the 308GTBi is both purposeful and good looking.

SMALL BOOT: behind the transverse-mounted 3-litre V8, Ferrari has provided a small, zip-up luggage compartment.

the mornings requires one to keep well clear of the throttle pedal lest the engine become flooded with fuel, but once the V8 has warmed up it will start at the first turn of the key. The Ferrari's five speed gearbox is controlled by the familiar notchy change running within the evocative open metal gate with which Maranello enthusiasts will be so familiar. First gear, away from the driver and dog-legged back towards the left, is a giant pain to engage, and the change from first to second gear needs a well-practised mixture of firmness and caution. If you hurry it unduly, second just doesn't seem to want to engage: if you're too gentle, then the spring-loading seems to take over and literally flicks the lever out of your hand as it comes out of first. But second to fifth, in a conventional "H" pattern, proves to be no problem at all. In fact, apart from moving away from standstill, first isn't really needed at all: the 308GTBi will pull from walking pace in second gear and the engine's tremendous flexibility enables one to dribble along in second and third in the middle of London's stifling rush hour traffic. One is aided in this task by a splendidly light and smooth clutch operation, contrasting with that on the Mondial 8 two years ago. In fact the pedals seemed all perfectly placed to my mind and heel-and-toeing was simplicity itself, even for those who like the lazy way and simply roll the side of their right foot off the edge of the brake and onto the throttle.

My colleague D.S.J. found the steering wheel too small for his taste and other short people may agree with him that it is angled too sharply. However, that's where being tall helped. From my point of view the wheel size and position was absolutely perfect, although the rim's proximity to the padded fascia cowling meant that my knuckles rubbed against that padding whilst swinging the wheel through the "twelve o'clock" position. Unlike the Mondial, the 308GTBi doesn't have a steering column instantly adjustable for length although Maranello Concessionaires tell me that the wheel is adjustable for rake — but this is a job for the workshop, not simply an instant adjustment for the owner. Although the instrumentation is comprehensive, the speedometer and matching rev. counter surrounded by oil pressure, water temperature and fuel contents gauges, they are not as clearly calibrated as one would like, particularly with the instrument lighting turned on. Oil pressure runs consistently at 85 lb./sq. in. and water temperature at 160 degrees. A word of warning, however: don't trust the fuel gauge. It was still flickering above the empty stop when the 308GTBi spluttered to a standstill in a traffic jam, its tank almost bone dry. Fortunately we were within sight of a garage which accepted the most convenient and appropriate credit card so this didn't turn into the drama it might have done.

This all independently suspended, four-wheel disc braked Ferrari, running on its Michelin TRX 205/55 VR390 radials, offers a well balanced ride which is just sufficiently taut to provide excellent handling without being harsh enough to impart a bumpy ride for the occupants. It's probably not *quite* as good as a Lotus Eclat, for example, but certainly doesn't have that characteristic "coarseness" by which most Porsche models can be identified. Like so many high performance cars it felt smoother and more responsive the harder it was driven. At low speed the steering feels somewhat dead and distant from the road wheels, but the faster one presses the 308GTBi, the more fluid and enjoyable the whole package becomes.

PRANCING HORSE
Continued

There is a reassuring touch of roll and although I was, surprisingly perhaps, aware of a trace too much understeer for my taste, this probably comes down to a matter of personal opinion. The steering is a little low geared for too much enthusiastic motoring round tight hairpins and the high level of rear end adhesion means you can't easily make it break away and slide round very slow tight turns unless the surface is loose or slippery. So the 308GTBi isn't really a "tight country lane" car, but on open, gently sweeping "B" and "C" routes it really comes into its own and a day's motoring romp round the Salisbury Plain area, in the surprisingly mild and sunny weather conditions which we enjoyed during January, proved to be joyful motoring experience the like of which we've not had for a long time.

Despite having to encompass several dubious first-to-second changes on that only adequate gearbox, the 0-60 m.p.h. time of 6.1 sec. reveals this Ferrari to be a highly impressive performer. And provided that you can change gear cleanly, 80 m.p.h. comes up in ten seconds from standstill and 100 m.p.h. in 15.8 sec. Top speeds seen in the gears were 43 m.p.h. (first), 60 m.p.h. (second), 86 m.p.h. (third) and 115 m.p.h. (fourth), in all cases, I have to confess, with the V8 fluttering against its electronic rev. limiter. The claimed top speed is fractionally over 150 m.p.h., although we didn't see more than 137 m.p.h. owing to the lack of room rather than the car's potential. But at that speed the 308GTBi was still pulling adequately in contrast, as we've mentioned before, to the old 2-valve V8 machine. Intermediate acceleration figures were equally impressive, particularly in fourth gear, and far more relevant than the car's ultimate speed. The feeling of security that this Ferrari's fourth gear overtaking performance imparts has to be sampled to be believed, and its ability to sprint from 60 m.p.h. to 80 m.p.h. in less than six seconds means that traffic queues and slow-moving lorries can be dealt with in the minimum possible time, and so with more safety. Of course, such performance tends to focus the mind extremely well, but since one can't guarantee that everybody else is paying so much attention to their driving, it's heartening to feel the large servo-assisted ventilated disc brakes (11" diameter front, 11.8" diameter rear) reacting with splendid precision time and again and demonstrating no signs of fade whatsoever throughout the duration of our test.

With not quite such a low overall gearing in top as the Mondial (21.14 m.p.h. per 1000 r.p.m. as opposed to 19.87 m.p.h. per 1000 r.p.m. on the slightly heavier two-plus-two coupe), the 308GTBi returned a very respectable overall fuel consumption of 20.1 m.p.g. during the course of 700 miles' motoring which encompassed City commuting, gentle top gear motorway cruising and enthusiastic third / fourth gear sprinting on secondary routes. It therefore isn't beyond the realms of possibility to envisage a consumption of 23/24 m.p.g. if the 308GTBi is handled with a degree of reasonable restraint. By any objective super-car standards, that is extremely impressive and certainly very efficient.

Secondary controls are mounted on the steering column stalks (flip-up headlights, windscreen wiper / washers and indicators) and onto an overcrowded console between the seats where you will find switches and levers controlling the electric windows, air conditioning, hazard flashers, three speed fan, variable speed control for the wipers and rear foglights. Although demisting isn't carried out with dramatic effectiveness in this Ferrari, I do particularly like the separate temperature and air direction controls for each passenger's footwell. Storage space is very restricted, with only a couple of pockets in the doors allied to a modest, zip-covered luggage compartment in the tail, behind the V8 engine.

The question of the reliability and durability of today's Ferrari range is something that can only be reported with the passing of time: by all accounts we've heard that the cars stand the test of time significantly better than equivalent models marketed ten years ago. Remember, these are thoughts for the essentially practical: nobody buys a Ferrari with longevity as his foremost criterion. But if today's £23,172 Ferrari 308GTBi survives the rigours of time in an impressive manner, then so much the better. So for the moment, just sit back and wind up the newly realised potential of that free-revving, four-valves per cylinder V8 with the rev. counter needle shooting round to beyond 6000 r.p.m. and, hopefully, the traffic evaporating in front of those menacing front wings, all's well with the World. There can be few other experiences like it. A.H.

ROAD TEST

Ferrari 308 Quattrovalvole

All the promise of a night on the town.

• Now that Triumph has pulled the plug on the Spitfire, and on itself for that matter, about the closest you can come to a real sports car is this Ferrari. It has all the stuff that has made real sports cars famous over the years: a noisy cockpit, a contortionist driving position, torturous entry and exit, quirky handling, an oven-temperature trunk, a thin-spread service network, and an extraordinary capacity to generate celebrity. This thing all but strews its own rose petals. You're a public figure in this car, a fit target for dropped jaws, searching stares, and unholy speculation. Every female assumes you must be somebody; every male just wants to be you. It is commonly understood, when you're arrowing across the landscape in this red dart, that you've got the world by the tail.

In other words, this is a fun car—flawed to be sure, but fun nonetheless, like every real sports car ever sold.

This is not like every other 308GTSi sold in the last six years, however, never mind that it looks almost the same. Now it's called 308 Quattrovalvole, in reference to the four-valve cylinder heads added to the 2.9-liter V-8. Everybody knows why four valves per cylinder are better than two: more power and more torque. Ferrari, along with most other manufacturers who have tried this configuration, says lower emissions and better fuel economy, too. But such talk of redeeming social value probably shouldn't even be allowed to come up in this discussion; it spoils the essential Ferrari flavor. The important thing is that horsepower is increased by 25, to 230 at 6800 rpm (up 200 rpm from the previous peak), and torque is enhanced seven pounds-feet, to 188 at 5500 rpm (up 500 rpm). Moreover, the engine is very well behaved. The power never "comes on"; it's just there all the time. It's stronger at the high end, of course, but there are no real weak spots and no balkiness anywhere. Just poke your foot into it and go. The fuel injection is much nicer in this regard than the carburetor setup of a few years back.

The same sort of "much nicer," in fact, is what you'd say about the whole car. Ferrari isn't claiming an arm's-length list of changes—we'll get to the what's-new details in a moment—but the whole car has apparently been rubbed on to minimize the irritants. For example, the steering seems easier now. Even though the system remains highly reversible—you are always aware of the wheel twitching left and right in response to road irregularities—it never seems to kick back as strongly as some

139

past models did, trying to wrench your thumbs off and head for the pasture. The clutch seems relatively easy too, and the shifter, though still notchy, is no longer a puzzle. The seat seems to have finally gotten to the right shape, and the interior noise level is merely high, not debilitating. To be sure, in absolute terms there is still plenty of room for improvement in all these categories. But on the other hand, this new car seems so much more refined than past examples that, given the choice between it at $59,500 and a several-year-old example for, say, $35,000, we'd certainly recommend the Quattrovalvole, if you've got the money.

Now, whether or not you should want such a car in the first place is another matter entirely. Never mind that this is a two-seater, with specifications at least superficially similar to the Corvette and several Porsches. In fact, the Ferrari is a whole lot more intense than the others—a real sports car, remember?—and is not really interchangeable with anything else on the market. A few will approach it because they want something Ferrari—not just something with the Ferrari name attached—and for them, the Quattrovalvole is a good call because, except for the lack of a V-12, it's faithful to the Ferrari way of doing things. The awkward, gated shifter is a good example. The steering, which despite its reversibility still has a great deal of friction damping, is another. A cockpit constantly invaded by engine noise is yet another. Many of these details are contrary to modern automotive engineering, but they are authentic Ferrari, trademarks of the old master, and are heartwarming when taken in that context.

Most buyers, however, have little notion of Ferrari, except that it's a famous name. They just want a car that makes everything else on the road look like a brown three-piece wool worsted. And

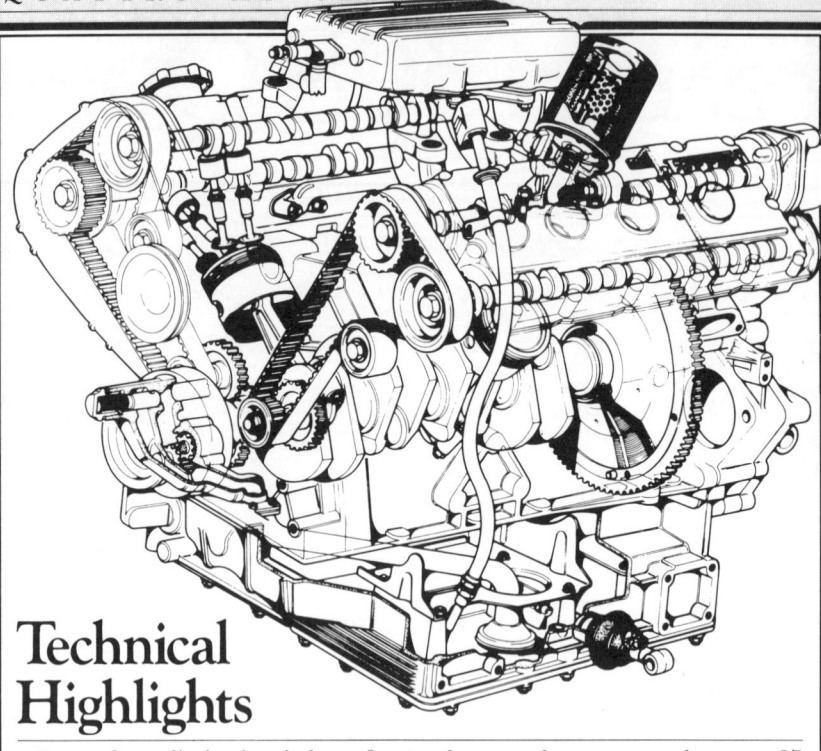

Technical Highlights

- Four-valve cylinder heads have for years been the standard approach to horsepower for the racetrack. Lately, engineers seeking more performance from regular production engines have also resorted to this layout. Datsun, Toyota, Lotus, and BMW each have at least one four-valve design on the market, and Mercedes-Benz, Volvo, Jaguar, and Saab have announced plans to join this elite club in the near future.

Doubling the number of intake and exhaust valves in an engine is costly, but four-valve conversions offer two significant benefits. First, volumetric efficiency is greatly enhanced because the intake charge has a much freer entrance to the combustion chamber. For comparison purposes, the important dimension to consider is the annular, or "curtain," area formed between a wide-open intake valve and its seat. In the case of Ferrari's new 2.9-liter four-valve engine, two 29mm intake valves produce an annular area 37 percent larger than that produced by the old engine's single 42mm intake valve. (Valve lift and cam timing have not been changed.)

Naturally, the annular area is larger on the exhaust side as well, but this is somewhat less important than the second advantage that justifies the *quattrovalvole* configuration: optimum spark-plug location. With two- or even three-valve designs, the spark plug is inevitably displaced to one corner of the combustion chamber; a four-valve layout, however, leaves plenty of space for an igniter in the geometric center of the cylinder. This is the ideal location, both for initiating the combustion process and for propagating the resulting flame front throughout the entire combustion chamber.

In user terms, the new four-valve head has increased the Ferrari's power output by twelve percent. (Unfortunately, the new heads can't be retrofitted to earlier engines.) The factory also attributes a fatter torque curve, lower exhaust emissions, and somewhat better fuel economy to this refinement. —*Don Sherman*

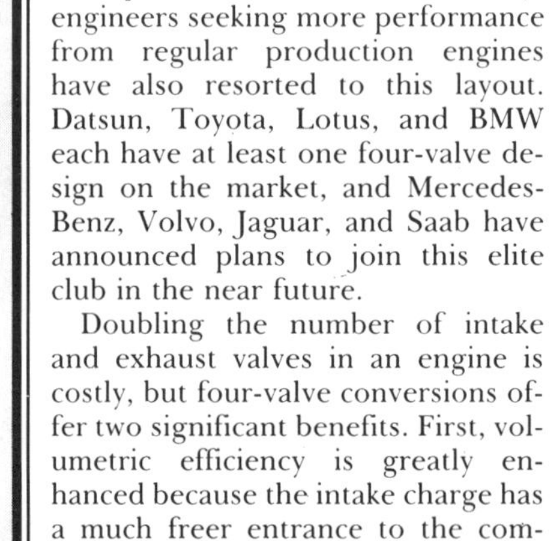

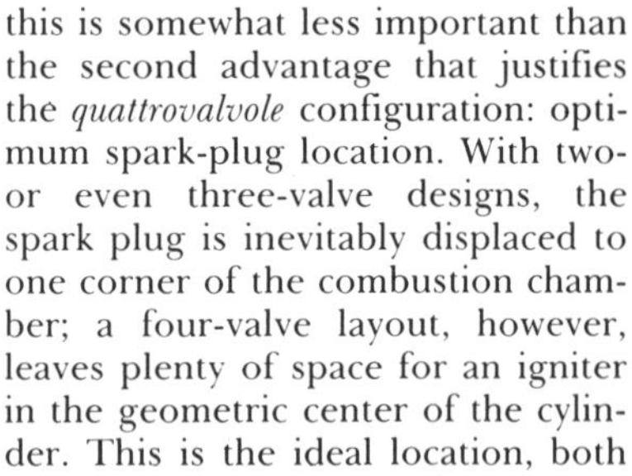

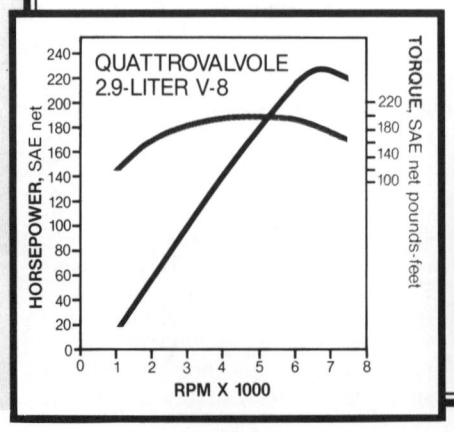

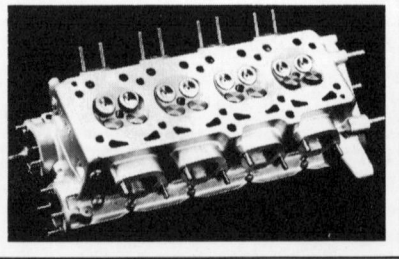

140

QUATTROVALVOLE

Vehicle type: mid-engine, rear-wheel-drive, 2-passenger, 2-door targa

Price as tested: $59,500

Options on test car: none

Sound system: none

ENGINE
Type .... V-8, aluminum block and heads
Bore x stroke .... 3.19 x 2.80 in, 81.0 x 71.0mm
Displacement .... 179 cu in, 2927cc
Compression ratio 8.6:1
Fuel system .... Bosch K-Jetronic fuel injection
Emissions controls .... two 3-way catalytic converters, EGR, auxiliary air pump
Valve gear .... gear- and belt-driven double overhead cams
Power (SAE net) .... 230 bhp @ 6800 rpm
Torque (SAE net) .... 188 lbs-ft @ 5500 rpm
Redline .... 7700 rpm

DRIVETRAIN
Transmission 5-speed
Primary-drive ratio 1.11:1
Final-drive ratio .... 4.06:1, limited slip

Gear	Ratio	Mph/1000 rpm	Max. test speed
I	3.08	5.2	40 mph (7700 rpm)
II	2.12	7.5	58 mph (7700 rpm)
III	1.52	10.5	81 mph (7700 rpm)
IV	1.12	14.3	110 mph (7700 rpm)
V	0.83	19.3	144 mph (7450 rpm)

DIMENSIONS AND CAPACITIES
Wheelbase .... 92.1 in
Track, F/R .... 57.5/57.5 in
Length .... 174.2 in
Width .... 67.7 in
Height .... 44.1 in
Ground clearance .... 4.6 in
Curb weight .... 3320 lbs
Weight distribution, F/R 40.6/59.4%
Fuel capacity .... 18.5 gal
Oil capacity .... 10.6 qt

CHASSIS/BODY
Type .... unit construction
Body material .... welded steel stampings, sheet aluminum stampings, fiberglass-reinforced plastic

INTERIOR
SAE volume, front seat .... 48 cu ft
trunk space .... 5 cu ft
Front seats .... bucket
Recliner type .... infinitely adjustable
General comfort .... poor fair **good** excellent
Fore-and-aft support .... poor fair **good** excellent
Lateral support .... poor fair **good** excellent

SUSPENSION
F: .... ind, unequal-length control arms, coil springs, anti-sway bar
R: .... ind, unequal-length control arms, coil springs, anti-sway bar

STEERING
Type .... rack-and-pinion
Turns lock-to-lock 3.3
Turning circle curb-to-curb .... 39.4 ft

BRAKES
F: .... 10.7 x 0.9-in vented disc
R: .... 10.9 x 0.8-in vented disc
Power assist .... vacuum

WHEELS AND TIRES
Wheel size .... 165 x 390mm, 6.5 x 15.4 in
Wheel type .... cast aluminum
Tire make and size .... Michelin TRX, 220/55VR-390
Test inflation pressures, F/R .... 33/33 psi

CAR AND DRIVER TEST RESULTS

ACCELERATION — Seconds
Zero to 30 mph 2.3
40 mph 3.5
50 mph 5.3
60 mph 7.4
70 mph 9.2
80 mph 11.5
90 mph 15.1
100 mph 18.5
110 mph 23.3
120 mph 32.0
130 mph 44.0
Top-gear passing time, 30–50 mph 8.7
50–70 mph 8.8
Standing ¼-mile .... 15.2 sec @ 92 mph
Top speed .... 144 mph

BRAKING
70–0 mph @ impending lockup .... 206 ft
Modulation .... poor fair good **excellent**
Fade .... **none** moderate heavy

Front-rear balance .... poor **fair** good

HANDLING
Roadholding, 282-ft-dia skidpad .... 0.80 g
Understeer .... **minimal** moderate excessive

COAST-DOWN MEASUREMENTS
Road horsepower @ 50 mph .... 15.0 hp
Friction and tire losses @ 50 mph .... 7.5 hp
Aerodynamic drag @ 50 mph .... 7.5 hp

FUEL ECONOMY
EPA city driving .... 11 mpg
EPA highway driving .... 19 mpg
EPA combined driving .... 14 mpg
C/D observed fuel economy .... 14 mpg

INTERIOR SOUND LEVEL
Idle .... 72 dBA
Full-throttle acceleration .... 89 dBA
70-mph cruising .... 81 dBA
70-mph coasting .... 77 dBA

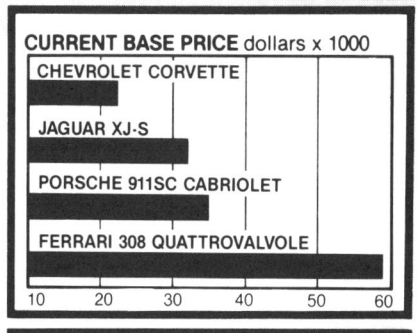

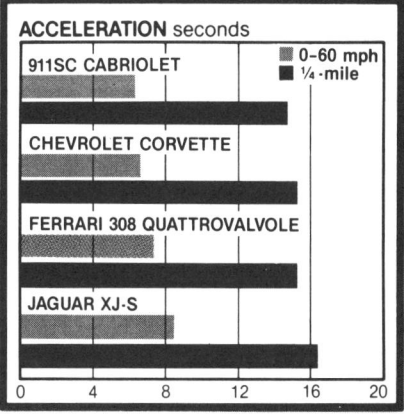

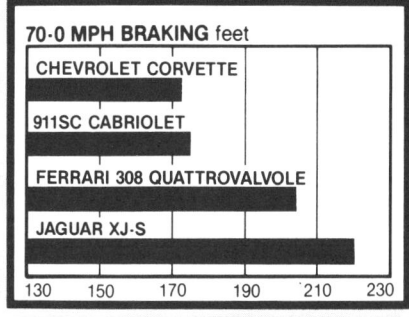

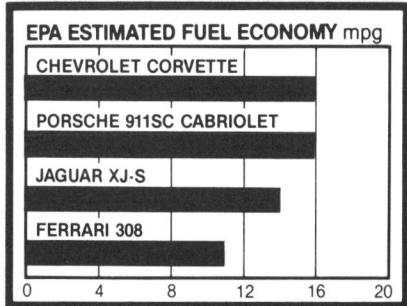

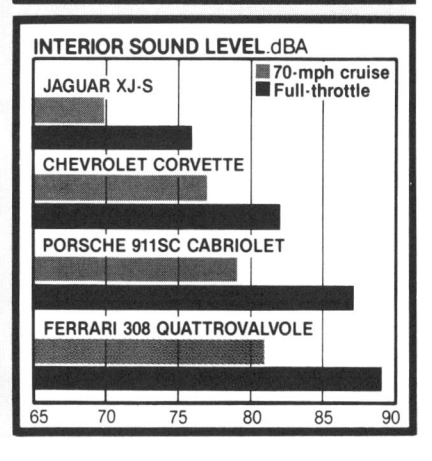

QUATTROVALVOLE

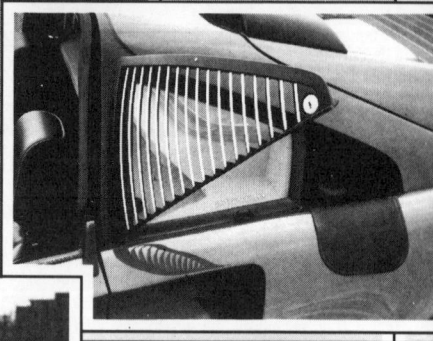

boy, the Quattrovalvole does that. If appearance is all that counts, this car is a winner. It helps a bit if you can appreciate that, to get the low roof and pointy nose, anything resembling space inside had to be pared to the bare minimum, and that applies whether you're speaking of room for your blow-dried or trunk accommodations for a clean pair of socks. This is the automotive equivalent of tight pants. Certain discomforts are part of the deal. If you can't live with them, don't wear such a tight car.

Now that we have that caveat out of the way, let's talk about what's new. Visually, the differences are small. Air vents have been cut into the hood, as on the Boxer, to aid airflow through the front-mounted radiator. The old vents atop the front fenders carry on, but they're black now—as are those on the hood—instead of body color. A basket-handle airfoil spans the gap between the roof sail panels, apparently for no good reason other than the American sales department's thinking it makes a nice decoration. And, of course, a new logo on the taillight panel proclaiming 308 Quattrovalvole instead of 308GTSi. A chrome "GTSi" still appears on the right side of the dash, however.

Our test car had 220/55VR-390 Michelin TRX tires on 6.5-by-15.4-inch wheels. By fall, there will be a change to Goodyear NCTs: 205/55VR-16 on 7.0-inch-wide wheels in front, 225/50VR-16 on 8.0-inchers in back.

In the powertrain, there is more to talk about than just twice as many valves, even though that change by itself was complicated enough to require new machining on the block, which means the old block and the new one are not interchangeable. The new one also has aluminum cylinder liners instead of iron ones, which is said to improve cooling.

The gearbox has been revised as well. Now it has its own oil pump. Also, the final-drive ratio has been shortened considerably, from 3.71:1 to 4.06:1, and first and fifth gear are slightly taller to compensate. Still, first gear is so short now that, on level ground or facing downhill, it's very tempting to start in second—partly because the car moves off just fine in second and partly because of the up-right-up notch the lever has to grate through between the first two gears. The transmission actually shifts really well for a Ferrari, certainly better than that of any other mid-engined version we've driven—once it has warmed up, that is—but it still has a crankiness unmatched by anything else produced in the Eighties.

The handling is cranky, too. If you drive only briskly, the Quattrovalvole gets around well enough, but then, so does everything else on the road. The problems show up when you're really cooking. As you approach the limit, the car understeers securely under power, but the lift-throttle characteristics are nasty: the tail comes right out, too far and too fast to be a handy course-correction aid. The relatively horizontal position of the steering wheel, which makes for a very long reach over the top, just compounds the problem of keeping up with the car when you back off the power. Probably the new Goodyear setup with wider wheels and tires in back will add stability, although that's just a guess at this point.

Regarding the steering-wheel position, it would seem a relatively easy thing to modify the upper column support to tilt the wheel down in a more vertical direction. The top U-joint is very close to the wheel, so there would be a big angle improvement for every increment of thigh clearance given away (there's plenty now). Of course, you'd say goodbye to the gas and water-temperature gauges, which would forever be blocked by the fat, leather-covered rim of the new black-spoked wheel—but that would be just one more trade-off in an already thoroughly compromised cockpit. This is a short-leg, long-leg car: the front wheel arch takes a big bite out of the driver's left-leg room, but if you slide the seat back to get more space, your right foot won't be able to do its job. A good footrest has been provided, so at least you have a flat and secure place to park the idle shoe; but it's so close to the clutch pedal, which itself is offset toward the center of the car, that it takes a while before you stop catching your foot on the side of it when you mean to stroke the pedal. Getting used to that takes longer than finding just the right slouch to keep your head from being in constant contact with the roof panel—that is, if you're under, say, six-foot-two. If you're over, maybe you should forget the whole idea.

Given the conspicuous appearance of this car, particularly when painted Ferrari red, we doubt that many people buy it for its speed. Nevertheless, now that it's called Quattrovalvole in deference to its nifty new engine, some measure of its newfound energy would be appropriate. Acceleration to 60 mph takes 7.4 seconds now; that's 0.2 second quicker than before, despite an extra shift caused by the shorter gearing. (Holding second gear to 60 mph, which requires exceeding the redline by 250 rpm, results in a 6.8-second clocking.) The standing-start quarter-mile is completed in 15.2 seconds at 92 mph, 0.4 second and 3 mph quicker. Top speed is up 4 mph, to 144, faster than the new Corvette, on par with the Porsche 928S. This Ferrari is a reasonably quick car, to be sure, and a constant threat to the record-of-violations side of your driver's license, but it's neither as quick as *quattrovalvole* sounds nor as fast as the new 180-mph speedometer (marked up from 85 mph, now that the government has deregulated the dial) looks.

So we assume that buyers will continue to come hither for reasons other than speed. And just to reassure them, we can report that those sixteen extra valves don't get in the way of any other pursuits.
—*Patrick Bedard*

COUNTERPOINT

• Hello, hello?... Mr. Truth, are you there? I'm a little confused, Mr. Truth, and can't seem to figure out exactly where you lie. Last fall found me in Modena and Maranello, you remember. The idea, supposedly in search of you, Mr. Truth, was that journalists should drive Mr. Ferrari's coming Quattrovalvole cars. Running rampant on the fabled Fiorano test track, the four-valve 308s felt absolutely wonderful. "Berserk with power" is the terminology, I believe. The four-valvers were utterly effortless horsepower rheostats, and we were told that those cars were U.S.-legal. Mr. Ferrari's organization plied us with still more vino and pasta and a day or two among Monaco's nobly nubile, and then put us on a Concorde flight back from Paris. Could it be, Mr. Truth, that you and the horsepower fell off the plane? Because the two of you don't seem to have arrived together in our genuinely legal Quattrovalvole test car. It runs okay, but not great. I hate to look a prancing gift horse in the mouth, but, hey, Mr. Truth . . .
—*Larry Griffin*

Saturday morning, a cloudburst. You have agreed to meet some friends for lunch in a town 90 miles away. Your radar detector is on the visor, but you won't need it, because policemen are not fond of standing in the rain writing tickets. The Ferrari blats along at a steady 80, the exhaust note seeming to warble in the whirl of rainwater sluicing by underneath. This is pretty close to heaven, if you ignore the steady drip of water that falls on your sleeve from the top of the windshield pillar. The *quattrovalvole* cylinder head is clearly the best thing that ever happened to the Ferrari 308. Suddenly it behaves just the way you always knew a Ferrari should. It is fast and crisp, and the Michelin TRX tires seem to have been designed for this car in the way they get its power onto the pavement, even wet pavement. It is a real Ferrari, for a change—maybe the first Ferrari equipped with something other than a V-12 that I've ever loved. At a price of $30,000, it would be an absolute worldbeater. At the price they're charging, it is a compelling sales pitch for a Porsche.
—*David E. Davis, Jr.*

What becomes a legend most? If it's a Ferrari, then the answer can only be an engine that makes your blood boil. The Ferraris of old rose to mythical proportions on the strength of their incredible twelve-cylinder powerplants and their blazing speed. Road tests of old Ferraris always talked about the passion with which the cars were built.

Well, the latest 308 doesn't seem like a passionate car anymore, and I'd say the problem lies right there in the engine compartment. It's not that this rendition lacks the credentials. Holy meatball, Guido, its power output and lofty redline are enough to make any engine designer drool.

Yet when I flatten the throttle, the 308 doesn't gush ahead like some sort of wild beast set free. It *sounds* thrilling, it revs as if there were no tomorrow, and the numbers tell me it's quick. But a religious experience? Uh-uh.

With all due respect to those who bow toward Modena once a day, the 308 commits too many sins *not* to be a rocket ship. If the legend is to continue, I submit that it's going to take even more power.
—*Rich Ceppos*

RoadTest

FERRARI 308 GTBi QUATTROVALVOLE

Fast and exceedingly beautiful, Ferrari's mid-engined 308 is also commendably practical and suprisingly economical

FERRARI'S PREMIER position among the Italian exoticar manufacturers remains as dominant as ever — and with good reason, considering their heritage of consistent success in the most rigorous of the racing formulae and the simultaneous production of some of the world's fastest and most beautiful sports cars.

Fuel crises and debilitating emissions regulations have caused the Maranello firm to falter, however, and still-traumatised Ferrari fans will recall how, in 1980, the adoption of fuel injection on Ferrari's quad-cam V8, to replace the bank of four Weber carburetters, dropped its power output from a once heady 255 to 214 bhp. As fitted to the then recently introduced Mondial the barely comprehensible result was a Ferrari hard pressed to exceed 135 mph.

It was Ferrari's racing experience that provided the answer and last year, by doubling the number of valves and upping the compression ratio to 9.2:1, power output leapt to an honest 240 bhp (DIN) at 7,000 rpm while torque was pushed up from 179 lb ft at 4,600 rpm to 192 lb ft at 5,000 rpm.

Thus the *quattrovalvole* Ferraris were born and the proof of the pudding in the Mondial tested by us last year was a top speed of 146.1 mph (measured in unfavourable conditions) and a 0-60 mph time of 6.4 sec. The smaller and lighter 308 GTBi and GTSi had retained respectable levels of performance even when fitted with the down-rated version of the V8, suggesting that in *quattrovalvole* form they would be real stormers . . .

Regarded by many as the most beautiful of the current supercars, the 308 GTBi, even more contentiously, can also lay claim to being one of the most attractive Ferraris ever. There is good historical reason for this, since it was styled by Pininfarina around the mechanicals of one of the less lovely Ferraris, the 308 GT4 2+2, while the deep air scoops along its flanks and its position in the range recall an all-time Ferrari classic, the Dino 246. The 308 GTBi *quattrovalvole* also wins the looks contest against its targa-topped derivative, the GTSi, with a more balanced appearance attributable to the window it has behind its doors rather than the extra black-painted louvres of the GTSi.

In other respects, though, the two cars are identical. They share the same 32-valve, 2,927cc all-alloy engine, which transformed the performance of the Mondial, and the classic supercar layout with the 90 degree V8 installed transversely ahead of the driven rear wheels. The limited slip differential is enclosed in unit with the five-speed gearbox which is mounted behind the engine. Suspension is by double wishbones, coil springs and an anti-roll bar front and rear with stopping power provided by four large servo-assisted ventilated disc brakes, and steering by an unassisted rack and pinion system. Though the 308 originally appeared with some glass-fibre bodywork its construction has now reverted to Maranello tradition with the well-protected all-steel body panels clothing a tubular steel chassis.

On specification and price the Lamborghini Jalpa 350 provides a natural rival for the Ferrari, the two trading punches all along the line with the Jalpa having the bigger and more powerful (250 against 240 bhp) V8 engine but also extra weight to hump around (26.6 against 25.6 cwt), while the relative costs are pretty close at £26,181 for the Ferrari and £26,423 for the Lamborghini. In comparison other quick cars such as the Porsche 911 Carrera Sport (£23,366), Lotus Esprit Turbo (£18,913), BMW 635 CSi (£23,995) and Audi Quattro (£17,722) almost qualify as bargains but, of course, few of them can provide the on-the-road charisma projected by the Italian cars.

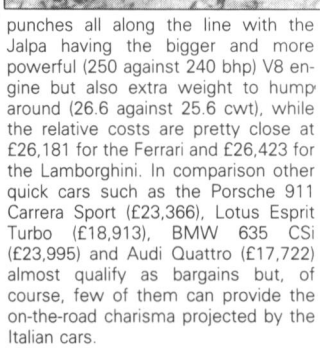

With the 308 GTBi *quattrovalvole* this image is backed up by true supercar performance. On a damp, blustery day, most unfavourable for high speed testing, it stormed around Millbrook's two-mile circumference bowl at a maximum average speed of 154.5 mph, suggesting that in more favourable conditions and on a straight level road it would comfortably pull maximum revs in fifth gear, which is 158 mph. This puts both the Jalpa (147.6 mph) and the lighter 911 Carrera (151.1 mph) in the shade and speaks volumes for the aerodynamic efficiency of Pininfarina's sensuous styling.

Off the line the Ferrari is similarly impressive with 0-60 and 0-100 mph times of 5.7 and 14.3 sec respectively, providing an interesting comparison with the Jalpa (5.8, 16.0 sec) with which it shares virtually identical times for each 10 mph increment up to 70 mph, but then, from 80 mph upwards, pulls steadily away from it. The Ferrari also bears comparison with other top grade standing-start merchants such as the Porsche 911 Carrera (0-60 mph, 5.3 sec; 0-100 mph, 13.6 sec) and Lotus Esprit Turbo (5.6, 15.9 sec).

This performance isn't accompanied by highly strung Italian temperament either, with the 32-valve engine's remarkable useable power band, extending from 800 to 7,700 rpm, illustrated by acceleration times in fifth gear (admittedly a relatively low ratio giving only 20.5 mph per 1,000 rpm) which vary between just 7.2 and 8.1 sec for every 20 mph increment between 20 mph and 110 mph, while in fourth gear the range is chopped down to 5.0 to 5.7 sec. The torquier Jalpa can match or better some of these times but it can't pull so effectively over such a wide speed range, while neither the

144

Though rather sparse-looking the leather trimmed seats are well-shaped and comfortable

911 Carrera nor the Espirt Turbo are quite in the same league with their higher upper ratios.

Even pottering around town leaves this Ferrari totally unflustered with the engine providing as little or as much urge as required for hours on end in slow moving traffic jams. Then as soon as there is a sniff of open road the engine will pull as clean as a whistle from any point within that astonishingly wide power band, with the really neck-snapping response coming from around 5,000 rpm upwards when a slightly harsh edge enters the engine note to set against its otherwise beautifully free-revving character.

As far as engine noise itself goes there is just too much of it to make prolonged high speed cruising on, say, an autobahn anything but a rather wearing experience. The *quality* of the engine note also lacks the magic often associated with well-bred power units although it is by no means unpleasant and at full chat is guaranteed to set the adrenalin racing in all but the most laid back of drivers.

The lack of temperament also extends to totally fuss-free starting and warm-up behaviour: and there is less of a penalty to pay for that scorching performance than might be expected at the petrol pumps. An overall fuel consumption of 19.6 mpg, the result of performance testing and hard road driving, is unlikely to impress the Friends of the Earth but it is an excellent result when set against the available performance and throws a rather harsh light on the 15.6 mpg notched up by the similarly quick Lamborghini Jalpa. With just a modicum of restraint most owners should find that around 23 mpg is attainable giving a theoretical range of 450 miles on each full 19.6 gallon tank of four star fuel.

This relatively restrained thirst is all the more commendable when the Ferrari's fairly hefty weight and low overall gearing are taken into account. Those closely stacked ratios do have their own advantages in give and take driving conditions, however, and, except from cold, they are satisfyingly pleasant to select with the lever moving swiftly and easily through the standard Ferrari open metal gate, albeit with the usual proviso that the gearchange responds more favourably to controlled force than carefully applied finesse. Our test car did have an uncharacteristically obstructive first to second dog leg shift, though, some compensation being provided by a moderately weighted, smooth acting clutch.

Generally, in fact, all the Ferrari's controls respond in a similar manner to the engine — requiring no more effort from the driver than any other car when motoring gently but offering the feel of a true thoroughbred when pressed to demonstrate its full capabilities. Thus the steering, though uncompromisingly heavy when parking and somewhat "dead" when running gently, is moderately weighted on the move and direct and accurate, with feel improving in direct proportion to the cornering force being generated.

And as would be expected, aided by its fat 220/55 VR 390 Michelin TRXs (the same front and rear), there is a great deal of grip available, this being retained to a fair degree in the wet. It is really only in the wet that the 308 can be provoked into biting back, with both terminal plough-on understeer and fairly abrupt lift off oversteer lying in wait for the unwary driver. In the dry the Ferrari is supremely forgiving but at the same time very rewarding to drive hard and fast.

In its basic behaviour the tendency is to mild understeer with even rapid cornering provoking no more than scarcely discernible body lean. In tighter bends the understeer can become strong but it is possible to balance the car's attitude on the throttle and really vicious provocation will kick the tail out. If a corner is entered too quickly then throttle lift off generally produces no more than a reassuring amount of tuck in; but even if the driver has completely over-cooked it, any lift-off oversteer is gentle and easy to control; in fact if he has sufficient mettle to allow it, it can be virtually self-correcting with the steering's castor action being just sufficient to gather everything up and get the car back on line.

Perhaps the best illustration of the Ferrari's high class breeding is that this beautifully balanced chassis is not achieved at the expense of ride comfort. At low speeds on rough surfaces the firm but well controlled suspension is shown in its least favourable light, really managing to do no more than round off the sharp edges of severe disturbances; but the small-bump ride is pretty reasonable even at town speeds, and with increasing speed the quality of the ride just gets better and better, demonstrating an outstanding ability to absorb bumps and ridges while still maintaining taut control.

As befits a car with the ability to achieve high speeds over the most difficult terrain the Ferrari's brakes work superbly. They may lack anti-lock control but the compensation is progressive, strong stopping power combined with excellent feel, moderate pedal weight and a total absence of fade during hard use.

To drive, then, the 308 GTBi *quattrovalvole* satisfies all the supercar criteria. And it backs this up by being surprisingly practical and habitable in other respects. The driving position for example, though slightly long-arm and short

MOTOR week ending October 29, 1983

MOTOR ROAD TEST No 50/83 ●
FERRARI 308 GTBi QV

Above: hardly a paragon of ergonomics, the 308 cockpit features plenty of instruments and a plethora of switchgear behind the open-gated gear lever

on headroom for tall drivers, is comfortable, with the offset pedals allowing easy heel-and-toeing and the leather-trimmed seat sufficiently well-shaped and supportive to keep the driver securely in place.

Less satisfactory is the switchgear, which is spread rather haphazardly around the centre console, and the indicator stalk which is too close to the lights stalk, creating the danger of the driver plunging himself into darkness when he meant to indicate right. The instruments also leave something to be desired, being commendably profuse in number but also prone to reflections and, particularly in the case of the speedometer, difficult to read.

On the other hand visibility is remarkably good for a mid-engined car with a not excessive three-quarter rear blind spot and good electrically-adjustable door mirrors. The lights, however, are poor in relation to the car's performance potential.

The heating system has separate controls for each side of the car and works fairly well with sufficient power, though it is slow to respond to temperature adjustment and the footwell vents (mounted underneath the facia at each end and identical in type to the facia top vents and thus adjustable for direction) have difficulty in projecting heated air right down to the base of the footwells. Ventilation is available (with the heater off) through the facia vents but for efficient cooling the optional air conditioner (costing £947) is really a must. Unusually, for a car so equipped, the air conditioning can be used simultaneously with the heater, to provide a proper bi-level temperature stratification.

Accommodation is strictly two-seater but, bearing this in mind, allows for a reasonable amount of luggage space with a surprisingly useful square-shaped compartment behind the engine (covered by a zip-up soft cover) with some odds and ends space behind the front seats, and a small locker between them.

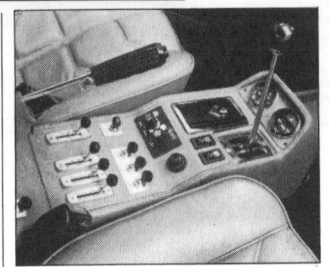

Beautiful though it undoubtedly is from the outside, some of our testers considered the interior aesthetics a bit bland, if very well executed with high standards of assembly and finish evident throughout the car. Also evident in almost indecent profusion throughout the GTBi were sufficient prancing horses to stock a Grand National field, the famous Ferrari insignia being applied to anything from the door mirrors to the ashtray lid.

If this indicates some sort of identity crisis within Ferrari then they should rest assured that their product is as exhilarating to drive and as worthy of the famous name as ever. Rather than being swamped by the onrush of advanced technology overtaking the car industry, Ferrari are right up there with the best of them. And even if their cars are easier to get along with, more comfortable and more practical than ever before, they certainly haven't lost that indefinable soul and charisma which is Ferrari.

Ferrari's magnificent four-valves-per-cylinder V8 engine is positioned transversely, just ahead of the rear wheels

PERFORMANCE

WEATHER CONDITIONS
Wind	5-15 mph
Temperature	52°F/11°C
Barometer	29.6 in Hg/1003 mbar
Surface	Dry tarmacadam

MAXIMUM SPEEDS
	mph	kph
Banked circuit	154.5	248.6
Best ¼ mile	157.4	253.3

Terminal Speeds:
at ¼ mile	100	161
at kilometre	126	203

Speeds in gears (at 7,700 rpm):
1st	42	68
2nd	62	100
3rd	85	137
4th	117	188

ACCELERATION FROM REST
mph	sec	mph	sec
0-30	2.1	0-40	1.8
0-40	3.0	0-60	2.8
0-50	4.5	0-80	4.5
0-60	5.7	0-100	6.1
0-70	7.5	0-120	8.2
0-80	9.2	0-140	10.5
0-90	11.6	0-160	14.3
0-100	14.3	0-180	18.1
0-110	17.2	0-200	23.1
0-120	20.5		
Standing ¼	14.2	Standing km	25.8

ACCELERATION IN TOP
mph	sec	kph	sec
20-40	8.1	40-60	4.7
30-50	7.4	60-80	4.5
40-60	7.3	80-100	4.4
50-70	7.2	100-120	4.8
60-80	7.3	120-140	4.9
70-90	7.4	140-160	4.9
80-100	7.4	160-180	6.0
90-110	8.0		

ACCELERATION IN 4TH
mph	sec	mph	sec
20-40	5.5	40-60	3.3
30-50	5.0	60-80	3.1
40-60	5.1	80-100	3.2
50-70	5.1	100-120	3.2
60-80	5.1	120-140	3.2

70-90	5.1	140-160	3.3
80-100	5.1	160-180	3.9
90-110	5.7		

FUEL CONSUMPTION
Overall	19.6 mpg
	14.4 litres/100 km
Govt tests	13.6 mpg (urban)
	31.5 mpg (56 mph)
	26.0 mpg (75 mph)
Fuel grade	97 octane
	4 star rating
Tank capacity	19.6 galls
	89 litres
Max range*	450 miles
	724 km
Test distance	1550 miles
	2494 km

*Based on an estimated touring consumption of 23 mpg

NOISE
	dBA	Motor rating*
30 mph	70	16
50 mph	74	21
70 mph	82	36
Maximum†	93	79

*A rating where 1=30 dBA and 100=96 dBA, and where double the number means double the loudness
†Peak noise level under full-throttle acceleration in 2nd.

SPEEDOMETER (mph)
Speedo	30	40	50	60	70	80	90	100
True mph	28	36	47	56	66	78	88	101

Distance recorder: 1.4 per cent fast

WEIGHT
	cwt	kg
Unladen weight*	25.6	1,300
Weight as tested	29.3	1,488

*with fuel for approx 50 miles

Performance tests carried out by *Motor's* staff at the Motor Industry Research Association proving ground, Lindley, and Vauxhall's Proving Ground, Millbrook.

Test Data: World Copyright reserved. No reproduction in whole or part without written permission.

GENERAL SPECIFICATION

ENGINE
Cylinders	V8
Capacity	2,927cc (178.6 cu in)
Bore/stroke	81/71mm (3.19/2.79 in)
Cooling	Water
Block	Aluminium alloy
Head	Aluminium alloy
Valves	Dohc per bank
Cam drive	Toothed belt
Compression	9.2:1
Fuel system	Bosch K-Jetronic ignition
Ignition	Marelli Digiplex electronic
Bearings	5 main
Max power	240 bhp (DIN) at 7,000 rpm
Max torque	192 lb ft (DIN) at 5,000 rpm

TRANSMISSION
Type	5 speed manual
Clutch dia	9.5 in
Actuation	Cable

Internal ratios and mph/1,000 rpm
Top	0.919:1	20.5
4th	1.244:1	15.2
3rd	1.693:1	11.1
2nd	2.353:1	8.0
1st	3.419:1	5.5
Rev	3.248:1	
Final drive	3.823:1	

BODY/CHASSIS
Construction	Tubular steel chassis, steel body panels
Protection	"Zincrox" protection for steel components, PVC underseal

SUSPENSION
Front	Independent by double wishbones, coil springs, anti-roll bar
Rear	Independent by double wishbones, coil springs, anti-roll bar

STEERING
Type	Rack and pinion
Assistance	None

BRAKES
Front	Ventilated discs 10.8in dia
Rear	Ventilated discs 11.0in dia
Park	On rear
Servo	Yes
Circuit	Dual, split front/rear
Rear valve	No
Adjustment	Automatic

WHEELS/TYRES
Type	Light alloy 165 TR 390
Tyres	Michelin TRX 220/55 VR 390
Pressures	33/33 psi F/R (normal)
	36/36 psi F/R (high speed)

ELECTRICAL
Battery	12V, 66Ah
Earth	Negative
Generator	Alternator, 80 A
Fuses	18
Headlights type	Twin halogen retractable
dip	110 W total
main	120 W total

Make: Ferrari **Model:** 308 GTBi Quattrovalvole
Maker: Ferrari Esercizio Fabbriche Automobili e Corse SpA, Viale Trento Trieste 31, 41100 Modena, Italy.
UK Importers: Maranello Sales Limited, Egham By-Pass (A30), Egham, Surrey TW20 0AX. Tel: Egham 36431
Price: £21,015.00 plus £1,751.39 Car Tax and £3,414.96 VAT equals £26,181.35. Extras fitted to test car: air conditioning, £946.83. Price as tested: £27,128.18.

The Rivals

Other possible rivals include the De Tomaso Pantera GTS (£22,730), Jaguar XJ-S HE (£20,693) and Mercedes 500 SL (£23,990).

FERRARI 308 GTBi Quattrovalvole — £26,181

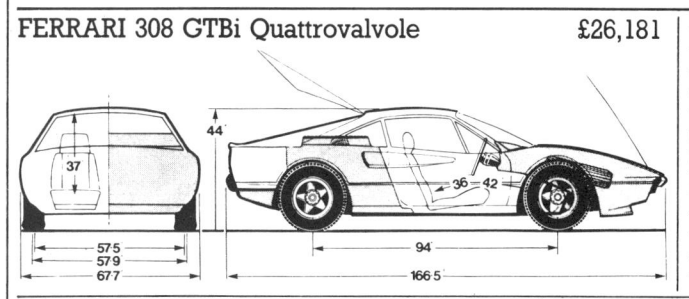

Power, bhp/rpm	240/7,000
Torque, lb ft/rpm	192/5,000
Tyres	220/55 VR 390
Weight, cwt	25.6
Max speed, mph	154.5
0-60 mph, sec	5.7
30-50 mph in 4th, sec	5.0
Overall mpg	19.6
Touring mpg	—
Fuel grade, stars	4
Boot capacity, cu ft	5.6
Test Date	October 29, 1983

Stunningly beautiful and capable with it, Ferrari's two-seater 308 GTB is a refreshingly practical supercar. The *Quattrovalvole* formula delivers scorching performance with respectable economy, while other virtues include safe, balanced handling, superb brakes, a comfortable ride and driving position and a generally satisfying gearchange. Visibility is fair, while instrumentation and switchgear leave something to be desired as do engine noise levels.

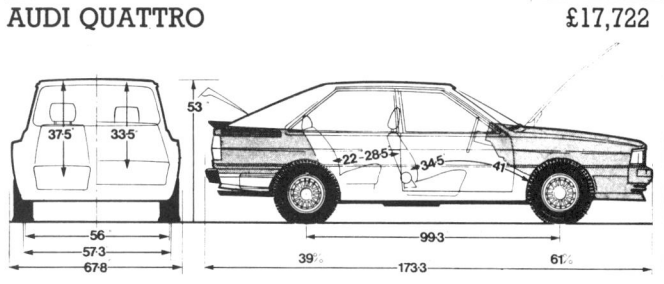

AUDI QUATTRO — £17,722

Power, bhp/rpm	200/5,500
Torque, lb ft/rpm	210/3,500
Tyres	205/60 VR 15
Weight, cwt	24.8
Max speed, mph	138e
0-60 mph, sec	6.5
30-50 mph in 4th, sec	8.2
Overall mpg	19.9
Touring mpg	—
Fuel grade, stars	4
Boot capacity, cu ft	7.6
Test Date	March 21, 1981

Now that it's looking as if its unique format — ultra-high performance and four-wheel drive — *has* started a new trend, the Quattro is more than ever a milestone in car design. It combines phenomenal roadholding and traction with performance, refinement, economy, comfort and accommodation in a way that has no equal, against which its weaknesses — poor ratios and slow shift, unprogressive heating, sparse instruments — are minor failings. Soon to be available with anti-lock braking.

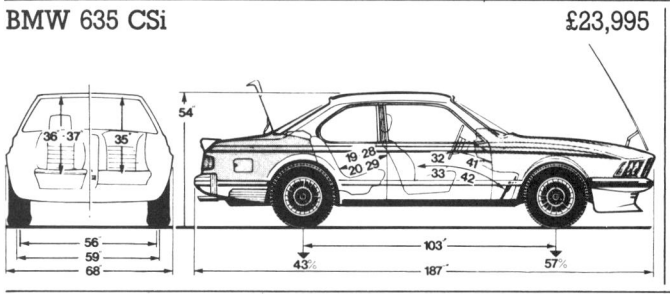

BMW 635 CSi — £23,995

Power, bhp/rpm	218/5,200
Torque, lb ft/rpm	229/4,000
Tyres	220/55 VR 390
Weight, cwt	28.0
Max speed, mph	137.1
0-60 mph, sec	6.9
30-50 mph in 4th, sec	7.2
Overall mpg	22.5
Touring mpg	—
Fuel grade, stars	4
Boot capacity, cu ft	12.5
Test Date	July 10, 1982

BMW's sleek flagship, the 635 CSi, is even more rapid and refined in its latest form, yet sets new standards of economy in the super coupé class. Handling and dry road grip are excellent, too, though in the wet traction is poor and the tail needs watching. Comfortable ride, superb gearchange, anti-lock brakes are further plus points, but rear accommodation is cramped and there is too much wind noise at speed, spoiling the otherwise good refinement. Well equipped, if fairly expensive.

LAMBORGHINI JALPA 350 — £26,423

Power, bhp/rpm	250/7,000
Torque, lb ft/rpm	235/3,250
Tyres	205/55 VR 16; 225/50 VR 16
Weight, cwt	26.6
Max speed, mph	147.6
0-60 mph, sec	5.8
30-50 mph in 4th, sec	4.3
Overall mpg	15.6
Touring mpg	—
Fuel grade, stars	4
Boot capacity, cu ft	—
Test Date	June 18, 1983

'Baby' of the two-car Lamborghini range, the targa-top Jalpa's ancestry runs back to the mid-engined Uracco of the early '70s. Magnificently vocal quad-cam V8 delivers stunning performance with massive mid-range punch, though economy is mediocre by today's standards. Very safe and ultimately forgiving handling married to reasonable ride. Fabulous brakes. He-man gearchange and poor visibility not so appealing, but practicality and finish better than of yore.

LOTUS ESPRIT TURBO — £18,913

Power, bhp/rpm	210/6-6,500
Torque, lb ft/rpm	200/4-4,500
Tyres	195/60 VR 15; 235/60 VR 15
Weight, cwt	22.6
Max speed, mph	152e
0-60 mph, sec	5.6
30-50 mph in 4th, sec	6.2
Overall mpg	18.5
Touring mpg	—
Fuel grade, stars	4
Boot capacity, cu ft	3.5
Test Date	March 28, 1981

Turbo power promotes Lotus towards the top of the first division in the supercar league. Stunning acceleration from smooth vice-less turbo 'four', allied to perfect ratios, strong roadholding, superb handling and tireless braking, all add up to a driver's car *par excellence*, with respectable economy and a comfortable ride as icing on the cake. But some shortcomings include poor heating, awkward visibility, poor pedal layout and lack of space for tall drivers.

PORSCHE 911 CARRERA SPORT — £23,366

Power, bhp/rpm	231/5,900
Torque, lb ft/rpm	209/4,800
Tyres	205/55 VR 16; 225/50 VR 16
Weight, cwt	23.0
Max speed, mph	151.1
0-60 mph, sec	5.3
30-50 mph in 4th, sec	5.6
Overall mpg	21.1
Touring mpg	—
Fuel grade, stars	4
Boot capacity, cu ft	9.8
Test Date	October 22, 1983

Little changed for 1984 except that it's quicker than ever — back at the top of the junior supercar acceleration league table — the 911 Carrera is also remarkably economical for its stunning performance. Still a great driving machine, with rewarding (though tricky *in extremis*) handling, potent brakes, superb ratios, good driving position and turbine-smooth engine. Gear change could be better, though, and remaining flaws include hard ride, poor heating/ventilation, and uncertain ergonomics.

Return Match

Back in 1978 we pitted a Porsche 911 against a Ferrari 308 in fast and furious twin test confrontation. The result was an honourable draw. Five years on, both models are running stronger than ever and are still natural rivals in the junior league supercar class. Does Latin flair still have the answer to Teutonic efficiency? There was only one way to find out....

LAST TIME it was a pair of Targas, and the day was a midsummer scorcher. This time the weather is autumn-dismal, so we don't mind that the cars are hardtops. What we do mind is the drizzle, which starts right on cue as we convoy through the gates of the Millbrook proving ground. If there is any truth in Porsche's and Ferrari's claims, we are going to lap Millbrook's high speed bowl at 150-plus for the first time, and a dry track would be desirable to say the least. For the acceleration contest later in the day at MIRA it will be essential. Now, it looks like the happenchance of a chance of dates which brought together our two exotics on an elusive, same-day, same-place showdown will be checkmated by the fickleness of the weather.

In truth, we hadn't even intended it as a twin-test at first: the logistics just weren't going to let it happen. But then fate forced a change of dates which suddenly meant that the Porsche and the Ferrari would overlap by a day, and created an opportunity that we were now loath to relinquish. For as much as they were in 1978, Ferrari's 308 and Porsche's 911 are still natural-born rivals.

Both cars have come a long way these past six years, even though superficially little has altered other than the adoption of broader Michelin TRX rubber for the 911's Pirelli P7 gumboots. It's under the engine covers that much has changed. Both cars have survived a fuel crisis and increasingly stringent emissions laws, emerging with high-tech engines that are fitter and stronger, cleaner and more fuel-efficient than ever before.

In the case of Ferrari, though, there was some faltering along the way. From the heady heights of 1978, when its classic quad-cam 2927cc V8 screamer was fuelled by a quartet of fat, gobbling twin-choke Webers and put out a claimed 255 bhp at 7,700 rpm, its outputs were forced down by environmental considerations to a nadir of just 214 bhp by 1981: a fall that owed less to the adoption of Bosch K-Jetronic injection and solid state Marelli Digiplex ignition than to drastic pruning of the engine's previously wild camshaft profiles. It took the adoption of new four-valve-per cylinder heads in late 1982 (plus a small compression lift from 8.8 to 9.2:1) to revive the 308's flagging outputs, which have now settled at 240 bhp at 7,000 rpm and 192 lb ft of torque peaking at 5,000 rpm: not quite, on paper, back to 1978's levels but, as we shall see, it seems a fair bet that 1984's horses are more honest than their 1978 equivalents.

By contrast, the Porsche's progress chart has climbed steadily upwards. A year after our '78 test its 2,994cc rear-mounted flat six received a mild boost from 180 to 188 bhp, followed by a more dramatic hike to 204 bhp a year later courtesy of new high-compression combustion chambers which also made it the first 911 ever to better 20 mpg on a Motor road test. Then towards the end of last year Porsche revived the Carrera name as part of an '84 model-year package of improvements which once again produced worthwhile gains in both performance and economy. A longer stroke pushed the capacity out to 3,164cc, new pistons and combustion chambers upped the compression from 9.8 to 10.3:1, new manifolding improved the breathing, and Bosch Motronic electronic engine management provided even tighter control of fuel and ignition settings. Maximum power was raised to 231 bhp at 5,900 rpm, maximum torque to 209.5 lb ft at 4,800 rpm, and revisions to the braking system helped to keep it all under control.

Minutiae apart, the rest reads much the same as in 1978. Both cars steer by unassisted rack and pinion, stop by large ventilated discs on all four corners, and are available only with a five-speed manual transmission. There the similarities end. Where the Porsche's ageless body (now in its twentieth year) is a galvanised steel monocoque (and carries a seven year anti-corrosion warranty), the Ferrari is more traditionally crafted with a tubular steel chassis clad in steel body panels, apart from the front and rear lids which are of aluminium alloy. Now coming on stream is Ferrari's unique Zincrox anti-corrosion treatment for all its steel body/chassis components.

As ever, the Porsche defies convention — and almost, it seems, the laws of the universe — with its air-cooled engine cantilevered out behind the rear wheels, in stark contrast with the Ferrari's state-of-the-art transverse and mid-engined configuration. In suspension design, too, they employ widely differing means to the same ends, the Ferrari's layout a classic wishbones and coil spring arrangement at each corner with an anti-roll bar at each end, the Porsche sprung by torsion bars all round with McPherson struts at the front and semi-trailing arms to support the rear, again with anti-roll bars at both ends. Ferrari road rubber takes the form of 220/55 VR 390 Michelin TRX's all round while the 911's Sport Equipment (which also includes front and rear spoilers and Sport dampers) means 205/55 VR 16 P7s at the front and 225/50 VR 16s at the rear.

Thus equipped, the Porsche weighs in at £23,366 (£21,464 without the Sport package), still with a useful price advantage over the £26,181 of the Ferrari. Not that a few thousand here or there is a make-or-break factor in this stratum of the market. Despite the price gap, the Porsche 911 Carrera and the Ferrari 308 GTB Quattrovalvole were born to compete.

At Millbrook the drizzle has dampened the track but not drenched it. We decide to give it a try, building up speed gradually, to see how it feels. We choose to do the Ferrari first, partly for its better balanced handling *in extremis* and partly because, on past performance, we suspect that its claimed 158 mph maximum is a lot more pie-in-the-sky than the Porsche's 152 mph — though given the weather conditions and the speed-sapping effect of cornering force through the banked turns, we don't expect either car to wind out all the way to its full potential. The comparative element, however, will be valid.

In the event there's no need for white knuckles. Despite the damp, grip is no problem, and there are no Italian histrionics. The Ferrari simply gets up there and does it: 154.5 mph for the lap, 157.4 mph over the fastest downwind quarter with the revs nudging the 7,700 rpm limit. The engine is howling but it's a healthy sound. Given a still day and a long enough straight, there's little doubt that it would pull the red line both ways, to equal the factory's 158 mph claim.

If anything, the Porsche is harder work. As we wind up towards max, suddenly the leading edge of the sunroof pops up a fraction and the wind noise is deafening. The steering, almost in the featherweight class at parking speeds, starts to centre powerfully as cornering force climbs and there's a trace of yaw when the wind is coming from the side. The 911 tops out at 151.1 mph for the lap, 154.7 mph for the best quarter — and with some relief I ease off the throttle. My arms are aching from the effort of steering it. The Porsche has come closer than the Ferrari did to equalling its makers claims and on a good day would probably also pull its red line (6,300 rpm and 153 mph) if not a shade faster still: but not faster enough to better the Ferrari. One up to Maranello.

Next stop is at MIRA's acceleration straights, and, when finally they dry out, a chance for the 911 to redeem itself. In 1978 the 911 had exploited superior traction and a 3.1 cwt weight advantage to more than offset a nominal 75 bhp power deficit; it outsprinted the 308 all the way to 100 mph, though the Italian car pulled away thereafter. This time the weight difference has narrowed slightly (23.0 cwt for the Porsche, 25.6 cwt for the Ferrari) and so has the power gap. According to the form books, the Porsche should retain its sprinting title (at least up to around 110 mph) while the Ferrari, by virtue of its substantially lower gearing (20.5 mph per 1,000 rpm vs 24.3 in fifth, 15.2 as against 19.2 in fourth) should take the honours in a high gear slogging contest.

In the event, an afternoon of laying rubber produces no upsets except that neither car is easily persuaded to deliver its ultimate. The Ferrari proves easy to power cleanly off the line with nicely controlled wheelspin, but uncharacteristic stiffness in the slotted-metal gearshift resists the usual enormously effective and satisfying slam-it-through technique. It takes perseverance and not a few missed shifts before we're satisfied we've wrung the best from it. The Porsche doesn't win any bouquets for its gearchange either. Its combination of cross-gate slop and notchy synchromesh is intrinsically less pleasurable than the Ferrari's ought to be — but on the day proves easier to master, provided you don't try to rush the first to second shift.

The tricky bit with the Porsche is getting it off the line: if you drop the clutch with too few revs on the clock you don't break traction, and the engine "bogs"; too many revs and excess wheelspin bounces the engine against the rev-limiter before you've even got moving. Get it right, though, and the Carrera claws off the line as if it has four-wheel drive. Thirty comes up in a blistering 1.9 seconds, sixty in 5.3, the ton just 13.6 seconds after blast-off.

In a test track drag race the Ferrari just can't quite cut it against the Porsche. By any other standards though, its figures are scintillating — 2.1 sec to 30 mph, 5.7 to 60, 14.3 to 100. And it's never far behind, eventually to close the gap and then forge away from the Porsche shortly after passing the kilometre post.

So much for the test track. On the road, driver differences would count for more than the ultimate abilities of the cars themselves, and more often than not the 308's more forgiving clutch would get it away quicker from the lights: in a fast getaway, over-enthusiastic slipping of the Porsche's more sudden clutch can make the friction plate wilt, while dropped-clutch wheelspin starts are inadvisable on the public road . . . And it's on the road, too, that the Ferrari scores with its almost unbelievably wide power band. What

Despite leather finish, the Ferrari's seats are grippy and comfortable. Legroom is generous but headroom less so, basic driving position mildly Italianate

Cloth trim fitted to the 911 test car is more slippery and but seat shaping is fine and rearward travel generous. Some drivers found the gearlever too far forward

149

Ferrari's black facia is rather plain but main instruments well placed. Column stalks apart, most controls are located on the centre console. Gearshift pattern places first on a dog-leg to the left

Instruments and minor controls are sprawled haphazardly across the Porsche's facia, with heater temp control on the floor between the front seats

other car will pull all the way from idle speed to 7,700 rpm without a stutter . . . in *top* gear? Subjectively you can detect a slight up-tempo shift as you pass 5,000 rpm, at which point just a hint of harshness starts to overlay the rising crescendo of engine noise. But just look at the figures: between 30 and 100 mph in fourth the rate of acceleration is almost completely constant, and barely less so over the same speed range in top. Throttle response is whistle-clean at any speed, starting and warm-up immaculate: what price Latin temperament? Only engine noise stays true to the cliches. It isn't even as nice a sound as it was when the 308 breathed through a quartet of carburetters, and it's sufficient in volume to render academic the relatively low levels of road and wind noise intrusion. At autobahn speeds you can forget about the radio.

On that score at least, the Porsche, despite its higher levels of tyre rumble and thump and a hint of wind hiss, is a notably more peaceful proposition on a long run. By any normal standard the Porsche is also a paragon of flexibility; but its altogether rangier gearing never quite lets it get on equal terms with the Ferrari, which takes 5.0 seconds for the fourth-gear 30-50 mph increment as against the Porsche's 5.6, and a full second less from 50 to 70 mph in top than the Carrera's 8.2 sec.

Yet for all that it's the 911 that somehow *feels* the faster car, as if by compressing its power delivery into a narrower rev band it packs a heftier wallop than the Ferrari's more extended and evenly spread power band. In the 911, there's a tangible increase in thrust as the tacho needle passes 4,000 rpm. and in its own way the Porsche feels perhaps even more exhilarating at 6,300 rpm than the Ferrari does at 7,700 — despite being a good deal quieter at the top end. At the other end, though, it's actually the Italian car that proves more amenable to traffic-queue trickling, where the Porsche, perhaps as a consequence of its overrun fuel shut-off, can get a bit snatchy.

On the day neither car's gearchange delivers the driver satisfaction that it ought, the Ferrari's snicking easily through its metal gate when you're taking it easy but with an untypical (from past experience) reluctance to be slammed through under hard use. In the Porsche's case, its newness tells against it — 911 shifts are slow to mature — though even at its best the 911's notchiness and its slightly imprecise (albeit conventionally arranged) gate require more concentration than the Ferrari's does on a good day. As was the case in 1978, the Ferrari stacks its ratios very closely together at the bottom end with wider gaps from third to fourth and fourth to top, while the 911's do precisely the opposite. In each case, the choice suits the car well.

Considering its greater weight and lower gearing, the Ferrari does well to return an overall consumption of 19.6 mpg; that speaks volumes for the inherent efficiency of its engine. Even so, the trimmer, longer-striding 911 takes the economy prize, with a remarkable 21.1 mpg overall. In neither car, though, need the cost of fuel be a make-or-break consideration.

Before we leave MIRA, a session on the skid pan affirms what experience has

150

already led us to expect: the potential treachery of the 911 when finally you push it too far. Snap the throttle shut when you're cornering near the limit in the 911 and strong understeer is exchanged for sharp tuck-in and then a relentless tail slide. It doesn't happen so quickly that a stab of corrective lock and re-application of power won't retrieve the situation, but your response has to be prompt and accurate. If not, the tail swing gains momentum and becomes unstoppable.

Up to a critical point, the Ferrari's handling is not that much different. It, too, is a natural understeerer — strongly so through tight bends, unless you're vicious enough with the throttle to break the tail away under power — and it too has strong lift-off tuck-in which can, under dire provocation, become a tail slide. The critical difference is that oversteer in the Ferrari is much less likely to be terminal: it's a gentler, less relentless slide that can be collected and corrected without greatly taxing the driver's skill. On a dry road, it's a difference that most people will never discover except by accident — but that's when it'll count the most.

On a wet surface, all the above still holds true, except that it happens at a lower threshold of grip: if it's greasy too, factors which in the dry are largely academic can suddenly, become a matter of pedestrian speeds, become a matter of dramatic reality. . . .

Yet on the road it's the Porsche which,

In the Porsche, no matter how you sit, some instruments will be obscured from view by the steering wheel and/or driver's hands. Oil level gauge is included, but there's no water temp gauge . . . .

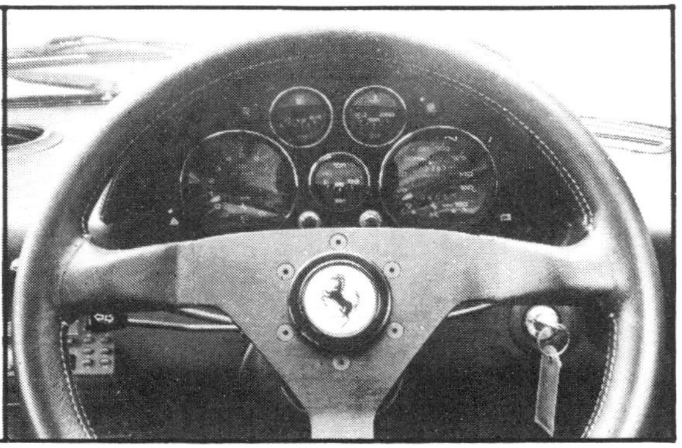

Although also reflective, most of the Ferrari's main display can be seen by most drivers, unless the wheel rake is set very low. Tachometer is red-lined at 7700 rpm and it's not just academic

even driven within its limits, is the superior adrenalin pump. Almost daintylight at parking speeds, its steering is sharp and pointy on the move, the effort building up in direct proportion to cornering speeds. Some might even find the kick-back excessive on bumpy roads, for there is probably no other car made that transmits so much pure feel: you feel every ripple, every change in surface; you feel the tyres knifing into the tarmac. Huge reserves of rear end grip let you power through tight curves without even a chirp from the tyres — though deliberate power oversteer, if you do manage to induce it, is a far more manageable proposition than the lift-off variety. As a pure driving machine, the 911 remains without peer.

By comparison the Ferrari, just as capable and ultimately more forgiving, simply doesn't provide the same driving sensations. It's a more low-key experience, perhaps too much so in the case of the steering which feels curiously dead in gentler motoring and only starts to communicate when you begin to load up the suspension with cornering force. But by the same token the Ferrari is less hard work, easier on the arms as you approach its limits, more manageable when you overstep them. And at very high speeds, in crosswinds and on uneven surfaces, the Ferrari tracks straight and true where the Porsche is starting to wander and twitch.

Where the Ferrari's suspension is at its most convincing is in terms of comfort,

PERFORMANCE

MAXIMUM SPEEDS	Ferrari		Porsche
	mph		mph
Banked circuit	154.5		151.1
Best ¼ mile	157.4		154.7
Terminal speeds:		
at ¼ mile	100		101
at kilometre	127		126
Speeds in gears:	at 7,700 rpm		at 6,300 rpm
1st	42		37
2nd	62		64
3rd	85		93
4th	117		121

ACCELERATION FROM REST

mph	sec		sec
0-30	2.1		1.9
0-40	3.0		3.0
0-50	4.5		4.1
0-60	5.7		5.3
0-70	7.5		7.2
0-80	9.2		9.2
0-90	11.6		11.2
0-100	14.3		13.6
0-110	17.2		17.0
0-120	20.5		21.5
Standing ¼ mile	14.2		14.0
Standing kilometre	25.8		25.5

ACCELERATION IN TOP

 | Ferrari | | Porsche
--- | --- | --- | ---
mph | sec | | sec
20-40 | 8.1 | | 8.0
30-50 | 7.4 | | 7.6
40-60 | 7.3 | | 7.7
50-70 | 7.2 | | 8.2
60-80 | 7.3 | | 8.5
70-90 | 7.4 | | 9.1
80-100 | 7.4 | | 9.5
90-110 | 8.0 | | 9.9

ACCELERATION IN FOURTH

mph	sec		sec
20-40	5.5		5.8
30-50	5.0		5.6
40-60	5.1		5.6
50-70	5.1		5.7
60-80	5.1		5.6
70-90	5.1		5.4
80-100	5.1		5.6
90-100	5.7		6.2

FUEL CONSUMPTION

Overall mpg 19.6 21.1
Tank capacity, galls 19.6 17.6

PORSCHE

ENGINE
Cylinders	Flat 6
Capacity	3,164cc (192.9 cu in)
Bore/stroke	95/74mm (3.74/2.93in)
Cooling	Air
Block	Light alloy
Head	Light alloy
Valves	Sohc per bank
Cam drive	Chain
Compression	10.3:1
Fuel system	Bosch Motronic injection
Ignition	Bosch Motronic solid state
Bearings	8 main
Max power	231 bhp (DIN) at 5,900 rpm
Max torque	209 lb ft (DIN) at 4,800 rpm

TRANSMISSION
Type	5-speed manual
Clutch dia	8.9in
Actuation	Cable
Internal ratios and mph/1,000 rpm	
Top	0.763:1/24.3
4th	0.966:1/19.2
3rd	1.261:1/14.7
2nd	1.833:1/10.1
1st	3.182:1/5.8
Rev	3.325:1
Final drive	3.875:1

BODY/CHASSIS
Construction	Galvanised steel, unitary
Protection	All cavities Tectyl treated; PVC underseal; 7 year anti-corrosion warranty

SUSPENSION
Front	Independent by McPherson struts; longitudinal torsion bars; anti-roll bar
Rear	Independent by semi-trailing arms; transverse torsion bars; anti-roll bar

STEERING
Type	Rack and pinion
Assistance	No

BRAKES
Front	Ventilated discs 11.1in dia
Rear	Ventilated discs 11.4in dia
Park	On rear by separate drums
Servo	Yes
Circuit	Split front/rear
Rear valve	Yes
Adjustment	Automatic

WHEELS/TYRES
Type	Alloy, 6J × 16in front; 7J × 16in rear
Tyres	205/55 VR 16 front; 225/50 VR 16 rear
Pressures	29/36 psi F/R

ELECTRICAL
Battery	12V, 66Ah
Earth	Negative
Generator	Alternator, 1,260 W
Fuses	24
Headlights	Twin halogen
type dip	110 W total
main	230 W total

FERRARI

ENGINE
Cylinders	V8
Capacity	2,927cc (178.6 cu in)
Bore/stroke	81/71mm (3.19/2.78in)
Cooling	Water
Block	Aluminium alloy
Head	Aluminium alloy
Valves	Dohc per bank
Cam drive	Toothed belt
Compression	9.2:1
Fuel system	Bosch K-Jetronic injection
Ignition	Marelli Digiplex electronic
Bearings	5 main
Max power	240 bhp (DIN) at 7,000 rpm
Max torque	192 lb ft (DIN) at 5,000 rpm

TRANSMISSION
Type	5-speed manual
Clutch dia	9.5in
Actuation	Cable
Internal ratios and mph/1,000 rpm	
Top	0.919:1/20.5
4th	1.244:1/15.2
3rd	1.693:1/11.1
2nd	2.353:1/8.0
1st	3.419:1/5.5
Rev	3.248:1
Final drive	3.823:1

BODY/CHASSIS
Construction	Tubular steel chassis, steel body panels
Protection	"Zincrox" protection for steel components, PVC underseal

SUSPENSION
Front	Independent by double wishbones; coil springs; anti-roll bar
Rear	Independent by double wishbones; coil springs; anti-roll bar

STEERING
Type	Rack and pinion
Assistance	None

BRAKES
Front	Ventilated discs 10.8in dia
Rear	Ventilated discs 11.0in dia
Park	On rear
Servo	Yes
Circuit	Dual, split front/rear
Rear valve	No
Adjustment	Automatic

WHEELS/TYRES
Type	Light alloy 165 TR 390
Tyres	Michelin TRX 220/55 VR 390
Pressures	33/33 psi F/R (normal) 36/36 psi F/R (high speed)

ELECTRICAL
Battery	12V, 66Ah
Earth	Negative
Generator	Alternator, 80A
Fuses	18
Headlights	Twin halogen retractable
type dip	110 W total
main	120 W total

Its ride is almost indecently good for a sporting machine, adequately pliant (severe potholes apart) even at town speeds and a masterful combination of suppleness and control as you go faster. The Porsche, too, is better at high speed than at low and is supremely well damped, but its ride is uncompromisingly sporty, hard and knobbly around town, remaining jerky at high speeds over surfaces you'd hardly notice in the Ferrari. Braking on both cars is marvellously potent and delivered courtesy of an ideally firm and progressive pedal action, albeit with a slight moan from the 911's discs under a severe pasting. Advantage Porsche, though, for its lesser tendency to lock its front wheels when braking hard on a wet surface.

By the time we get back to North London the evening traffic is starting to clot. Sitting in jams becomes an opportunity to consider practicalities and creature comforts. With its tall, comparatively narrow cabin and slim pillars the 911 provides an exceptionally good view out, though for a mid-engined car the Ferrari is pretty good too, its rear pillars creating only a mild obstruction. Wipers and door mirrors (electrically adjustable in both cars) are excellent in each case, but the Ferrari's pop-up headlamps are weak. In each car the instrumentation is an old-fashioned collection of individual, reflection-prone glass covered dials, though the Ferrari's closely-grouped collection is easier to see — visible to most drivers through the centre of the wheel — than the Porsche's facia-wide sprawl. In both cases switchgear, and heating and ventilation controls, are scattered and confusing. As ever, the Porsche's heater, drawing heat from around the exhaust manifolds of the air-cooled engine, is ultimately powerful but almost impossible to regulate accurately. A powerful blast of independent fresh air ventilation is available, though only with the fan on its noisy fast setting, and the siting of the small centre-facia vents favours the passenger at the expense of the driver. In the case of the Ferrari the (extra-cost) air conditioning can be used to supply face-level cool air separately from the heater, though the supply of warm air to the footwells is not overgenerous and response is slow to adjustment of the temperature.

Seat shaping is excellent in both cars, with the Ferrari locating you well despite its potentially slippery leather upholstery. For our two mid-size testers the Ferrari's slightly long-arm driving position is marginally better-liked, especially as its steering wheel is rake-adjustable (though it's a spanner job), besides which the Porsche's gearlever can be set a shade too far forward. There's a vast amount of legroom for tall drivers in both cars, but headroom is limited in the Ferrari. The 911 also scores on the flexibility of its accommodation, with two tiny rear seats that can individually, or together be folded flat to create an extra luggage platform, and a front luggage compartment that'll take almost twice as much as the Ferrari's rear mounted one — though in the Ferrari, too, it's possible to cram some stuff behind the rear seats if they aren't set too far back. Neither car's interior finish is exactly plush — the Porsche's facia has scarcely changed in 20 years, and the 308's plain black leathercloth dash trimming is decidedly dated. To the eye and to the ear, however, Ferrari build integrity is now every bit the equal of the Porsche's.

At day's end we rendezvous back at *Motor's* offices. It's been a long haul. Now, there comes the dawning realisation of something so unexpected that, at first, we're reluctant to admit it to each other. Our enthusiasm for the Porsche remains undiminished … and yet. The feeling sneaks up on us at first, and then takes hold with growing conviction. We *like* the Ferrari better.

This takes some getting used to. After all, everyone knows about the crushing efficiency and logic of German engineering; and we all know about the patchy brilliance of Italian exotica, fast and beautiful yet temperamental and riddled with impracticalities. Yet here the cliche is turned on its head. It's the Porsche that emerges as the idiosyncratic design of uneven ability, the Ferrari as the less flawed and more rounded design.

It's not an easy thing to rationalise — but then you don't buy cars like these for rational reasons. You buy them for the way they make you feel. On that score, the Porsche is still the thrill-machine. It may not be any faster than the Ferrari in practical terms, but it feels as if it is; it may not corner any quicker, but the experience is a more vivid one. On a more mundane level, the Porsche is quieter on a motorway. But then there are the flaws. The hard ride and the road noise. The instruments that you can't control. The clutch and gearchange that demand concentration. And the potentially treacherous handling on the limit.

For the chance to own a Porsche we could live with all those things; but with the Ferrari we wouldn't have to. The Maranello car's flaws are fewer, its virtues more subtle. But what finally clinched it is this: when we parted that evening after 12 long hours of driving and testing, the car in which we both wanted to drive home was the Ferrari.

Jeremy Sinek

Due in Australia in the coming months, Ferrari's 308 Quattrovalvole is as fast as a Ferrari should be. Steve Cropley reports

There is no better car to show that an early 1980s crisis at Ferrari has passed than the latest 308GTB, the *Quattrovalvole* fuel injection. A couple of years ago, the Maranello men, led by their Fiat masters, were forced by Europe's tightening clean-air laws, plus Fiat's designs on the important US sports car market, to adopt fuel-injection for their quad-cam, 3.0 litre V8, then running two valves per cylinder.

The move slashed the power of the ballsy V8 from 191 kW to 160, and the GTB, the most spirited user of the Ferrari transverse V8 package, suffered greatest harm to its reputation. Gone, along with much of the acceleration kick-in-the-back, was the Webers' characteristic roar, and the car was diminished by that, too. The seriousness of the problem is shown up by the fact that whereas early fuel-injection V8s are in easy supply in the UK at low prices, carburettor cars are much more keenly sought and rarely available in good condition.

Then, 18 months ago, Ferrari unveiled a four valves per cylinder 3.0 litre V8 and this boosted power to 180 kW at 7000 rpm accompanied by 265 Nm of torque at 5000 rpm. In reality, the engine's outputs are very similar to those of the carburettor cars, since in pre-injection days the Italians didn't have the same interest in telling the truth about power figures as is forced on them now, principally by the German transport department. Thus the 308GTB QV has all the Weber-fed car's flashing performance, plus far better fuel economy and torque spread.

Through the changes, the 308 has kept its looks. Most people say it's the most beautiful exotic of them all, which is a bonus when you consider the GTB is far from being the most expensive. But under the skin, the changes have been considerable. When the company introduced the two-valve injection engine, Ferrari updated the chassis to forestall any impression that the car was aging in its handling, compared with cars like the Lotus Esprit and Lamborghini Jalpa. This required considerable suspension adjustment, since the Pirelli P7 and Michelin TRX tyres which took over from the faithful old Michelin XWXs were fatter in section but had a smaller overall diameter. The car's ride height had to be adjusted, spring rates and characteristics were altered, there were modifications to the inner body to accommodate the bulkier tyres on lock and, as with so many cars which swap to low profile Michelins and Pirellis, much work was needed on the suspension bushes to minimise bump-thump and 'tune' their compliance.

The result is a car which can be fairly called a 1984 Ferrari. The looks remain magnificent; they're only improved by the chunkier tyres. The engine is a magnificence. It lacks the roar of the Webers, but its elasticity-with-power more than makes up for that. Besides, the lightning responses are the kind which aren't available in cars less specialised than this one, and the sounds are still that blood-curdling combination of whizzes, whines, barks and grumbles that go to make a noise which is unique to the multi-cylindered Italian. Perhaps in print it seems a little banal to attach importance to looks and noises, but the Ferrari's sounds are as implicit in its overall function as the wheels. The noise, and to an extent the sheer grace of the body's curves, is the reason I'd buy a 308 over a Lotus Turbo, even knowing the Ferrari is 20 per cent more expensive (in the UK) and that it wouldn't go as hard as the car from Hethel.

But the key to the Ferrari, of course, is the way it goes. There's enough power — in European tune — to put away standing 400m times below 15 seconds; standstill to 100 km/h is consumed in 5.8 sec (only about 0.3 sec slower than a Boxer) and, with its 32-valve power, the 308's a genuine 240 km/h car again. The engine is smooth but the throttle is faithful to every heartbeat. If you want the car to proceed smoothly, you have to drive it that way. Muffed gearchanges and untidiness with the power are far more noticeable in Ferraris than others, even Lambos. So important are sensitive controls to the Italians that the 308's throttle pedal has a small bevel gear set at its base to transmit the fore/aft movement of your foot into the horizontal push-pull action of the throttle cable.

That kind of integrity is applied to

Four-valve flash

the workings of all the controls. The steering feels sharp but lifeless at low speeds but truly comes alive when the car passes 80 km/h, the steering wheel angled farther away from the vertical than in most exotics as has been the Ferrari fashion for a couple of decades. The clutch, though not light, has an utterly predictable throw and a firm takeup. The gearchange, a small lever sprouting from an open gate, can be hideous to use when cold and is never actually foolproof, but it has a terrific metal-to-metal action, and when you concentrate properly it'll change ratios as quickly as you can move the lever.

The 308 is a fairly difficult car to

Four-valve injected V8 (below right) restores power to healthy (and genuine) 180 kW, doesn't sound as nice as Webered engine but is fabulously flexible. Cockpit (left) is workmanlike as ever, controls need firm yet sensitive driver for GTB to offer up all its considerable charms

drive, which is why many first-time owners are disappointed. The GTB needs sure hands to cope with its firm and precise controls. It needs concentration to drive smoothly and fast. It needs a sensitive driver to sort out the myriad of messages that are imparted to driver through the hands, the seat, the pedals; perhaps merely the vibration of the toeboard. It even needs a driver with rather good eyesight, to pick early the characteristics of corners seen from the lower-than-normal seating position. The 308 Quattrovalvole is a fast car, and it tends to arrive at difficult corners sooner than most. As far as handling balance goes, the Ferrari is approaching the foolproof, now that it has the huge cornering grip of the TRX/P7 tyres to depend on. The tail still snaps out eventually, and without much warning, but it happens only when road conditions are very bad or in the dry when the driver's contemplating suicide. Apart from a tendency to go straight on in tight corners (or on slippery roads) under brakes, the car has no real vices. It rarely displays attitude at all. Some body roll, evident in photographs, is never felt.

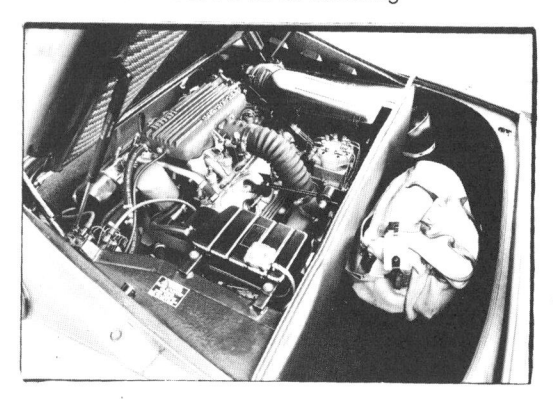

The 308 is not perfect, of course. It is cramped in the cockpit, especially for headroom, and the instrument and dashboard layout is unimpressive for something that costs as much as two Mercedes-Benzes. On the other hand, the Ferrari is not a luxury car (another reason for some owners' disenchantment) and there's no reason why the decor should say different. In its latest guise the 308 is well-finished, with the last word in rustproofing through much use of zinc-coated metal. For such a sophisticated, beautiful and fast car, it's well priced among supercars. And it's comparatively well-known at service garages. Were I on satisfactory terms with my bank, mine would be red.

SCW

FERRARI MONDIAL CABRIOLET QUATTROVALVOLE

The one to be seen in

WHEN WE FIRST tested the Ferrari Mondial 8 more than three years ago, it was less than fast and more than a little frumpy. Trying to move its 3640 lb, even the excellent 308 engine—then still in 2-valve form—was at best uninspiring. The car had Ferrari's typically good gearbox, brakes and handling, but we wondered who the intended customer really was. Who would choose the Mondial—not even a good 2+2 let alone a real 4-seater—over the quicker, more nimble and much prettier 308GTBi or GTSi?

If you're an affluent open-air driving enthusiast, there's now a pretty good reason: a true folding soft-top version, the first regular production cabriolet from Ferrari since the 330/365 GTS models of—good grief!—the early Seventies. Add the Quattrovalvole (4-valve) cylinder heads and 25 more *cavalli* and you have a Mondial with some spirit. Not the same car at all.

It's even lighter (though not light) at 3545 lb. Whether Pininfarina saved some weight in converting it from the coupe (the rear glass of which would be a significant amount) or whether our 1981 test coupe was an excessively heavy early production example, this one is nearly 100 lb leaner. It even looks leaner.

ROAD & TRACK ROAD TEST

312 PB AND 330 P3/4 SPYDERS COURTESY HARLEY CLUXTON

Mondial Quattrovalvole Cabriolet flanked by 330 P3/4 and 312 PB spyders.

and Pininfarina has done a good job on the top, both esthetically and functionally. Up, the black fabric top gives the Mondial a rakish line; down, it makes the car festive and inviting. This is the car for the boulevardier, the Monte Carlo or Newport Beach sportsman hard at play.

We were pretty skeptical about the top, having heard stories of a diabolically difficult mechanism and wondering where, between rear seat and engine, they could have found a place to stow it. It's definitely a 2-man top, but if you read the instruction manual and make sure that one of the crucial bows is held at the right angle as the top goes down, it's a piece of cake. It does protrude above the rear deck when down, and with the Mondial's low seating you really must depend on the mirror for rearward vision. What Pininfarina has done is leave part of each rear sail in place; this shape is duplicated by the protective boot, helping to minimize the apparent height of the folded top. Another good touch is the provision of retractable quarter windows, which can be lowered electrically even with the top up. There is a certain amount of drumming from the top driving with it up; this becomes obtrusive at about 85 mph, then surpris-

AT A GLANCE	Ferrari Mondial Cabriolet	Aston Martin Volante	Porsche 911 Cabrio
Curb weight, lb	3545	4330	2750
Engine/drive	V-8/rwd	V-8/rwd	flat-6/rwd
Transmission	5-sp M	5-sp M	5-sp M
0–60 mph, sec	7.6	8.9	7.0
Standing ¼ mi, sec @ mph	16.0 @ 87.0	16.8 @ 84.5	15.5 @ 88.0
Stopping distance from 60 mph, ft	153	165	146
Interior noise at 50 mph, dBA	76	70	73
Lateral acceleration, g	0.808	0.667	0.803
Slalom speed, mph	60.3	na	59.8
Fuel economy, mpg	13.5	13.0	23.5

Mondial Cabriolet: A stronger 4-valve with open-air *raison d'être*.
Volante: Classic extravagance for the very few
911 Cabrio: Porsche performance with the lid off

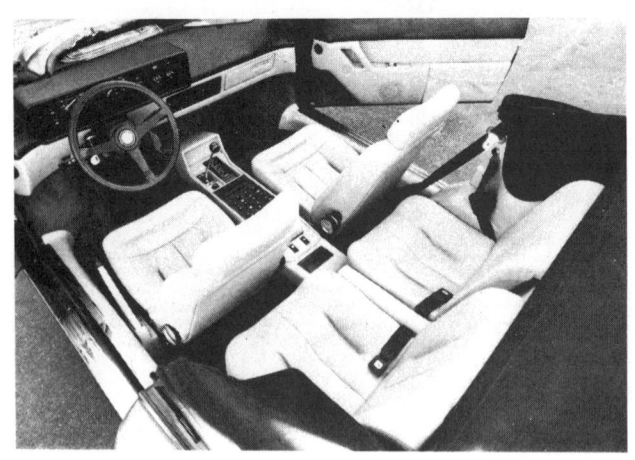

ingly diminishes and gradually builds up again as the car approaches its maximum speed.

What kind of use will the Mondial Cabriolet get? We still don't see it as a 4-seater—even children aren't well accommodated in the back—but as a 2-seater with very occasional short-run rear seating, such as a blast from cabaña to cafe on a sunny day with the top down. The interior is still not especially luxurious for a $65,000-plus car, nor particularly well arranged. To get the seating package within the wheelbase (quite a bit of which is used by the engine, even though transverse), the front seats have been shoved forward, very near to the large front wheelhouses, giving the driver and front passenger a shoehorned feeling that is made worse by the low cushion height, which reduces the view all around. If you're driving fast, only looking down the road, it's not a problem, but maneuvering in dense traffic requires a less than graceful amount of neck craning.

Fortunately, all the controls are where they should be; the steering wheel rim is of just the right thickness and when you drop your right hand, that very positive shift lever is right there. Not so good is the view of the instruments; for some drivers the tachometer is obscured by the wheel. The seats and door panels are covered in simple but high-quality tan leather, the lighter color making the interior design less forbidding than on previous Mondials in black. We were also pleased to see that the formerly protruding mesh speakers (which could scrape the knuckles of the left hand when the brake lever was being used) have been replaced by nicely integrated grilles in the door panels. But the ventilation system is mediocre for such an expensive car; the center vents put out a fair volume of air but the lack of side vents limits the effectiveness on a really hot day. But then, the top ought to be down, right?

The Quattrovalvole, as we said in our more recent 308GTBi test, is a superb engine, growly and eager, giving out a wonderful howl at its maximum of 7700 rpm. The Mondial gearing has been changed to suit the Quattrovalvole's characteristics; the final drive is numerically higher (4.06:1 versus 3.71), 1st and 5th are slightly lower, and the middle three ratios remain the same, the result being moderately shorter gearing throughout. The 4-valve engine gives vastly improved performance off the line, as Americans are wont to enjoy, much stronger acceleration all the way up, and an increase in top speed, now 138 mph at 6800 rpm in 5th. Cruising for long distances can be unpleasantly loud, more so than with the steel top; the problem is not so much engine noise as structure resonance. Frankly, the top end of the performance spectrum would be better enjoyed in a GTB or GTS, but the Mondial Cabriolet will *feel* just as fast because of its higher sensory inputs.

Using a positive but not excessively heavy clutch, the Mondial can be eased through the gears in a relaxed manner or driven fiercely for all it's worth. The shifting, within the beautiful, no-nonsense gate, needs to be done with absolute assurance; you can't get it into the next gear with the fingertips but when you move it forcefully it goes in with absolute directness, telling you in a very mechanical way that, yes, by God, that's 3rd all right. Even if you don't use the gears religiously, the engine's flexibility lets you burble through traffic in a leisurely way.

The steering has a direct, positive feel that some might find a little heavy. It's a bit slow for low-speed maneuvering, and the turning circle is rather large, but for fast road work it really does the job. There is a tendency toward understeer that increases as you go faster. This means you use a bit of muscle controlling the car but the big Michelin 240/55VR-390 tires have more grip than you're ever likely to use in normal spirited driving. There is oversteer at the very limit, as in our skidpad test (0.808g). Throwing the car from side to side in the slalom also makes the tail come out, and you have to keep the power on. The suspension is supple but noisy over sharp irregularities; you also get noticeable bump-steer from anything really protruding from the surface. The Cabriolet's structure is less rigid than the coupe's, transmitting some flexing and shaking.

We now feel that Ferrari has a Mondial with real *raison d'être*; faster, better looking, with wind-in-the-hair driving and all the attention from the sidelines you can handle. The Cabriolet was genuinely admired by most observers; drive it, and you will not be ignored. Forget the back seat, or put a Doberman in it as a guard dog. Two seats are enough, and the Ferrari has the performance to provide the most exhilarating open-air driving you could want.

ROAD TEST
FERRARI MONDIAL CABRIOLET QUATTROVALVOLE

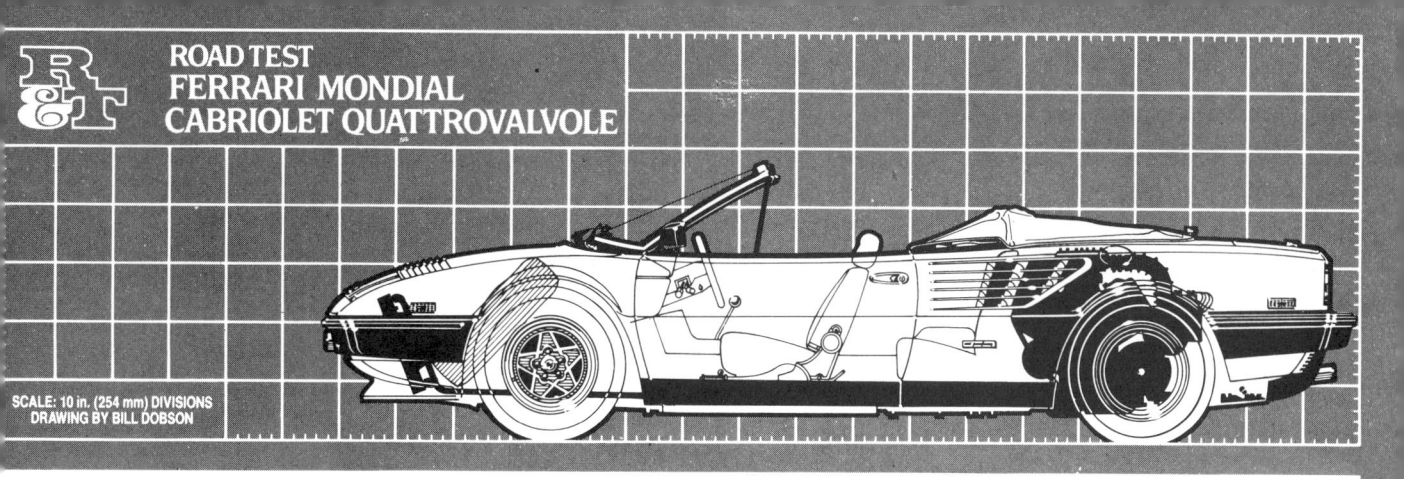

SCALE: 10 in. (254 mm) DIVISIONS
DRAWING BY BILL DOBSON

PRICE
List price, all POE $65,000
Price as tested $66,180
Price as tested includes std equip. (leather interior, elect. window lifts), metallic paint ($830), dist prep ($350)

IMPORTER
Ferrari North America, 777 Terrace Ave, Hasbrouck Heights, N.J. 07604

GENERAL
Curb weight, lb/kg	3545	1609
Test weight	3680	1671
Weight dist (with driver), f/r, %		44/56
Wheelbase, in./mm	104.3	2650
Track, front/rear	59.6/60.4	1513/1535
Length	182.7	4640
Width	70.5	1790
Height	49.6	1260
Ground clearance	4.6	117
Overhang, f/r	39.7/38.7	1008/983
Trunk space, cu ft/liters	5.0	142
Fuel capacity, U.S. gal./liters	18.5	70

ACCOMMODATION
Seating capacity, persons		2+2
Head room, f/r, in./mm	38.5/33.5	978/851
Seat width, f/r	2 x 18.0/2 x 18.5	2 x 457/2 x 470
Seatback adjustment, deg		45

ENGINE
Type		dohc V-8
Bore x stroke, in./mm	3.19 x 2.80	81.0 x 71.0
Displacement, cu in./cc	179	2927
Compression ratio		8.6:1
Bhp @ rpm, SAE net/kW		230/169 @ 6800
Equivalent mph / km/h		139/224
Torque @ rpm, lb-ft/Nm		188/255 @ 5500
Equivalent mph / km/h		92/148
Fuel injection		Bosch K-Jetronic
Fuel requirement		unleaded, 91-oct

Exhaust-emission control equipment: dual 3-way catalytic converters, air injection, exhaust-gas recirculation

DRIVETRAIN
Transmission		5-sp manual
Gear ratios: 5th (0.92)		3.74:1
4th (1.24)		5.03:1
3rd (1.69)		6.86:1
2nd (2.35)		9.54:1
1st (3.41)		13.84:1
Final drive ratio		4.06:1

INSTRUMENTATION
Instruments: 180-mph speedo, 10,000-rpm tach, 999,999 odo, 999.9 trip odo, oil press., oil temp, coolant temp, fuel level

Warning lights: oil press., alternator, brake sys, handbrake, converter overheat, coolant level, trans oil level, fuel level, rear side glass retraction blocked, door ajar, hood ajar, washer level, a/c freon low, service, light failure, lights on, rear-window heat, seatbelts, hazard, high beam, directionals

CHASSIS & BODY
Layout		transverse mid engine/rear drive
Body/frame		unit steel
Brake system		11.4-in. (290-mm) vented discs front & rear; vacuum assisted
Swept area, sq in./sq cm	424	2736
Wheels		cast alloy, 180TR390
Tires		Michelin TRX, 240/55VR-390
Steering type		rack & pinion
Overall ratio		na
Turns, lock-to-lock		3.5
Turning circle, ft/m	41.0	12.5

Front suspension: unequal-length A-arms, coil springs, tube shocks, anti-roll bar
Rear suspension: unequal-length A-arms, coil springs, tube shocks, anti-roll bar

MAINTENANCE
Service intervals, mi:	
Oil/filter change	7500/7500
Chassis lube	15,000
Tuneup	15,000
Warranty, mo/mi	12/unlimited

CALCULATED DATA
Lb/bhp (test weight)	16.0
Mph/1000 rpm (5th gear)	20.3
Engine revs/mi (60 mph)	2950
Piston travel, ft/mi	1380
R&T steering index	1.44
Brake swept area, sq in./ton	251

ROAD TEST RESULTS

ACCELERATION
Time to distance, sec:
- 0–100 ft 3.5
- 0–500 ft 9.3
- 0–1320 ft (¼ mi) 16.0
- Speed at end of ¼ mi, mph .. 87.0

Time to speed, sec:
- 0–30 mph 2.7
- 0–50 mph 5.5
- 0–60 mph 7.6
- 0–70 mph 9.8
- 0–80 mph 12.8
- 0–100 mph 21.5

SPEEDS IN GEARS
- 5th gear (6800 rpm) 138
- 4th (7700) 112
- 3rd (7700) 84
- 2nd (7700) 60
- 1st (7700) 41

FUEL ECONOMY
- Normal driving, mpg 13.5
- Cruising range, mi (1-gal. res) ... 235

HANDLING
- Lateral accel, 100-ft radius, g 0.808
- Speed thru 700-ft slalom, mph 60.3

BRAKES
Minimum stopping distances, ft:
- From 60 mph 153
- From 80 mph 258
- Control in panic stop ...... excellent
- Pedal effort for 0.5g stop, lb 17
- Fade: percent increase in pedal effort to maintain 0.5g deceleration in 6 stops from 60 mph .... nil
- Overall brake rating .... excellent

INTERIOR NOISE
- Idle in neutral, dBA 73
- Maximum, 1st gear 87
- Constant 30 mph 74
- 50 mph 76
- 70 mph 79

ACCELERATION

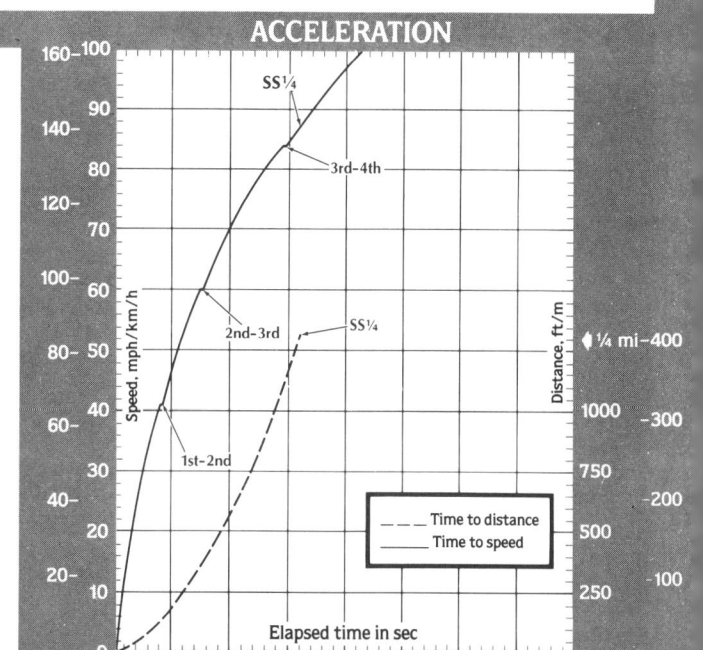

Ferrari 308 GTB QV Supertest

Ferrari 308 GTB QV Supertest

Now, remarkably, approaching a decade in production, the 308 carries its age well, and remains in most people's eyes the most beautiful of Ferrari's current production cars. Perhaps not so strong on practicality, this two-seater scores highly in all dynamic departments, with handling to match its 150mph-plus top speed.
Photographs by Tim Wren

IT IS conceivable that someone who had never driven a Ferrari, even a fairly experienced driver, would be a little disappointed by first acquaintance with the 308.

You have to slide down into that beautiful Pininfarina body, and when you've achieved that mildly athletic feat, you might just be a bit troubled by the sound that greets you when you turn the ignition key. At rest, and at low revs, the transverse V8 mounted just behind your shoulders doesn't sound anything special; loud, certainly, but not special, though it will only take a couple of seconds for the smoothness of the throttle response to impress itself upon you.

Move the lever away from you in the open gate and pull it back into first gear, and take off. So far so good, as the clutch is not too heavy and is progressive in action. But you may find that, from cold, not all the gears are too keen on being selected. Also, if you make the error of trying to shift gear slowly when the 'box has warmed up, the gearchange may not live up to your high expectations.

Drive over a bumpy road at low speed, and you may not find the ride quality very absorbent, while the steering wheel has a tendency to kick whenever the front wheel encounters a protuberance or depression in the surface, in a manner not common in modern high-performance cars. At low speeds, the steering feels dead as well as low-geared, and its lock is poor.

However, once you've put this learning period behind you, the sheer brilliance of this car can only lead you to love its dynamic qualities as much as its appearance. The way the power goes down on to the road, the speed of the lever as you flick up through the gears, the speed at which you can enter and leave bends, the power and smoothness of the brakes, the astonishing flexibility of the engine and its keenness to rev right up to its red line at 7,700rpm, and, above all, the Transylvanian howl of the exhaust when you really get moving . . . If you don't love this car by now, you'd better lead the rest of your motoring life in sensible saloons.

In comparative terms, the 308 GTB QV is the poor man's Ferrari. Though it costs less than half the price of the recently announced Testarossa, you'll still need 36 pence change on top of £29,100 to pay for it, so don't expect a lot of sympathy for your financial predicament. That's a few hundred pounds less than a Lamborghini Jalpa, but £6,500 more than the Ferrari's most direct rival, the Porsche 911 Carrera (without the "Sport" pack).

The Targa-top version of the 308, the GTS, weighs in at £30,399.77. In all other respects, the two cars are identical.

What you get for your money is probably the most attractive of all the current roadgoing Ferraris. The all-steel, Scaglietti-built body (glass fibre was used only for the first year of production), despite the fact that it is due to clock up a decade of production at the end of this year, remains pleasing from any angle. No stronger evidence for its continued appeal is needed than the fact that the GTO competition car, introduced recently, is in appearance a stretched GTB, whereas Ferrari and Pininfarina could easily have chosen an entirely unrelated design.

The 2,927cc V8 is now in its finest form, since the adoption of *quattrovalvole* (four valves per cylinder). With the aid of Bosch K-jetronic fuel injection and Marelli Digiplex ignition, the all-alloy unit produces 240bhp at 7,000rpm, and 192lb ft of torque at 5,000rpm. The twin overhead camshafts per bank are driven by toothed belts. This results in less mechanical clatter than existed in earlier Ferraris with chain-driven cams but, as we have mentioned, it does not result in Jaguar levels of quietness nor in quite the nerve-tingling Cosworth-like scream of the Jalpa.

When you lap Millbrook's five-lane banked bowl at an average speed of over 150mph, various factors take on a greater significance than they might have at lower speeds: wind direction and speed, any moisture that might lie on the surface, and the small bump in the top lane as the 1-mile post is approached.

On the day we tested the 308, the wind was gusting quite fiercely. As the bump was approached we were gathering speed, towards our fastest ¼-mile (156.1mph). The bump unsettled the car slightly, and just then – on every lap – the wind attacked the front nearside of the car from an angle of about 45 degrees, and a small degree of extra lock was needed to counteract the elements' intention of creaming this fine machine into the Armco.

Some lightness of the front end was noticeable at speeds of more than 120mph. Almost all 308s sold in the UK – unlike our test car – are ordered with the optional front spoiler (£300.24, tax paid) and although this clips a few mph off the top speed, it apparently eradicates this minor instability. Purchasers may also specify a 'rear aerofoil' for an extra £170.68, but the purpose of this is apparently more cosmetic than dynamic.

Given the unhelpful conditions which prevailed during our test, it is not unreasonable to accept Ferrari's claimed top speed for the car of 158.4mph, for we narrowly beat the suggested standing 400 metre time and were within 0.3sec of the standing km time.

Despite the considerable rearward weight bias, breaking traction for a good standing start in the 308 is no problem. It reached 60mph in a mean of 6.3sec, 100 in 15.8 and 120 in 23.8, figures which place it near the top of what could be termed the junior supercar league. With calmer weather, the statistics would be marginally improved, without a doubt. As things stand, the 308 has performance which is almost exactly on a par with that of the Porsche 911 Carrera. By that we mean *real* performance – it's impossible for one to pull away from the other on a straight road.

The spread of available power is the best thing about this engine. We did not subject it to pulling from below 20mph in 5th (there seemed little point), but we feel sure that it would have done so without complaint, despite the fact that that corresponds to fewer than 1,000rpm. Peak torque is achieved at 5,000rpm, but the curve is unusually flat, and there's plenty of pulling power from 2,500 onwards to the 7,700rpm line.

With such a wide band of useable power, it's possible to drive the 308 in a manner that suits your mood at the time. You don't *have* to drive it like the Number Two *Scuderia* driver attempting to justify his continued presence in the team (though it's very rewarding when you do!) – you can go shopping in it, without having to blip the throttle every time the traffic halts your progress. You can cruise along in top or 4th, with no more likelihood of the plugs fouling than is the case with a Granada.

Always there is the knowledge that you can snick down a gear or two, press the throttle pedal to the floor, and leave almost anything else standing. The point is that if you buy a car with this potential, you'll want to use it where appropriate, but the 308 is not a thinly disguised racing car. It's a road car and, in dynamic terms at least, a very practical one.

At motorway speeds, the engine noise is at an acceptable level, and it's possible to hear the radio or carry on a conversation with ease at 70mph or more. Indeed, as the speed rises, some of the noise is left behind, but we wouldn't necessarily recommend such an argument for a plea in mitigation. Not that it makes much difference, there's virtually no wind roar at high speeds, while road noises are restricted to tyre rumble over some surfaces and occasional bump-thump.

We didn't spend much of our test mileage exploiting the superb flexibility of the 308's engine, and the result was 16.5mpg overall. It doesn't require severe restraint, however, to push that figure closer to 20mpg, and the 19.6-gallon fuel tank allows a better range than that of many saloons. The fuel filler, in a neat Pininfarina touch, is hidden behind what looks like a vent extract on the offside.

The gearbox, like the engine, of the 308 is identical to that of the Mondial we tested recently, but the linkage is slightly different. It may be that, or mere variation in tolerances, but we found this one a little looser and easier to change smoothly with. Even so, it's no index finger and thumb job – some applied violence is required. Positive movements of hand and foot (it's essential for the clutch pedal to be fully depressed) yield satisfying results, and once you've got the knack of it, you can slide up and down the ratios almost unconsciously, leaving you free to concentrate fully on the road ahead.

While the ratios are well chosen to exploit the engine's power band to the full, with speeds of 42, 62, 85 and 117mph attainable in the intermediates) we did feel that the overall gearing could be just a shade higher: if *your* 308 could equal the factory's maximum speed claim, the rev needle would be 7rpm beyond the red line. Granted that there isn't anywhere – even a German autobahn – where such a speed may be maintained for long, it is nevertheless uncomfortable to put the engine into the yellow warning sector (7,000rpm onwards) for more than a brief period. Perhaps a six-speed gearbox with a real cruising top ratio is the answer. At 100mph, the engine is spinning at just under 4,900rpm in 5th, and it is only beyond that point it takes on its manic howl.

The 308's engine slightly overhangs the rear wheel centre-line. Directly beneath it and to the offside is the transmission, which incorporates a limited-slip differential. LSDs can be a problem in road cars, suddenly locking up and making unexpected demands of the driver, but the 308's does not "bite back" in this way, instead doing the job of helping the driver to transmit as much of the power to the road as his brain considers prudent in the circumstances. Traction, even on a slippery surface, is one one of the 308's strongest suits.

The suspension too, is supremely well sorted; it consists of the classic racing set-up – double wishbones all round, with coaxial coil spring/dampers and an anti-roll bar at each end. We've already referred to the 'deadness' of the rack and pinion steering at low speeds. Once you're on the move, it takes on a different character. Apart from occasional kickback, it doesn't quite have the animal tendencies of the 911's steering, but it is superbly precise, building up effort as cornering forces increase, and progressively lightening as the front Michelin TRXs eventually go beyond their very high dry-surface limits of adhesion.

By modern standards, the spaceframe chassis is 'yesterday's technology', but it doesn't feel like it. It is admirably rigid and free from any detectable flexing under load. It is the suspension that does the job when you corner hard, and it does that job splendidly, with taut damping control that prevents float, keeps roll to a minimum, but which smooths out progress over bumps or potholes, except at urban or suburban speeds.

In almost all circumstances, the 308 will corner with a trace of understeer, even when charging through curves under full power. If the front end slide shows signs of excess, easing pressure on the throttle simply brings the car back on to the chosen line. A tail slide will only occur in the dry under extreme provocation and when it does, it is easy to hold. Quite simply, this is one of the easiest cars to drive quickly in good conditions. A degree of extra caution is required on wet roads, but the same holds true of many relatively high-powered cars fitted with wide tyres.

The massive ventilated discs of the 308 (mounted outboard all round) are more than enough to cope with the car's performance. On a damp surface, it is possible to lock up the fronts, but such is the progression of the system that this is easy to control. We found no tendency whatever for the brakes to fade when used vigorously in the dry, and we found the weighting of the pedal close to ideal. The handbrake, in contrast, was far less satisfactory. The lever's siting makes its fulcrum too high and too far back, with the result that it takes two hands to apply it when the car is on a slope. Even then we were unsure enough of its effectiveness to leave the car in gear every time.

The interior of the 308 shows its age much more than any other area of the car, the switchgear and instruments being similar to those of the 400 rather than the Mondial. The simple, traditional dials may not be "state of the art" but they are easy to read, and that's really what counts. If the switches are a bit scattered around, they aren't too dificult to memorise, particularly since the centre console is cleverly lit at night by a light (perhaps excessively bright – in contrast to the feeble instrument illumination) in the underside of the handbrake lever. The worst item of swichgear is the headlamp stalk: if mistaken for the indicator stalk at night, it can supply total darkness, an undesirable condition in such a fast car.

Not everyone will find the driving position ideal. There's not quite enough legroom for tall drivers, and certainly not enough head clearance, while the steering wheel's lack of reach adjustment imposes a stretched-arms posture on most drivers. The pedals are superb, large, perfectly spaced and allowing room for the left foot to rest comfortably. As for the seat itself, it's bound in leather and reasonably well shaped, but its cushion is really too hard for comfort on a long journey, especially if the driver is not personally well padded in that area.

The surprising aspect of the car (especially in view of its age) is its low waistline, which imparts a much more light and airy feeling to the interior than is the case in most mid-engined cars, whose cockpits are all to often claustrophobic. All-round visibility too is much better than might be expected, the pronounced humps of the front wings giving a good clue to where the front end is.

Though relatively small, the door mirrors allow a good field of view, aided by an electrical adjustment device of BMW origin. On main beam, the headlights are powerful enough, but when dipped they are disappointingly weak, discouraging rapid night driving, especially on unknown roads.

Most 308 owners will also have at least one other car, so will have taken into account its poor luggage capacity. There's no room under the front lid, which is occupied by the radiator, full-size spare wheel, battery and servos. The rear lid, hinged at the roofline, uncovers the engine and a neatly trimmed but small boot (large enough to accept a couple of squashy bags). Inside, oddments space is strictly limited – there is a small locker at the rear of the centre console, shallow door pockets, a bit of room behind the seats and a coin tray.

Italian cars are justifiably lacking in worldwide renown for the efficiency of their heating and ventilation systems, but the 308 (at least in the form we tested it, with the £1,100.08 option of air conditioning fitted) is a welcome exception. Heater volume is adequate if not perfect in its directional ability (the footwells take a while to warm up as a result), while the air conditioning allows a refreshing (and these days increasingly rare) combination of cool air to the face and warm air in the lower half of the cabin. Demisting is rapid, and the fan is not too noisy on its lower setting.

Both inside and out, the finish is very tidy indeed, indicating a more fastidious attention to detail than was sometimes the case with Ferraris of earlier times. Curiously, the tiny exterior door catches seem far more modern than those of the Mondial (which are borrowed from the 400).

For your £29,100, you get – as you have a right to expect – a very high specification for your 308. This includes electric operation of the windows, door mirrors and central locking, leather upholstery, and an electric aerial and twin speakers. Maranello do not fit a radio/cassette player, reasoning that customers will wish to select their own.

There's not much else, apart from the front spoiler, that can be specified. You can have the car fitted with Pirelli P7 tyres (unlike the Michelins, these are of unequal size, front to rear – 205/55 VR 16 on the front 7J rims and 225/50 VR 16 on the 8J rears), air conditioning (well worth having, even if expensive) and a metallic paint finish, which sounds like mere extravagance.

Although the 308 continues to require considerably more loving care and attention than, for example, a Porsche 911, it can now boast of genuine 6,000-mile service intervals (which is what Porsche dealers, as opposed to the factory, recommend for their cars), and it has a robust feel to it which suggests that you *could* use it every day if you so chose. It would be a bit of a shame to do so, though. While Ferrari's intimate connection with the Fiat empire has led to a large degree of "productionising", this has not been allowed to emasculate the cars which carry Barracca's Prancing Horse emblem (almost everywhere – even on the ashtray lid!) The greatest strength of the 308 is in the sense of liberation and excitement it generates when you jump into it after driving other cars.

PERFORMANCE

Tests carried out at Millbrook Test track, Bedfordshire.

Maximum speed (lap of banked circuit) **151.1mph**
Fastest ¼-mile **156.1mph**

Acceleration through gears:
0-30mph	2.3sec
0-40	3.5
0-50	4.9
0-60	6.3
0-70	8.0
0-80	10.1
0-90	12.8
0-100	15.8
0-110	19.5
0-120	23.8

Standing ¼-mile 14.5sec/97mph
Standing km 26.5sec/125mph

Acceleration in single gear:
	5th	4th
0-40mph	–	5.6sec
0-50	8.3	5.5
0-60	8.1	5.5
0-70	7.9	5.4
0-80	8.4	5.2
70-90	9.3	5.1
80-100	9.7	5.5
90-110	10.7	6.3

ECONOMY
Overall consumption 16.5mpg
Composite* 21.2mpg
Test distance 539 miles
Tank capacity 19.6 gallons
Maximum Range 415 miles
Based on government test figures (one half of Urban figure plus one quarter of each of the steady-speed figures, 56/75mph).

WEIGHT
Unladen (with fuel for 50 miles) 25.5cwt

ENGINE
Twin overhead camshafts per bank (driven by toothed belts), water-cooled V8, transversely mid-mounted, 2,927cc (bore/stroke 81/71mm). Five main bearings. Compression ratio 9.2:1. Aluminium/silicon alloy cylinder heads, aluminium alloy engine block. Nikasil-coated aluminium liners. Four valves per cylinder. Marelli Digiplex electronic ignition. Induction by Bosch K-jetronic fuel injection. Maximum power 240bhp at 7,000rpm. Maximum torque 192lb/ft 5,000rpm.

TRANSMISSION
Rear-wheel drive, five-speed all-synchromesh gearbox, cable-operated single-plate clutch. Limited-slip differential. Internal ratios and mph/1,000rpm:
Top	0.919:1	20.5
4th	1.244:1	15.2
3rd	1.693:1	11.1
2nd	2.353:1	8.0
1st	3.419:1	5.5
Reverse	3.248:1	
Final drive	3.823:1	

BODY/CHASSIS/SUSPENSION/STEERING
Tubular steel frame, steel body panels. Front and rear suspension: independent, by double wishbones, coil springs, anti-roll bar.

TYRES/WHEELS/BRAKES
Michelin TRX, 220/55 VR 390 on 165 TR390 cast light alloy rims. Brakes: servo-assisted ventilated discs, 10.8 in front, 11.0 in rear dual circuits, split front/rear. Parking brakes operates rear di

DIMENSIONS
Overall length 166.5in, width 70.5in, height 44.1in, front/rear track 57.5/57.5in, wheelbase 92.1in.

ELECTRICAL
1120W Alternator, 12V 66 Ah battery, twin retractable halogen headlights, 110/120W total, 18 fuses.

FERRARI 308 GTB QV
Maker: Ferrari Esercizio Fabbriche Automobili e Corse SpA, Viale Trento Trieste 31, 41100 Modena, Italy. Importer: Maranello Concessionaires Ltd, Egham Bypass, Egham, Surrey TW20 0AX (0784 36431). Price: £23,358.00 plus £1,946.66 Car Tax plus £3,795.70 VAT equals £29,100.36. Price as tested with £1,100.08 air conditioning: £30,200.44

THE SPECIALTY FILE

Twin-Turbo Ferrari 308

An ersatz GTO for Ferraristi who can't wait for the real thing.

PHOTOGRAPHY GEORGE LEPP

• Ever since the rebirth of the GTO, Ferrari fanatics have been pushing and shoving to queue up for the new supercar. None that we know of have actually succeeded in laying hands on one as yet. The process seems to be a lot like waiting for campaign promises to come true: the factory in Maranello is known for talking big and delivering later, if at all.

An alternative worth considering is the injection of some of the GTO's twin-turbo magic into a current Ferrari Quattrovalvole (308GTSi). This requires doing without the GTO's other exotic performance accouterments, but the prospect of old-fashioned twelve-cylinder muscle in a car wearing the rampant-stallion insignia is usually enough to make the faithful salivate in anticipation.

We recently tested such a double-blown car, a 308 Quattrovalvole supplied by Prancing Horse, Inc., a Ferrari service and high-performance emporium in Campbell, California (408-559-1757). Although Prancing Horse supplies and installs its own kits, this particular development was undertaken jointly with now defunct Pfaff Turbo, in nearby San Jose.

As turbo kits go, this is one of the most straightforward installations we've ever seen. Each bank of the V-8 feeds an IHI RHB52 turbocharger via a special exhaust manifold. Each compressor mouth is protected by a K&N air filter, and the twin streams of compressed air produced by the turbos are gathered and then routed through the K-Jetronic fuel-injection system's metering unit on the way to the original intake manifold. Peak boost pressure is 7.0 psi, regulated by the turbo's integral waste gates. Minor modifications to the fuel-pressure regulator provide a slightly richer mixture whenever manifold pressure rises above atmospheric. On the exhaust side, Prancing Horse has fitted a European, catalyst-free system to minimize back pressure. There are no internal changes to the 32-valve engine, no major intake-system revisions, and no complicated engine-control systems.

According to Rick Brady, the proprietor of Prancing Horse, this simplicity is made possible by the inherent stoutness of the Ferrari V-8. The good combustion and detonation resistance inherent in a four-valve combination chamber are also helpful, and the modest, 8.6:1 compression ratio doesn't hurt.

There's no denying that this turbo installation really brings the 308 engine to life. As a matter of fact, it transforms the mid-engined machine into one of the fastest road rockets going. The twin-turbo 308 sprints from a standing start to 60 mph in just 5.6 seconds, to 100 mph in 12.3 seconds, and then claws to 130 mph in 23.0 seconds. In the process, it devours a quarter-mile in 13.5 seconds, achieving 107 mph through the traps. Top speed is limited to 147 mph by gearing and the engine's redline, but with an estimated 350 horsepower, it doesn't take long to get there.

Any 308 driver should be able to appreciate the benefits of a major boost in horsepower. The turbos trim nearly two seconds from zero-to-sixty and quarter-mile times, so the homemade GTO should never have to sneak around, fearing encounters with Porsche 911s or 928s, Chevrolet Corvettes, or the current domestic pony cars. In fact, the double-blown 308 reminded us of the much revered Ferrari Daytona. Except in top speed, its performance almost perfectly matches the older twelve-cylinder's, putting the 308 at the overachieving end of the speed spectrum, exactly where Ferraris belong.

In exchange for this vast improvement, the modified engine extracts little penalty in everyday, nonfrenetic driving. One reason is the essentially stock intake system, which keeps low-speed response respectable. In top gear, the modified 308 goes from 30 to 50 mph in 9.5 seconds and from 50 to 70 mph in 8.1 seconds—versus 8.7 and 8.8 seconds, respectively, for the standard car. Obviously, in the 1500-to-3500-rpm range used in this test, there's not much boost available; but once the engine is turning 3000 rpm, the boost gauge is on the rise and there's a full 4000 rpm worth of engine operation left.

The engine's sound is also thoroughly refined. At the upper end of its rev range, the well-known Ferrari shriek is very much in evidence. At the low end, however, the twin-turbo V-8 is surprisingly silent and docile. Indeed, at a steady 70 mph, we measured a sound level of 78 dBA, 3 dBA lower than a standard car. Our impressive C/D fuel economy of 17 mpg is another indication that this car is a capable cruiser.

The 308's chassis accepts the extra power with eagerness. Our test car, equipped with Goodyear NCT tires, retained the basic combination of initial understeer and terminal oversteer, but the transition could be prompted a bit sooner with the stronger engine. The 308's handling characteristics are still quite manageable, as long as the driver doesn't get carried away with the extra horsepower.

Obviously, this twin-turbo installation eliminates all the emissions controls, and it's unlikely to enhance the engine's longevity. Still, the system did withstand the rigors of our performance tests, and Brady says he's seen excellent reliability in several similar installations. The complete package costs $6000, but you can save a grand if you bolt the hardware on yourself. Considering that some dealers are demanding $10,000 for a highly dubious GTO "reservation," $6000 for a pair of turbos to tide you over sounds entirely reasonable.

—*Csaba Csere*

Vehicle type: mid-engine, rear-wheel-drive, 2-passenger, 2-door targa
Kit price: $6000 (installed)
Engine type: twin-turbocharged V-8, aluminum block and heads, Bosch K-Jetronic fuel injection
Displacement . . . . 179 cu in, 2927cc
Power (C/D estimate) . . . . 350 bhp @ 6500 rpm
Transmission 5-speed
Wheelbase . . . . 92.1 in
Length . . . . 174.2 in
Curb weight . . . . 3350 lb
Zero to 60 mph . . . . 5.6 sec
Zero to 100 mph . . . . 12.3 sec
Zero to 130 mph . . . . 23.0 sec
Standing ¼-mile . . . . 13.5 sec @ 107 mph
Top speed . . . . 147 mph
C/D observed fuel economy . . . . 17 mpg

FERRARI 308GTB

Mark Hughes looks at the affordable supercar; on the following pages we celebrate the marque with stories on landmark cars and personalities

PRICES

£10,000-£15,000 The bottom range populated by cars which are best avoided except by super-competent DIY owners. Severe structural decay, tired mechanicals and lack of history characterise this price bracket.

£15,000-£20,000 Moving towards decent early cars, but perils abound. A poor glassfibre car would start here, typically 25 per cent dearer than early steel GTB – but would still need lots of work. Avoid short-lived, modest-performing GTBi unless image is all.

£20,000-£25,000 First-time buyers should look here. Middling-to-good 308GTBs with decent history are available, but any QV likely to be in fairly ordinary condition. GTS models also common in this price region, but cost 15-20 per cent more than parallel GTB.

£25,000-£30,000 Good 308GTB QV or exceptional examples of earlier models – although the very best glassfibre cars could be higher. With careful research, you shouldn't go wrong and depreciation-free ownership should be surprisingly inexpensive, long-term.

£30,000-£35,000 Solid territory for 328GTB/GTS, age and history having more bearing on price for these post-1985 cars. Earlier models would have to be near-perfect glassfibre, GTS or QV versions to climb this high.

Over £35,000 The best, most recent, low mileage 328s, rising a whisker above £40,000 for the handful of brilliant specimens.

GTS models are far more numerous in the Uk, but carry a 15-20 per cent price tag over the GTB version

Have you ever wanted to own a Ferrari? Did you, during the boom, watch with gloom as so many motoring icons, none more so than Ferraris, soared out of sight to price levels beyond any realistic ambition? After all the turbulence, you will have seen the figures in the price guides and the ads tumble again to the point where they were six or seven years ago. You may have more pressing concerns than wondering whether or not a car purchase of such unashamed indulgence is feasible, but one certainty has returned. If they have the desire, a great many people can once again entertain the idea of making sacrifices to buy and run a Ferrari.

There will not be a better moment to buy your first Ferrari. A reasonable 308GTB can be bought for £20,000, a baseline figure for presentable cars which is unlikely to drop any further, and which is already showing signs of starting to drift up again. Other Ferrari models can be bought for a similar figure, but a 308GTB – or any of the various GTB/GTS versions – has decisive advantages as the best to drive, the most reliable and the least expensive to maintain.

With mind made up, you would then have to find one. These Ferraris are rarer than you might expect, a fact which might just add to their attraction. Of 19,555 GTB and GTS cars built between 1975 and 1989, only 1637 were sold in the UK in right-hand drive form. Put another way, Ferrari took 15 years to build as many cars as Porsche was capable of turning out in five months at its peak a few years ago, and the total UK stock could have been manufactured by Porsche in less than a fortnight.

It should go without saying that you need to do lots of research to learn about these cars, but then most Ferrari buyers do. Contrary to public perceptions, most Ferraris are owned by devoted enthusiasts, not posers. Look at better cars than you can afford so that you can

Wheels are porous on early cars, easily kerbed on later cars using low-profile tyres. Retail price £300–400 inc VAT, independent specialists usually cheapest

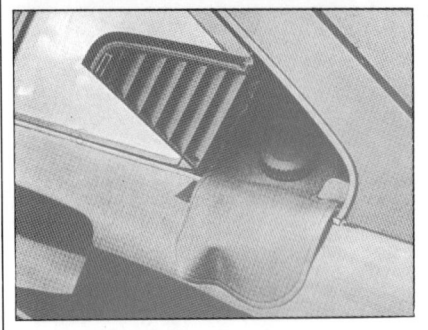

Above: neatly hidden fuel filler. Below: early cars have plated minor controls, later one have round black knobs. Take any seat work to a specialist trimmer

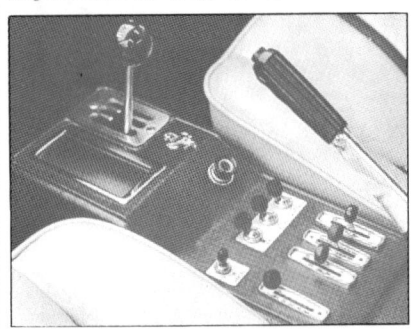

Below: only one pair of headlamps on 308, but most cars have extra driving lamps set under the front bumper. Pop-up actuation is by electric motors

This early car is a glassfibre 308 GTB, made from 1975 to '77. It wears larger wheels and lower-profile tyres than standard

WHICH IS WHICH?

308GTB (glassfibre) 1975-77. Lighter than later steel version by around 200lbs. 2926cc V8 has 8.8:1 compression, four twin-choke Weber 40DCNF carbs. Max power, 250bhp at 7700rpm. Max torque, 210lb ft at 5000rpm. Sought-after for rarity, more durable body, strong driver appeal. *Production: 712 (154 UK RHD).*

308GTB (steel) 1977-80. Change to steel body coincides with US launch and addition of GTS (with detachable roof panel) as alternative version. Car is heavier at 2784lbs (GTB) but mechanical spec is unchanged. Extra weight slightly detracts from performance and handling, but these cars are cheaper. *Production: GTB, 2185 (211 UK RHD); GTS, 3219 (184 UK RHD).*

308GTBi 1980-82. Interim model with Bosch K-Jetronic fuel injection lasted only 18 months, sold while Ferrari developed four-valve heads. Black sheep of this family, as performance noticeably inferior; majority of models are GTSi. Max power, 214bhp at 6600rpm. Max torque, 179lb ft at 4600rpm. This is, and always will be, the cheapest way to buy a GTB. *Production: GTBi, 494 (42 UK RHD); GTSi, 1743 (67 UK RHD).*

308GTB QV 1982-85. Performance fully restored and manufacturing quality improves dramatically. Engine still 2926cc, but now has four-valve heads (Quattrovalvole) and 9.2:1 compression. Max power, 240bhp at 7000rpm. Max torque, 192lb ft at 5000rpm. *Production: GTB, 748 (74 UK RHD); GTS, 3042 (233 UK RHD).*

328GTB 1985-89. Most complete of GTB family, with sharpest performance and handling, option of ABS, best build quality. Engine, still 32-valve, up to 3185cc (bore and stroke up 2mm), compression is 9.8:1. Max power, 270bhp at 7000rpm. Max torque, 224lb ft at 5500rpm. *Production: GTB, 1344 (130 UK RHD); GTS, 6068 (542 UK RHD).*

---

form a sharper view of those towards the cheaper end of the spectrum. Try dealers and private vendors, tap the experience of other Ferrari owners to guide you to reputable specialists, and enjoy the hunt. Check out the insurance, which needn't be absurdly expensive if you can keep your usage within a limited-mileage policy. And before you actually buy a car in the right condition and with good history, have it professionally inspected by a Ferrari expert.

To find out more, I visited Bob Houghton in Northleach, Gloucestershire. Now one of the most highly regarded independent specialists for road and race cars, Bob has been working with Ferraris since he left school, his 28 years in the business having included co-founding Graypaul Motors with David Clarke and running his own franchised Ferrari dealership in Cirencester. He agrees with our judgement that a 308GTB is the ideal first-time buy – one of these was his first Ferrari.

"This is the Ferrari which will give you fewest headaches," says Bob. "Provided they're carefully maintained, they're very reliable – Ferraris only ever get a bad name because they're not looked after properly. And prices at this moment are amazing. You can buy a good 308GTB for not much more than £20,000, although I think spending less than this would end up being a false economy."

Body and chassis

The greatest worry, as ever, concerns the structure of a GTB. Early glassfibre cars excepted, these steel-bodied cars will become ravaged by rust as quickly as any mass-produced alternative of the same age. A car can look wonderful at 20 paces, but be distinctly ropey when you're close up. As a general rule, the older cars corrode more rapidly not only because of their age, but also because rust prevention was tackled seriously by Ferrari only after zinc-coated steel – known at Maranello as 'zincrox' – was gradually introduced from 1982. Any 328

"I OWNED A 308 AND LIVED" – STEVE CROPLEY

Cropley's Ferrari had to work for its living, including shopping runs (left). In the mid-'80s he became nervous of the car's rising value and it was sold for a tidy profit

In the middle of the '80s, just before the prices went silly, I owned a Ferrari 308 GTB. It wasn't the ordinary steel-bodied version but the early, lightweight, glassfibre version which was made for about a year after the car was launched in 1975. My car, a late '76 version, came with just under 18,000 miles on the clock, and went just under two years later with 30,000 miles up.

Over the years, there has been much foolish talk about 'glass 308s. At first they were seen as undesirable 'plastic' things. Then, when rust began to bubble through the unprotected steel bodies, they became more attractive, especially when it was remembered that the standard of body finish was exemplary. Finally, when 308s began to appear here and there in club racing, the weight saving of about 200lbs over all-metal cars (the honest and actual figure, like so many Ferrari technical specs, is shrouded in mystery) made the plastic 308 seem glamorous.

It went well, of course. The original carb-fed 3-litre V8, burning leaded four-star and lacking anti-pollution plumbing, was supposed to produce 255 bhp, though these were the days of the Prancing Horse's think-of-a-nice-round-number power figures. In view of the realistic 215 bhp claimed for the later, cleaner 3.0-litre injection models, I'd say an honest output for my car was around 225 bhp. It was enough.

I never used the 7700 rpm rev limit (reduced a bit on later models) but I did see 7000 now and then. The performance was definitely in the junior supercar league: 0-60mph in seven seconds, perhaps less. An honest 130mph would come up fairly readily, a thing I discovered on a journey through Europe visiting various motor museums. But the joy of that F1-derived V8 was its instant throttle response and the breadth of its power band.

Over 12,000 miles I expected niggling problems but my 308 didn't even blow a light bulb. It did need one rear wheel bearing replacing, and a coolant pipe running through the valley of the V8 sprang a slight leak. But, considering that it did more miles in two years than many Ferraris do in 20, it was reassuringly reliable.

I sold my car because, frankly, its rising value scared me. People started putting quite ordinary things into garages rather than using them, and I didn't want to be one of those. The £12,500 I paid for it had seemed a fortune at the time, but it went for £19,000, I understand the owner after me sold for £22,000, the one after that caught the boom and sold to a dealer for £35-£40,000. The one who bought it from him still owns the car. Now, it's probably worth £25-30k, though latest market contacts say the market for Ferraris is firming again.

I loved my car, although if truth be told I set out at first to buy a Porsche 911, but changed my mind after a friend volunteered the following profound thought: "The man who has a chance to own a Ferrari and doesn't take it," he declared, "needs shooting." I think he was right.

Above: dash was clearly laid out and quite sensible for a '70s Italian car, although cabin noise levels are high in comparison with cars of the '90s

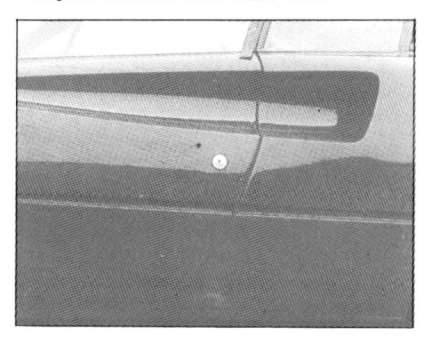

Above: first 712 cars were glassfibre-bodied and quality is good, although some have minor crazing. Door handles neatly tucked away at bottom of window frame

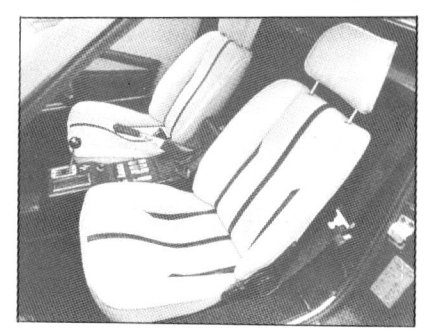

Above: body-hugging Ferrari seats give good lateral support. Below: doors don't usually suffer badly from rot as drainage is good. Check they fit snugly though

167

should be largely sound, whereas a mid-priced carburettor 308, built from much poorer steel, is likely to be fraying at the edges.

Fortunately, it is relatively straightforward to assess how far the dreaded tin-worm has nibbled because the method of construction – steel panelling over a tubular frame – means that most corrosion is visible, and difficult for a bodger to conceal from a trained eye. As well as checking through the typical tell-tale areas of decay, you need to form a view about the quality of any partial restoration attention, respray work or accident repairs.

All four wings are vulnerable because mud can accumulate in the pronounced lips around the wheelarches and the recesses at the top. Although deterioration here can be kept at bay by regularly washing out the wheelarches and coating them with Waxoyl, not all cars have enjoyed such diligence. A classic early sign is bubbling on the lower rear corners of the front wings, just ahead of the A-posts, but the rear wings can suffer similarly ahead of the wheels.

The steel sills are box sections which are attacked so quickly by condensation from within that earlier cars are likely to have had them replaced – a fairly simple job. The doors seem to survive quite well because they contain efficient drainage arrangements, but look for an even and snug fit all round their apertures.

Eruptions of rust surrounding the bonded windscreen are common because the rubberised seal around the glass hardens and cracks with age, letting water in. Resealing prevents this if caught in time, but rot will work from the inside outwards once water has penetrated. A similar condition can develop where the join between the top of the rear wing and the roof buttress is filled with black mastic, which can split and cause a fan of rust to grow down the wing. A filled and painted join here is a revealing pointer to third-rate remedial attention.

Apart from frequently being victims of accident damage, the nose and tail also have corrosion peculiarities. The front valance panel above the bumper tends to rot because paint often didn't reach its underside on earlier cars, while the lip at the top of the vertical tail panel can become so flaky that the metal will give way with finger pressure. You also need to examine all the car's inner recesses, like engine bay, headlamp pod cavities, front spare wheel well and rear luggage compartment. Don't forget the alloy wheels, for those on early cars have a record for porosity, causing the lacquered surface to blister when the alloy oxidises underneath.

Luggage and spare wheel bays are neatly covered by zip-up covers

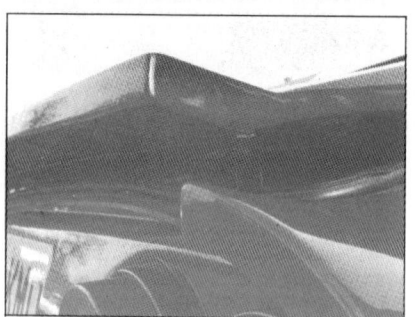

Above: underside of engine cover's trailing edge can rust and bubble, and cracks can appear at kick-up at front edge of lip if repeatedly slammed hard

Above: leading edge of door shuts collect stone chips. Below: bonnet stay (there's only one) is liable to seizure. Forcing front lid down then distorts the bonnet skin

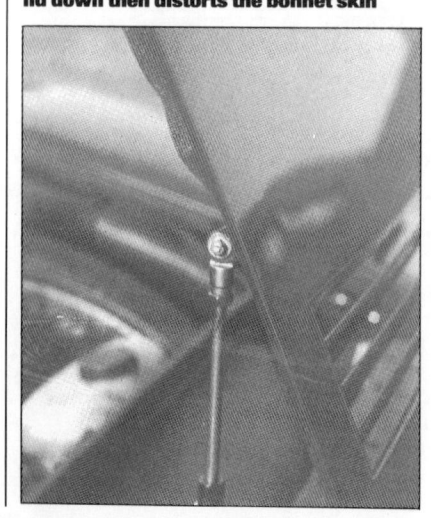

THE CLUB

The Ferrari Owners' Club is at 35 Market Place, Snettisham, King's Lynn, Norfolk PE31 7LR; tel: 0485 544500.

On the chassis side, mercifully, GTBs are largely bullet-proof. The oval-section main tubing, characteristic of all Ferraris, is so stout that rot is unknown, but outriggers at front and rear can decay. Those at the front, which support the radiator and bodywork, suffer the worst, but are difficult to inspect because the glassfibre front bumper assembly needs to be removed. A car which needs work here, however, will already have revealed enough bodywork rot to point you elsewhere.

Although a GTB doesn't rust as severely as the earlier 308GT4 or Dino 246GT, very careful inspection is vital because structural restoration of an indifferent early car will be inevitable one day. Bob Houghton reckons that sorting out a worst-case car bodily would swallow around £18,000 at a reputable professional, "although you could spend half the money for half a job". Even for an expert DIY owner, factory panel prices would add up to a fair chunk of the cost, a few examples being £455 for a tail panel, £254 for a bonnet skin and £2089 for an entire front end.

It's wise, clearly, to avoid bottom-money cars – or go for a glassfibre GTB. There was a time when the idea of a plastic Ferrari seemed like heresy, but the tables have turned to the extent that these cars command a hefty price premium because they're both rare and durable. Scaglietti, Ferrari's body builder, did a fine job during its brief acquaintance with resins and matting, producing a thick, strong and smooth structure of the highest quality.

A few stress cracks and 'starbursts' may have crept into the surface over the years, but nothing alarming. Remember, though, that these cars still have vulnerable steel sills and chassis outriggers. Paint also tends to fade more quickly, so most cars have received a respray by now. Producing a lasting finish on glassfibre is notoriously labour-intensive, so good quality work backed up by bills would be reassuring. Apart from its cost, the big problem with a glassfibre GTB could be finding one: the surviving stock in the UK is probably around 125 cars; only a handful are on the market at any one time.

Engine

The all-alloy V8 engine, a free-revving, flexible, raucous unit, is one of the sublime joys of owning one of these Ferraris. It's very reliable and promises the least expensive maintenance – by Maranello's standards – of any Ferrari. Bob Houghton, whose procedures are so thorough that he requires a car for three days to carry out a major service, esti-

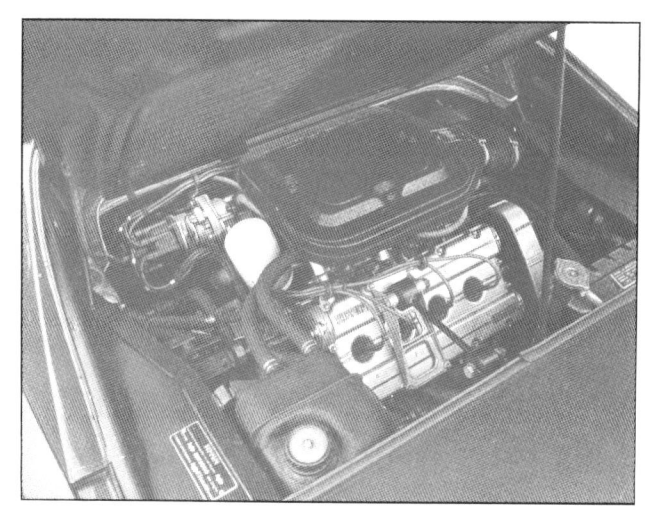

Engine, a V8, is bulletproof if cared for: timing belts need changing every 20,000 miles, accessed through cover in right-hand wheel arch. 2926cc until 1985, 3185cc after, with four-valve heads added 1982

Pininfarina's lines have stood the test of time well, and the shape of the cabin section was borrowed for the F40

SPECIALISTS

Franchised dealers
Maranello Concessionaires Ltd, Thorpe Ind. Est., Egham, Surrey; tel: 0784 436222.
Maranello Sales Ltd, Tower Garage, Egham Bypass, Surrey; tel: 0784 436431.
Graypaul Motors Ltd, Nottingham Road, Loughborough, Leics ; tel: 0509 232233.
Nigel Mansell Sports Cars Ltd, Salisbury Road, Blandford, Dorset; tel: 0258 451211.
HR Owen Ltd, 27 Old Brompton Road, London SW7; tel: 071 584 8451.
Lancaster Garages Ltd, Auto Way, Ipswich Road, Colchester, Essex; tel: 0206 855500.
Evans Halshaw, Monaco House, Bristol Street, Birmingham B5; tel: 021 666 6999.
HA Fox, Torquay, Devon; tel: 0803 294321.
Melbourne Garage Ltd, Route des Issues, St John, Jersey; tel: 0534 62709.
JCT 600 Ltd, Tordoff House, Sticker Lane, Bradford BD4 8QG; tel 0274 668241.
Stratton, Altrincham Road, Wilmslow, Cheshire; tel: 0625 522222.
Reg Vardy, Stoneygate, Houghton-le-Spring, Tyne & Wear; tel: 0915 120101.
Glenvarigill Company Ltd, 300 Colinton Road, Edinburgh; tel: 031 441 6850.
Cissbury Garage Ltd, 150 Findon Road, Worthing, Sussex; tel: 0903 830447.

Independent specialists
Bob Houghton Ltd, Midwinter Road, Northleach, Gloucester; tel: 0451 860794.
Nick Cartwright Specialist Cars, Butterley Reservoir Farm, Butterley Lane, Ashover, Derbyshire; tel: 0629 56999.
Terry Hoyle, Unit 6, Heybridge House Industrial Estate, Maldon, Essex; tel: 0621 55391.
Rardley Motors, The Avenue, Headley Road, Grayshott, Surrey; tel: 0428 606606.
DK Engineering, Unit D, 200 Rickmansworth Road, Watford, Herts; tel: 0923 255246.
Moto Technique Ltd, 141 Molesey Avenue, West Molesey, Surrey; tel: 081-941 3510.
Talacrest Ltd, Station Road, Egham, Surrey; tel: 0784 439797.
RTR, Unit 2, 41 Millers Road, Warwick; tel: 0926 403973.
Kent High Performance Cars, Unit 1, Target Business Centre, Parkwood Industrial Estate, Maidstone, Kent; tel:0622 663308.
Euro Tec, Winfrith Newburgh, Dorset; tel: 0305 852896
Forza 288, Highcliffe, Dorset; tel: 0425 273682
Euro Spares, 8 Rosemary Lane, Halstead, Essex; tel: 0787 473678.

ALASTAIR FERGUSON ON EVERYDAY MOTORING IN A FERRARI

Alastair Ferguson has owned his glassfibre, dry-sumped 308 GTB since July 1991, its purchase representing the achievement of a boyhood ambition, and since then he reports the car has been a joy to own and use.

"I decided early on that I wanted a glassfibre car because, in addition to its inability to rust, it is about 200lbs lighter than the steel version. I found it privately advertised after checking out various specialised dealers and being unhappy with both the prices asked and the condition of some of the alleged 'one owner, low mileage, full service history' examples I was offered." Alastair describes his car as a sound and honest machine in good rather than excellent condition but has no hesitation about using it as it was intended, in all types of driving conditions, including trips to Tesco's.

"Within a month of buying it I took it on a tour of France and northern Italy – the trip through the Alps was wonderful and memories of if will live with me forever. Some of my friends thought I was mad, but I had no qualms and the car really does inspire confidence."

On a day-to-day basis the car is much more practical than most pub 'know-alls' would have you believe, says Alastair: "The Webers do not go out of tune – they physically cannot do so – and it doesn't overheat, even in mid-summer in St Tropez. With its glassfibre body, my car can be used all year round without worrying about the dreaded biodegrading that Italian cars are famous for and even though it uses contact breaker points and carbs, it is an instant starter hot or cold."

The heart of any Ferrari is its engine and, as Alastair explains, the 308's motor is a remarkable piece of engineering. "It will run from 1000rpm in top gear with complete flexibility and will also pull right up to the 7500rpm, although I haven't tried this in top gear yet. On a decent run I get 23mpg and the only demerit points from the day-to-day driveability viewpoint are the heavy clutch and quite high noise levels by '90s standards."

Ferraris are very strong cars and ones of this age are of comparitively simple design which makes DIY maintenance a real possibility for the skilled amateur. There's no need to pay official dealer labour rates, in any case, as there are many skilled specialists doing excellent work for around half the official dealer's labour rate. Alastair uses Colin Clark in Acton, although there are several others.

"Spare parts are no problem because Ferrari, like Lotus, makes it a matter of policy to keep its obsolete models running and virtually everything is available. And there are a few independent specialists such as Eurospares in Halstead, Essex which can supply parts, new and second-hand, at cheaper prices.

"The Ferrari 308 has always been compared with the Porsche 911. I believe it is just as practical a proposition as a Porsche of equivalent age. There are probably more myths propagated about Ferraris than any other marque but my experience is that the car can be used reliably, provided it is properly maintained, and turns every journey into an occasion. If you have the chance to buy one and choose not to take it, you will probably always regret it."

Ferguson: "I'd take it on the Monte if I could."

---

mates that routine work for a GTB covering, say, 6000 miles a year will cost around £1500 a year for one complete 'nut-and-bolt' service and one or two intermediate visits. This level of maintenance is more diligent than the factory schedule, which attempted to ape Porsche by specifying 9000-mile intervals for early 308s and late 328s, and 6000-mile visits for all models in between.

It's important to replace the rubber timing belts every 20,000 miles or two years, while camshaft wear – a potentially horrendous expense at £1650 for each of the four camshafts – can be avoided by changing tappet pads before their hardness wears through.

A complete service history and an encouraging picture of past usage are crucial to engine life. A cylinder head overhaul of valves, seats and springs, and possibly new pistons and liners, is the typical requirement at around 90,000 miles on a carefully maintained and sympathetically driven car. One which has been used regularly around town might be suffering by 60,000 miles, while an abused engine – perhaps one which has been thrashed from cold – could need a rebuild at 40,000 miles, costing £4500 to £6500

Signs of advancing wear are exhaust smoke, low oil pressure and piston slap. Oil pressure should be around 85psi at 5000rpm when hot, but having a specialist check it with a master gauge is a good precaution because the car's instrument can be inaccurate. Piston slap can develop on a 3-litre engine when pistons and steel liners become worn, but the 3.2-litre engine's Nikasil alloy liners are much more durable.

Apart from regular servicing, most of the mechanical repairs on these cars concern the exhaust system and the clutch. Exhausts appear to have a short lifespan, partly because these cars tend to run hard and hot when they're used, then they sit for long periods. As many classic car enthusiasts know, infrequent use can exacerbate the corrosive effects of condensation within the system. A 328's stainless steel system lasts well, but exhaust replacement on the earlier models – normally involving a new mild steel main silencer section costing £439 – can be expected every couple of years.

The exhaust manifolds on 3-litre cars are also vulnerable, largely because they do not have enough strength designed into them to resist cracking through vibration. Replacement is expensive because a pair of manifolds costs £751 and fitting the front one, which involves removing the right-hand fuel tank and partially dismantling the suspension, is time-consuming. The 328s have much more durable stainless steel manifolds with additional fluting for strength, but you will notice that with these, the engine note is slightly dulled.

Noses often get bent on 308s – panels for a complete front end cost about £2500 including VAT

170

THE ALTERNATIVES

There are other candidates as a first Ferrari for around the same money as one of the 308/328GTB family. But what are their pros and cons?

400i Long-lived four-seater with sharp-edged Pininfarina lines, shape starting as 365GT 2+2 in 1972 and evolving into 412 which ceased production in 1989. Big, long-legged grand tourer, usually fitted with automatic transmission. Not the most rewarding car to drive, but it has special charisma as the only V12 Ferrari from this period, and front-engined. Good examples are available at £20,000, but these are risky for first-time buyer. Whatever you do, don't try to run one on a modest budget, as complexity means routine maintenance costs around three times the level of a 308GTB – and restoration of rough bodywork can lead to horrendous bills. Mechanically very strong, as owners of cars on 200,000-plus miles can testify. *Production (all models 1972-89): 2911 (518 UK RHD).*

308GT4 Unfairly maligned – and priced accordingly – because its Bertone styling isn't generally liked, but this mid-engined car has fine virtues. V8 with Weber carbs shares performance gusto of early GTBs and genuine two-plus-two seating is a practical bonus – the GT4 has more room in the back than a Jaguar XJ-S. Handling is superb, close to a GTB's in precision and balance, but poor steering lock is tiresome. The main drawback is rust-prone bodywork, for deterioration is quicker than GTBs of comparable age. Appearance and structural doubts keep prices depressed to comfortably below £10,000 for poor cars, a shade over £20,000 for the best. *Production (1973-80): 2826 (547 UK RHD).*

246GT Maranello's definitive mid-engined car, with 2418cc V6 engine and luscious Pininfarina shape. Exquisite to drive, generally less durable mechanically and structurally than GTB. Engine seems to demand more servicing, because it has to be revved hard; bodywork needs to be pampered. Perhaps too ambitious a choice as a first Ferrari, but a good target for the second buy. Collectibility well-established, so prices are higher than GTB at £35,000-£45,000 for the finest examples – but they've tumbled from six-figure boom levels. As an alternative for the cost of a good early GTB, an indifferent Dino ought to be avoided. *Production: 246GT (1969-74), 2487 (488 UK RHD); 246GTS (1972-74), 1274 (235 UK RHD).*

Mondial The modern interpretation of the two-plus-two 308GT4, but this time with Pininfarina styling and even more room inside – but less driver appeal. A masterpiece of mid-engined packaging, but far-forward driving position feels odd until you get used to it. Mechanically similar to GTB range but weight is 350lb more and handling is less responsive. Maintenance costs should be on a par with a GTB. Prices tend to run at 60-70 per cent of a GTB's, so typical range for an early one is £8000-£20,000. Cabriolets, available from 1983, much more desirable. *Production: Mondial 8 (1980-82), 703 (74 UK RHD); Mondial QV (1982-85), 1145 (216 UK RHD); Mondial Cabriolet (1983-85), 629 (24 UK RHD); 3.2 Mondial (1985-89), 987 (91 UK RHD); 3.2 Mondial Cabriolet (1985-89), 810 (57 UK RHD).*

Transmission

The common transmission problem is short clutch life, which is a result more of bad driving than weak componentry. A sensitive driver might manage 40,000 miles on a clutch, but someone who habitually slips the clutch, even momentarily, during take-off and gearchanging can wear it out within as little as 6000 miles. A worn clutch plate can also generate enough heat to cause surface cracking on the flywheel. For parts and labour, replacing a clutch will cost around £500.

Anyone familiar with Ferraris knows that second gear synchromesh is always stiff when the gearbox is cold. This is deliberate: second gear is used more than any other when you let a Ferrari engine sing, so its synchro ring is stronger to cope. It's adviseable, therefore, to make first-to-third changes until the oil is warm, but not all owners do this. If second gear is loose when the car is cold, it's inevitably knackered when it has warmed up. That said, gearboxes are very reliable, and when a rebuild does become due the work is usually just a case of replacing bearings and baulk rings.

Suspension, steering

Provided it has received proper maintenance, a GTB's suspension and steering should cause no problems. But a surprising number of cars are running with tired dampers and creaking suspension joints because their owners haven't noticed the gradual deterioration – which rather defeats the point of driving a Ferrari. So much of a GTB's pleasure comes from its ability to wrap itself around the driver, instilling a rare sense of supremely agile responses and razor-sharp precision, that a car which floats and pitches is under-performing.

There are a few maintenance observations about a GTB's underpinnings. Dampers, neglected by Ferrari owners as much as everyone else, are Koni adjustables which should be checked regularly for bump and rebound, tightened as necessary, and overhauled when all the adjustment has been taken up. Rear outer suspension bushes can seize, or become noisy if neglected. Early cars are prone to rear hub bearing failure because water can seep in.

Conclusion

The rationale of weighing maintenance costs against depreciation – the bottom line of the financial equation – really does hold water. Now that others have suffered the sting of plummeting values, you could step in and enjoy the privilege of Ferrari ownership; just buy the right car with your eyes open. All you need is the £20,000 to start with...

Four-pipe exhaust is mild steel on 308s and gives a lovely sound, but can have short lifespan. Manifolds can crack too. Later models have more long-lived stainless systems

OWNER'S VIEW

Derek Collins explains why his enthusiasm for the 308GT4 has grown

"When I'm standing outside looking at my car I want a GTB, but when I'm sitting inside I'm glad I've got a GT4"

As far as Derek Collins, a retired marketing director living in the Cotswolds, is concerned, a car has to be Italian to be interesting. His stable of four cars contains a Ferrari 166/195MM *Barchetta* with a known history which includes having been driven on the 1950 Targa Florio by Luigi Villoresi, a 1930 supercharged 1750 Alfa Romeo, a Fiat Panda . . . and a Ferrari 308GT4.

"I really stumbled across my GT4," explains Derek, "although I had been thinking about buying a 308, either a GTB or a GT4. This was three years ago, and I was running a 250GTE at the time. That was a nice car with its wonderful V12 engine, but somehow I could always find a good reason not to take it out — it was very large, and felt large to drive. People had said that I would enjoy a mid-engined 308, and one day while I was staying in Brighton with a friend he told me that there was a GT4 for sale locally. I looked at it and liked it, and made a deal to trade it for my 250GTE plus £3500."

Somewhat neglected

The car is a 1978 model, one of the last series before electronic ignition was introduced. It had covered 33,000 miles when Derek bought it, and since then he has added another 17,000, using it frequently and in all weathers. But it had been somewhat neglected in its short lifetime.

"The engine was fine," says Derek, "but there was some rust, and the rear anti-roll bar, a piece of metal as thick as your thumb, had snapped like a carrot. I bought a new one from Maranello Concessionaires and fitted it myself quite easily. I took the car for a check-over to my nearest Ferrari specialist, Bob Houghton in Cirencester, and he agreed with me that the engine was excellent — that had been my top priority when buying the car. For bodywork and sill repairs I took the car to a good tin-basher and squirter I know in Cheltenham called Pember & Miles. One of the unusual things about the GT4 is that the sills come in two halves, and they made new rear sections for each side for a very reasonable sum. The main area of bodywork rust was around the edges of the near-side rear wing — I had a go at this myself, but by the time I'd finished with it the metal was like lace, so Pember & Miles welded in new metal and leaded the area, respraying with the original acrylic paint from Maranello. The repair is now undetectable. A few months later it became obvious that the rear under-valence was also rusting through, so I had that done as well."

Since then Derek has carried out all regular maintenance himself, following Ferrari's servicing guidance. Some people say that changing the plugs on the mid-engined 308s is a headache, but all that's necessary, he says, is a suitable short socket attached to a long arm to reach the four plugs facing the cockpit bulkhead. With one of the rear wheels removed for easy access, oil changing is simple, and the drain plugs are the Allen key type. The oil filter is conveniently sited within the vee of the engine. The only regular item which Derek leaves to Bob Houghton is the distributor, which is really (on this model) two distributors in one casing with a pair of contact breakers and coils. The drive is a multi-spline arrangement which has to be accurately re-assembled and set up on an electronic gadget by a specialist. Changing the toothed camshaft drive belts every 25,000 miles is the other job best left to a specialist, as it could be all too easy to fit the new ones inaccurately, with dire consequences for the valve gear and pistons. Derek's only departure from Ferrari servicing recommendations has been to replace the normal Champion spark plugs with Japanese NGK BP7ES plugs, which he finds far better for varied driving thanks to their superior heat range.

Derek has also taken the precaution of treating the whole car with Waxoyl, and drilling all the cavities. The only corrosion to have materialised since then is a small area on each front wing where mud accumulates within the arch just in front of the doors. Little can be done about this short of cleaning within the arches after every run, but patch repairing is quite simple.

"Apart from a wheel bearing failure," Derek continues, "I've had only one major mechanical problem, and that was something very unusual. I am a born instrument watcher — I think flying with the RAF in the war taught me this — and one day I was alarmed to see the oil pressure plummeting. It turned out that one of the two jack-shaft bearings had collapsed, an oil seal had been destroyed as a result, and the engine had pumped out all its oil in about 20 miles. I coasted to a stop just a quarter of a mile from a garage, bought a gallon of oil, and managed to get to Bob Houghton's. He replaced both bearings, a job which required engine removal. If you're wondering what jack-shaft bearings are, they support the shafts on which pulleys carry the camshaft drive belts."

What does Derek particularly like about his GT4?

"Well, Ferraris are all about performance, and in this respect it's not bad. Friends who race GT4s have told me that the quoted 155mph maximum can be reached, and I saw 127mph at the end of a standing kilometre at a Ferrari Owners Club Sprint at Keevil. I finished third in class at that event, which sounds wonderful until I add that there were only three cars in the class! My excuse is that I wasn't prepared to drop the clutch at 6000rpm.

Civilised

"I'm always curious about fuel consumption on my cars, even though Ferrari owners aren't supposed to want to know. I regularly return 20-21mpg driving as fast as I feel safe — the worst I've seen is 18.5, and the best 23. The car is quite civilised and comfortable, and feels very stable at 110-120mph. It's not particularly noisy, and my wife, who never liked the 250GTE, enjoys travelling in it. My car doesn't have air conditioning, which as far as I'm concerned is a good thing since it is one less servicing worry.

"People say it has a rotten gearbox, but I like it. It's not at all heavy when you make nice, smooth, quick changes, just dipping the clutch a fraction. One odd feature — and this has been the case with all three 308s I've driven — is that it's advisable to avoid second gear until the gearbox has warmed up. It's very important to treat the engine gently until it's warm: I never go above 3000rpm until the oil temperature has reached 60 degrees centigrade. Some owners have complained about the rear tyres wearing very quickly: I've heard stories of tyres lasting as few as 4000 miles, and as many as 15,000. My experience has been around the average of that range.

"I wouldn't pretend that there's nothing I don't like about my GT4. All Ferraris have terrible steering lock, which makes them really cumbersome when parking in town. There's no way you can carry adults around in the rear seats, although a 6ft 2ins friend of mine did just manage to squeeze himself into the back for a short demonstration run. The view immediately behind the car in the mirror is terrible, and when you're reversing you find the door is so far from the edge of the seat that you can't get your head out of the window. But compared with the 308GTB, the GT4 is more practical: it's not quite as low, there's a little more room inside because the wheelbase is longer, and I prefer the driving position."

GT4 owners must be fed up with people knocking these cars, largely on account of their appearance. What does Derek think about all the old criticisms?

"I used to feel the same about the GT4, but most Ferrari enthusiasts now recognise that the car is under-rated, and values are climbing accordingly. I can sum up my feelings like this: when I'm standing outside looking at my car I want a GTB, but when I'm sitting inside I'm glad I've got a GT4."

172